I'm Pregnant!

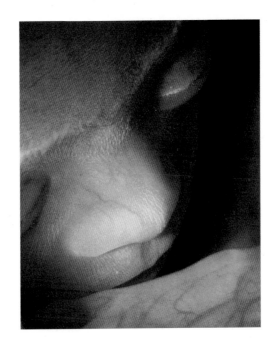

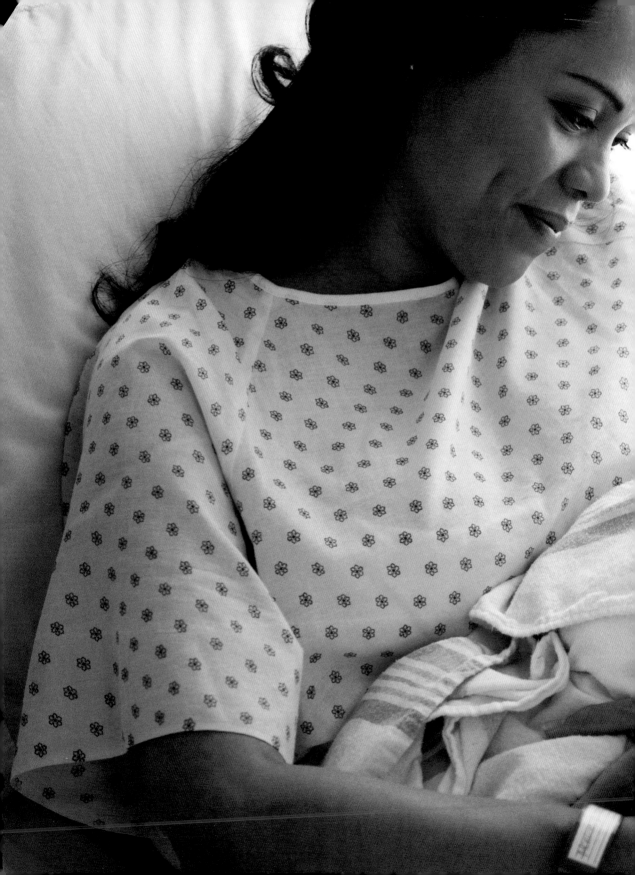

I'm Pregnant!

A week-by-week guide from conception to birth

LESLEY REGAN, MD

US medical consultant Paula Amato, MD
Foreword by Joe Leigh Simpson, MD

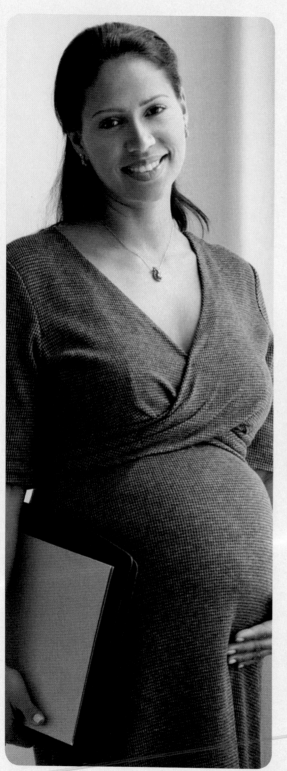

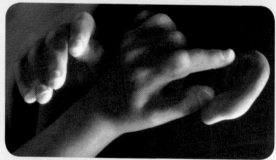
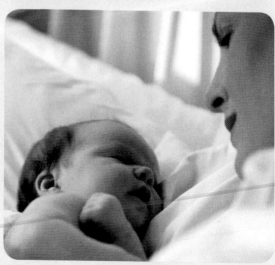

Contents

Your journey through pregnancy

Labor and birth

Life after birth

Concerns and complications

Foreword

Professor Lesley Regan clearly understands the dilemma faced by readers seeking a book on pregnancy. With so many to choose from, why this one? Yet this book truly is different, written for the discerning pregnant woman and her family who really want to know what they will experience.

I'm Pregnant! offers valuable detail in every chapter. Nothing important is overlooked. We learn how conception occurs and what may impede it. We learn the importance of our parent's genes and how birth defects arise. We are led through pregnancy, week-by-week. Expected side effects are discussed, not only famous ones like nausea but late pregnancy nuisances such as heartburn and backaches. There are answers to questions you might hesitate to ask: Intercourse? For how long, and will I feel differently? Why isn't my husband as excited as I am? Not only are the normal mechanisms of labor discussed but also forceps, vacuum, and cesarean delivery. Options for pain relief during labor are detailed.

A unique feature of I'm Pregnant! is that Professor Regan has been there. And, it shows. The text is peppered with personal anecdotes from her own pregnancy, bestowing credibility. From start to finish, I'm Pregnant! is a gem. I recommend this book not only to every inquisitive pregnant woman but also to obstetricians, whether trainees or experienced. It is packed with up-to-date facts, and glued together with wisdom and empathy.

Joe Leigh Simpson

66 *Pregnancy is one of the most important journeys that you will ever embark upon...* **99**

Introduction

There are plenty of pregnancy books on the market so why should I embark on writing another one? The reason is simple—the women I care for keep telling me that they would like more detailed answers to their questions about pregnancy and childbirth. Furthermore, they want a book that provides them with clear, comprehensive information without any prescriptive or personal agenda.

I understand and sympathize with the request since, when I was expecting my twins, I remember feeling both astonished and intimidated by books that seem to suggest that there are right and wrong approaches to pregnancy and childbirth. I have no argument with the many different childbirth philosophies, but I do have a problem when they result in pregnant women feeling that they have failed in some way if they do not follow the advice to the last letter, or their pregnancy does not follow a textbook pattern.

So the agenda here is very simple—knowledge is key. My goal is to offer you a level of information about your prenatal care that is not always readily available. I want to ensure that you are fully aware of the extraordinary events in your baby's development and the remarkable adaptations your body will make. My belief is that the only way to be confident about the choices and decisions that you will need to make in pregnancy is to have a clear understanding of everything that may happen to you. I also believe that this is the best way to achieve the happiest outcome—a healthy mother and a beautiful take-home baby.

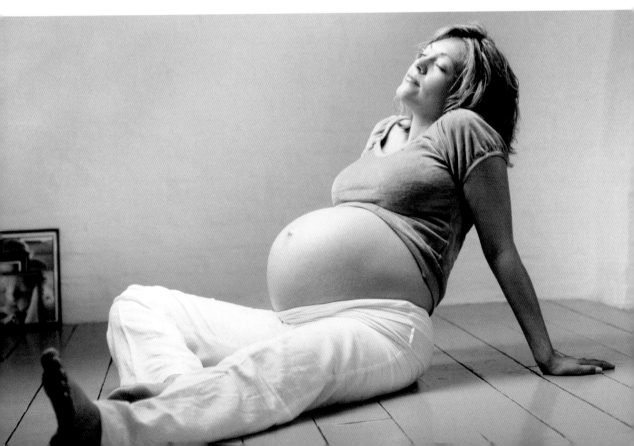

> 66 *The first trimester is the crucial period when all the organs, muscles, and bones of your baby are formed.* 99

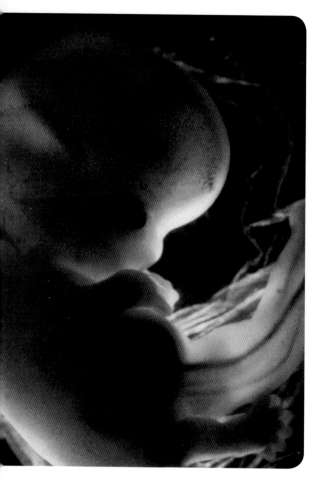

About this book

Pregnancy is one of the most important journeys that you will ever embark upon. To help you understand as much as possible about this eventful and exciting period of your life, this book is chronological. It takes you from the moment you conceive, through each week of your pregnancy to the delivery and then provides all the information you need to give birth and care for yourself and your baby afterward. This makes it easier to find your way around the book. I hope that you will be able to find the answers to your questions quickly and easily. Above all, the clear, comprehensive, and up-to-date information will help you understand the medical jargon and the experiences you are likely to encounter in the next few months.

The "journey" section is divided into the three trimesters of pregnancy. Everyone seems to have her own idea about exactly which weeks fall in each trimester; the only important thing is that each one corresponds to an important and distinct phase of your baby's development. At the beginning of each trimester, there is a broad overview of the major milestones that occur, then it is broken down into three more detailed

THE JOURNEY TIMELINE ··

The three trimesters are divided into 4–6 week sections offering detailed information that relates to your exact stage of pregnancy.

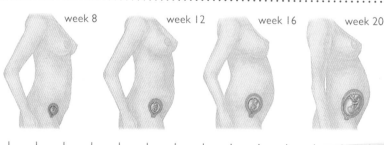

week 8 week 12 week 16 week 20

| 1 | 2 | 3 | 4 | 5 | 6 | 7 | 8 | 9 | 10 | 11 | 12 | 13 | 14 | 15 | 16 | 17 | 18 | 19 | 20 |

▶ WEEKS 0–6   ▶ WEEKS 6–10   ▶ WEEKS 10–13 ▶ WEEKS 13–17   ▶ WEEKS 17–21

▶ FIRST TRIMESTER   ▶ SECOND TRIMESTER

"through the weeks" guides. Each of these covers what routinely happens in pregnancy, including a description of your baby's development, your own body changes, how you may be feeling both physically and emotionally, with a section on the prenatal care you can expect and some of the common concerns linked to that particular time. To keep things simple I define the age of your pregnancy and your baby as the number of weeks from your last menstrual period.

In this book, I have defined the first trimester as 0–13 weeks of pregnancy. Generally speaking, the first trimester is the crucial period when all the organs, muscles, and bones are formed. During the first eight weeks, we refer to the baby as an embryo; subsequently, it is called a fetus. In the second trimester, all the basic structures that have developed are consolidated. During this period, the fetus grows rapidly and starts to make facial expressions, hear sounds, and can be felt kicking. In the third trimester the baby undergoes an important final phase of growth and maturation. Everything in the third trimester is based on the building blocks laid months before.

Although the vast majority of pregnancies are relatively problem-free, not every pregnancy is an entirely smooth journey. Whenever less common problems arise, you

will be referred to the last section of the book for further explanation.

No two labors are alike. The summary of how normal labor progresses and the various pain-relief options will be all most readers need, but for those who may need more specialized treatment there is detailed information and advice on topics such as cesarean birth and premature delivery.

The Life after Birth chapter looks at the highs and lows you may experience after your baby is born. Delight in your new baby will be mingled with anxiety about minor problems and your ability to cope in your new role. I hope my advice will help ease you and your baby through these first important weeks together.

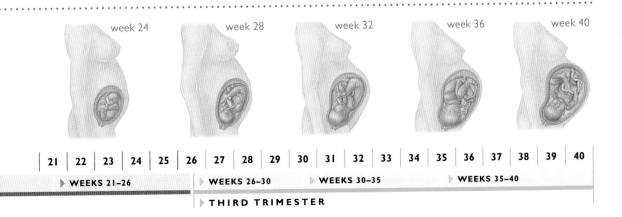

| week 24 | week 28 | week 32 | week 36 | week 40 |

| 21 | 22 | 23 | 24 | 25 | 26 | 27 | 28 | 29 | 30 | 31 | 32 | 33 | 34 | 35 | 36 | 37 | 38 | 39 | 40 |

▶ WEEKS 21–26   ▶ WEEKS 26–30   ▶ WEEKS 30–35   ▶ WEEKS 35–40

▶ **THIRD TRIMESTER**

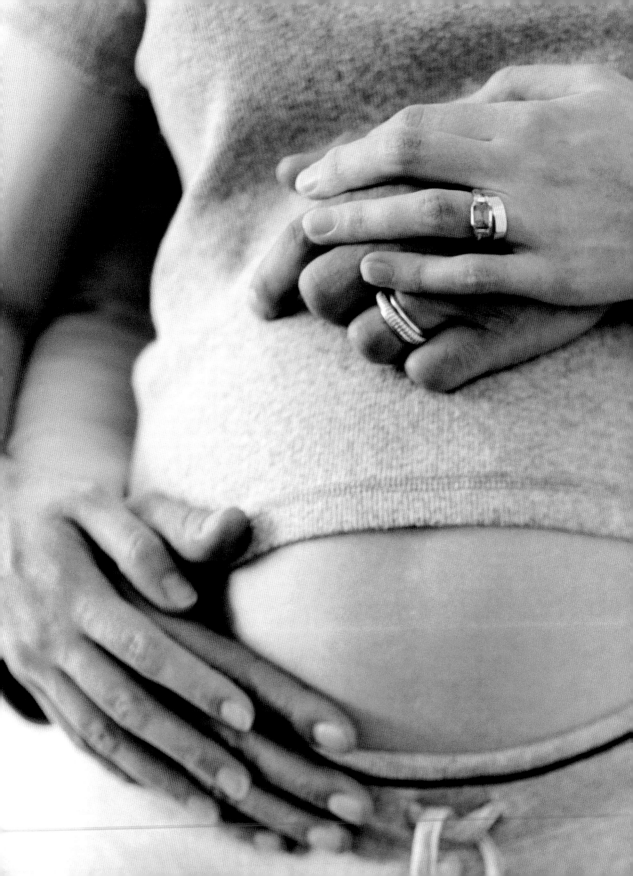

Your Journey through Pregnancy

The very beginning

Whether you are pregnant already or have just made the decision to have a baby soon, this is the beginning of one of life's most exciting and, occasionally, daunting experiences. This section helps you lay the foundation for an enjoyable pregnancy—offering insights into conception and answering questions about what is safe in pregnancy; the best foods to eat; how to stay fit over the coming weeks, and how to negotiate your rights and benefits as a parent. These are your guidelines for the journey ahead….

CONTENTS

The origins of life

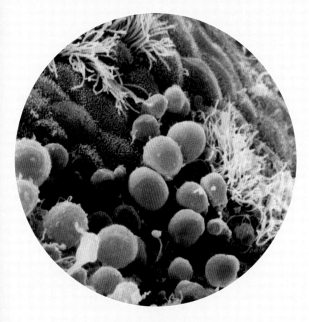

INSIDE THE UTERUS
The lining is perfectly primed for a life to begin. The granules of mucus (yellow) will provide nutrients for a newly fertilized egg.

MATURING SPERM
As sperm pass slowly along the epididymis—a coiled tube that lies behind the testes—they mature until they are ready to be ejaculated.

JOURNEY'S END
A swarm of surviving sperm lock onto the thick, inviting surface of the egg after their long journey up the fallopian tube.

> *"A mature egg drifts in the fallopian tube... conditions are perfect for fertilization."*

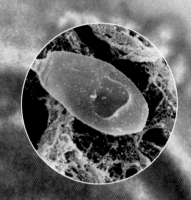

THE WINNER
Just one sperm penetrates the ovum's thick outer layer, and fertilization occurs.

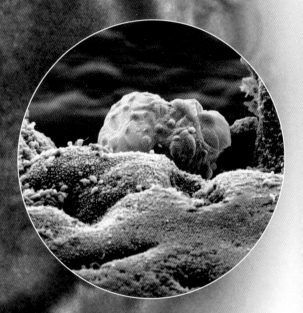

SIX DAYS OLD
The tiny cluster of cells, now known as a blastocyst, embeds in the wall of the uterus— a pregnancy begins.

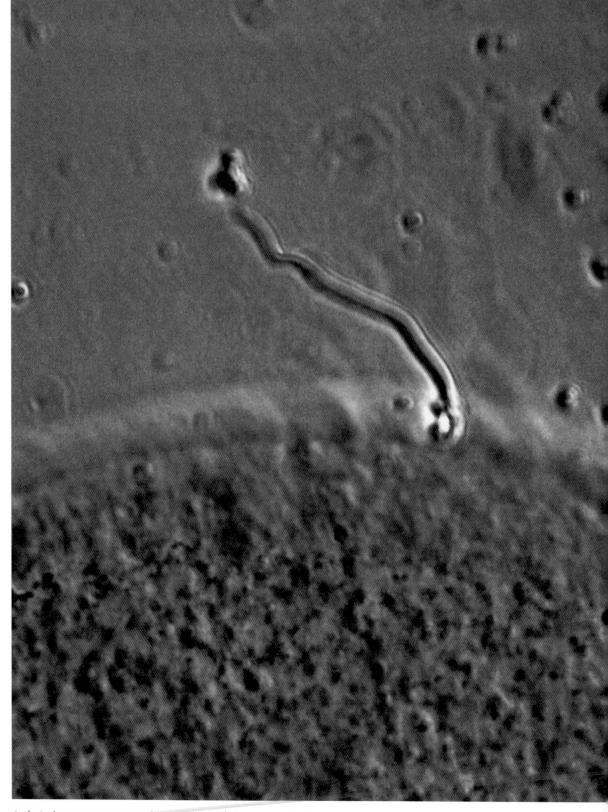

▲ A single sperm penetrates the outer layer of the comparatively huge ovum.

Conceiving a baby

WHEN YOU CONSIDER the complex sequence of hormonal events that are necessary for you to become pregnant, and the hoops that your partner's sperm have to go through to fertilize an egg, you will begin to realize why the expression "the miracle of conception" is not an overstatement.

The series of events that occur during your menstrual cycle need to be carefully orchestrated if a pregnancy is to occur. Conception is a bit like a jigsaw puzzle—just one piece missing and the puzzle can't be completed.

As soon as your period ends, a hormone called follicle stimulating hormone (FSH) is secreted into your bloodstream by the pituitary gland, a small bean-sized gland that lies just behind your eyes in your brain. This hormone acts on your ovaries, which sit at the end of your fallopian tubes and contain many thousands of eggs. You were born with about three million eggs, but by puberty these will have degenerated to about 400,000. All of these eggs are exposed to FSH, and it is not clear why only a few are recruited to start developing. On average, a woman will release only 400 or fewer mature eggs during her reproductive lifespan.

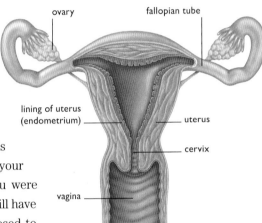

FEMALE REPRODUCTIVE SYSTEM The ovaries store and release eggs, which travel along the fallopian tube into the uterus. The narrow cervix connects the uterus to the vagina.

Ovulation

Each ovary releases one egg (or occasionally more than one) in alternate menstrual cycles, so ovulation takes place from only one ovary at a time. Eggs that are selected to start to mature do so within a fluid-filled bubble called a follicle, which starts to enlarge under the influence of FSH. Every month about 20 eggs start this process, but usually only one "dominant" follicle becomes fully mature and ovulates; the others shrivel, and the eggs within them are lost. The egg grows to one side of the follicle, surrounded by special cells called granulosa cells, which feed it with nutrients and also produce estrogen. This hormone stimulates the growth of the lining of the uterus (endometrium) as well as breast tissues, which is why breast tenderness is a common premenstrual symptom.

> *Conception is a bit like a jigsaw puzzle—just one piece missing and the puzzle can't be completed.*

As the level of estrogen in your blood rises, it feeds back a message to the hypothalamus (a control center in the brain) telling it that the follicle is mature and ready to ovulate. In response, the hypothalamus informs the pituitary gland to release a short sharp burst of luteinizing hormone (LH), called a pulse, which triggers the release of the egg about 36 hours later. The egg bursts from the follicle, which has grown to about the size of a quarter coin. This is called ovulation and it usually occurs around the 14th day of the menstrual cycle.

The mature egg has developed several important features. It contains chromosomes (which carry genetic information) at the right stage for further development and it is capable of taking in a single sperm while blocking the entry of any other sperm surrounding it. The egg is swept into the fallopian tube by delicate wandlike projections called fimbria, which resemble the fronds of a sea anemone. Tiny hairlike strands, called cilia, line the fallopian tube and these help the newly released egg to pass down the tube toward the uterus.

Meanwhile, the remaining cells in the ruptured follicle form a swelling in the ovary called the corpus luteum, which starts to produce the hormone progesterone. Like estrogen, progesterone has an effect on the uterus, breasts, and the hypothalamus and pituitary glands in your brain. In the uterus, progesterone makes the cells receptive to pregnancy by producing the nutrients needed to support a developing embryo and by thickening the lining of the uterus.

If the egg is not fertilized after ovulation, your production of luteinizing hormone begins to fall and the corpus luteum withers away. When the levels of both estrogen and progesterone have fallen below the threshold needed to maintain the lining of the uterus in a receptive state for embryo implantation,

THE RACE TO THE EGG ···

Ovulation occurs mid-menstrual cycle when the mature ovum (egg) bursts from its follicle.

Cohorts of sperm stream through the narrow opening of the cervix and into the uterus.

the blood-filled lining starts to disintegrate and your menstrual period begins. In a normal cycle, this usually occurs around 14 days after ovulation. The onset of your menstrual period signals that a new cycle of follicle growth can begin again.

Your partner's role in conception

This may seem to be the simpler part of the proposition, but the statistical probability of your partner's sperm meeting your egg is minute. On average a man ejaculates about 5ml (a teaspoonful) of semen containing 100 to 300 million sperm. Less than 100,000 make their way through the cervix; a mere 200 survive the journey up into the fallopian tubes to fertilize the egg.

Boys are not born with a full complement of sperm. Production starts around puberty and from this time onward, sperm are manufactured in the testes regularly at a rate of 1,500 per second. Each sperm has a lifespan of about 72 days. The testes deliver the sperm into the epididymis (a long coiled tube sitting at the top of each testis) and over the next two or three weeks they become capable of moving on their own and fertilizing an egg. From here the sperm move into the vas deferens. These tubes contract during male orgasm and transport the sperm out of the scrotum, past the seminal vesicles and the prostate gland (from which they collect seminal fluid) and into the urethra—the tube that travels between the bladder and the penis. During ejaculation, the opening to the bladder is shut off and the sperm are rapidly transported into the penis ready for their journey into the vagina.

Arriving in the vagina is not the end of the sperm's extraordinary obstacle race; they still need to navigate a considerable distance before they have an opportunity to fertilize an egg. The environment is fairly hostile to sperm:

A sperm negotiates the frond-filled lining of the fallopian tube on its way to the ovum.

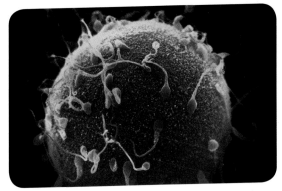

Mission accomplished—the successful surviving sperm swarm over the mature ovum.

66 *It is amazing to reflect that, even before you realize you are pregnant, the design of your baby's body is being programmed.* **99**

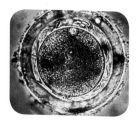

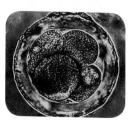

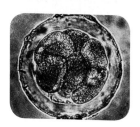

FIRST DIVISIONS After fertilization, the zygote divides rapidly. Within 36 hours it is made up of 12 separate cells.

vaginal secretions are quite acidic to prevent bacteria and other organisms from reaching the uterus and fallopian tubes and causing damaging infections. However, once in the vagina, the semen coagulates rapidly and this helps keep the sperm in the right place and protect them from vaginal fluids.

Within 5 to 10 minutes after ejaculation some sperm have entered the uterus and are heading toward the fallopian tubes. During this journey the sperm become hyperactive and attain full fertilizing capacity so that when they get close to the egg they are able to shed their cap (acrosome) and fuse with the egg. For the next 72 hours, more sperm from the pool at the cervix continue to enter the uterus. Once in the fallopian tubes, the remaining sperm (by now reduced to about 200) swim upward aided by muscular contractions of the uterus and tubes. This in itself is an extraordinary feat, since at the same time the egg is being propelled down toward the uterine cavity.

Fertilization

The process during which the sperm enters the egg, fuses with it, and the egg starts dividing takes about 24 hours to complete and usually takes place while the egg is still traveling down the fallopian tube.

Only the strongest cohort of sperm reaches the egg but it seems that the "winner" in the race is entirely random. Several sperm may bind to the surface of the egg and this stimulates them to lose their caps, exposing enzymes that digest their way through the outer membrane (the zona). However, only one sperm penetrates the oocyte, the innermost part of the egg, and fertilization occurs. The sperm tail, which has been so vital in propelling it to this point, is left outside and eventually disintegrates. The newly formed single cell that results is called a zygote and it now forms a thick wall around itself to prevent penetration by any other sperm. Your pregnancy has begun!

The zygote now begins to divide into more cells, called blastomeres, which number about a dozen by the third day. This tiny cluster of new life then takes about 60 hours to make its way to the uterus, by which time it is made up of about 50–60 cells and is called a blastocyst.

Already there are two distinct cell types—an outer layer of trophoblast cells, which will develop into the placenta, and an inner cell mass, which will eventually form the fetus. Two to three days later (about one week after fertilization) the blastocyst embeds itself in the lining of the uterus. It has now subdivided into approximately 100 cells and starts to produce the hormone human chorionic gonadotropin (HCG), which sends a signal to the corpus luteum to carry on producing progesterone. Having failed to do so, the lining of the uterus would begin to break down and menstrual bleeding would begin.

In the second week after conception, the trophoblast cells continue to enter the uterine lining and the inner cell mass develops into an embryo. It is only a dot, but has already started to differentiate into three different cell layers, called the germ layers, which will each become a different part of the baby's body. It is amazing to reflect that, even before you realize you are pregnant, the design of your baby's body is being programed.

HOW TWINS ARE CONCEIVED

Twins and triplets are conceived in two different ways:
▶ When two or more eggs are released and fertilized, the result is nonidentical twins.
▶ When one egg is fertilized by one sperm and then divides into two separate zygotes, the result is two separate embryos. These share identical genetic structures and will therefore become identical twins.

With both types of twin each baby develops surrounded by amniotic fluid, within its own amniotic sac. But because nonidentical twins are conceived from two separate eggs, each has its own placenta. Identical twins share one placenta, but each twin has a separate umbilical cord.

The number of twin pregnancies has doubled compared to a generation ago. Twins now make up around two percent of all

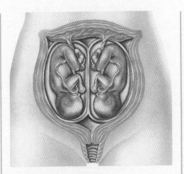

IDENTICAL TWINS share a placenta.

one fertilized egg divides

pregnancies—partly because of medical advances, such as in vitro fertilization (IVF) and fertility drugs (both of which carry an increased risk of multiple pregnancy), and partly because women are having babies later in life. Women over 35

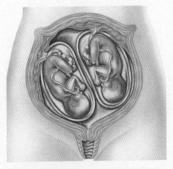

NONIDENTICAL TWINS have two placentas.

two separate eggs are fertilized

have a greater chance of conceiving nonidentical twins because they are more likely to release more than one egg per cycle. Nonidentical twins also run in families. However, there are no factors that increase the chance of conceiving identical twins.

Genes and inheritance

Genes control the growth and repair of our bodies and are also the code by which we pass on physical and mental characteristics to our children. At conception, your baby inherits a unique package of genes that will make him or her different from any other person.

The 30–40,000 or so genes that make up the human genome are arranged in pairs along chromosomes, long strands of genetic material that are found in the nucleus of virtually all body cells. Individual genes are single units of information inherited from the parents and occupy a specific position on the chromosome. Genes contain numerous small segments of DNA (the genetic blueprint), which provide codes for specific traits such as blood type and also dictate the specialized function of cells. In some cases, the presence or absence of a gene can predispose a person to a disease or protect against it. Genes are also either dominant or recessive. In a pair made up of one dominant and one recessive gene, the dominant gene prevails and this has effects on inherited traits such as eye color (see opposite) and also in the development of some genetic diseases (see p.144).

When a baby is conceived, both the mother's egg and the father's sperm contribute a single set of 23 chromosomes to the embryo, making a total complement of 46. Each egg and sperm carries a different mix of genes, which is why, except for identical twins (see p.21), every baby inherits a unique selection. However, because all cells are derived from this single fertilized egg, the same genetic material is duplicated in every cell of a baby's body.

BOY OR GIRL?

When an egg is fertilized by a sperm, the embryo that results has 23 pairs of chromosomes. The sex of your baby is determined by just one pair: chromosomes 45 and 46 (known as pair 23) are the sex chromosomes.
▶ The sex chromosomes are labeled the X (female) and Y (male) chromosomes. All eggs carry a single X chromosome, while sperm carry either a single X or a single Y in equal numbers. It is therefore the sperm that determines the sex of a child.
▶ When a sperm carrying the X chromosome fertilizes an egg, it forms an XX pair of chromosomes and the result is a girl.
▶ When a sperm carrying the Y chromosome fertilizes an egg, this results in an XY pair—a boy.

▶ Methods that claim to tip the balance in favor of a boy or girl rely on the fact that the Y-bearing sperm swim a little faster than the X-bearing sperm, but the latter survive a little longer. However, theories on the timing of conception to produce a boy or girl vary considerably. In spite of these methods, the ratio of boys to girls remains reassuringly normal.

BROWN EYES OR BLUE?

One of the most easily understood examples of inherited traits involving dominant and recessive genes concerns eye color. Because the brown-eye gene is dominant and the blue-eye gene is recessive, the gene for brown eyes always prevails.

Both you and your partner carry a pair of genes for eye color and this offers four possible combinations of genes in your own children. To work out your baby's chance of having blue or brown eyes you need to look at the eye-color genes you have inherited from your own parents. Even if you both have brown eyes, you may each have a blue-eyed parent and therefore be carrying recessive blue-eye genes. If these recessive genes combine, your baby will have blue eyes.

If you both have blue eyes, neither of you possess dominant brown-eye genes, so together you cannot have a child with brown eyes. (The brown-eye gene also covers hazel eyes, and the blue-eye gene includes gray and light-green eyes.)

Two brown-eyed parents In this example, the dominant brown-eye gene masks the recessive blue-eye gene so all children have brown eyes.

 PARENTS BR/BL + BR/BR

CHILD

- BR+BR child will have brown eyes
- BR+BR child will have brown eyes
- BL+BR child will have brown eyes
- BL+BR child will have brown eyes

One brown-eyed parent; one blue-eyed parent
The dominant brown-eye gene will prevail or two recessive blue-eye genes will combine to produce blue eyes.

 PARENTS BR/BL + BL/BL

CHILD

- BR+BL child will have brown eyes
- BR+BL child will have brown eyes
- BL+BL child will have blue eyes
- BL+BL child will have blue eyes

Two brown-eyed parents In this example, both parents carry a recessive blue-eye gene from their own parents so have a 1 in 4 chance of a blue-eyed child.

 PARENTS BR/BL + BR/BL

CHILD

- BR+BR child will have brown eyes
- BR+BL child will have brown eyes
- BL+BR child will have brown eyes
- BL+BL child will have blue eyes

Two blue-eyed parents Blue-eyed parents both carry two copies of the recessive blue-eye gene so cannot have a brown-eyed child.

 PARENTS BL/BL + BL/BL

CHILD

- BL+BL child will have blue eyes
- BL+BL child will have blue eyes
- BL+BL child will have blue eyes
- BL+BL child will have blue eyes

HOME PREGNANCY TEST Most kits are reliable and very easy to use.

Confirming that you are pregnant

As soon as you begin to suspect that you might be pregnant, you can find out with certainty. Now, testing is quick and easy and can give a reliable answer as soon as you have missed a menstrual period.

Most women opt for a urine test, which measures rising levels of the hormone human chorionic gonadotropin (HCG), produced by the blastocyst about a week after fertilization has taken place. Your gynecologist or family planning clinic can perform the test for you or you can use a home pregnancy testing kit, available from your local pharmacy. These are accurate and simple to perform and offer the great advantages of speed, privacy, and convenience. The test strip or wand is packaged in a plastic container that looks like a tampon applicator; before you use it, check the sell-by date and read the instructions carefully. Most kits advise you to perform the test several days after your first missed period, although it may be positive earlier. The result may be negative if you take the test too early and there is too little HCG in your urine.

Using a pregnancy testing kit

To use a kit all you have to do is use the toilet in order to pass fresh urine over the strip. The concentration of HCG is always highest in your early morning urine but newer test kits are sensitive enough to be used later in the day. When the HCG in your urine comes into contact with the test strip it causes a color change. At first a blue line or pink circle appears in the test window to confirm that the test kit is working. In a matter of minutes, a second blue line or a pink circle in the results window will indicate whether or not you are pregnant. The kits always provide a second test strip, so it is a good idea to check a borderline positive result with a follow-up test a week or so later. Sadly, some positive tests become negative later on because the embryo has not been able to implant successfully and will be followed shortly by your menstrual period.

Other ways of testing

In some circumstances, a blood test may be used to detect the exact level of HCG in your body within the first month of pregnancy, before it is detectable in your urine. If you have been undergoing fertility treatment you may want to know if it has been successful even before your menstrual period is due. Accurate measuring of HCG is also needed if an ectopic pregnancy (a pregnancy that develops outside the uterus) is suspected (see p.81).

Another method of diagnosing a pregnancy is using an ultrasound scan. Scans are not performed routinely in very early pregnancy, but they can be useful if you are unsure of your dates, have a history of miscarriage, or if you have symptoms or signs to suggest that you may have an ectopic pregnancy. Although there is little to see on the scan until about 10 days after your missed menstrual period, after this time it is usually possible to see the pregnancy sac in the uterus with a tiny fetal pole (a definite rectangular blob within the sac) and sometimes even a beating fetal heart.

Before the days of home pregnancy testing kits, most women had their pregnancy confirmed with an internal examination performed by a gynecologist after missing one or two periods. An experienced doctor can diagnose a pregnancy by noting the bluish tinge of the vaginal skin and cervix, the fact that the uterus and cervix are softer than normal and, by six weeks, that the uterus is slightly larger. These changes are due to the increased blood supply to all the pelvic organs. Internal examinations are not performed routinely, but every now and again an unsuspected pregnancy is diagnosed this way.

YOUR EMOTIONAL RESPONSE

I have had many conversations with women who have just discovered they are pregnant and their reactions range from elation to mild panic. Here is a selection of the feelings that are expressed most frequently:
▶ I don't believe it.
▶ It's wonderful.
▶ I'm ecstatic.
▶ Help—I didn't think it would happen this quickly.
▶ What have I gotten myself into?
▶ Can I afford a baby?
▶ I shouldn't have had that glass of wine after work.
▶ Will my job be safe?

▶ Why didn't I stop smoking last month, like I planned to?
▶ Where will I have my baby?
▶ What can I do to help my baby?
▶ Will my baby be normal?

Not all of these are positive reactions and this is entirely normal, so don't feel guilty about any negative thoughts that you may be having. Even if you have been planning your pregnancy for some time, you may feel daunted as the consequences begin to sink in. Added to the reaction is a cocktail of pregnancy hormones coursing through your bloodstream that is

enough to make anyone feel emotionally unpredictable. There is no doubt that being pregnant and bringing a baby into the world will be one of the most eventful and unpredictable periods of your life and there will be times when the sheer magnitude of it leaves you feeling overwhelmed.

However, like every other important life event, pregnancy will be much more enjoyable if you feel confident and in control of the situation. The only way to achieve this is to find out everything you can about it—I hope this book will help you do just that.

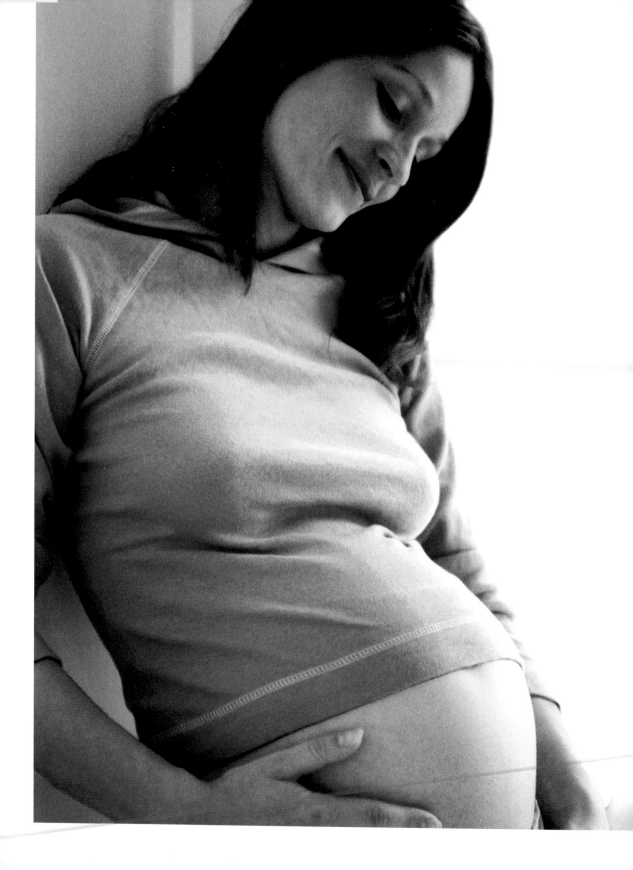

Staying safe in pregnancy

EVERY WORRYING PREGNANCY STATISTIC and media horror story comes into sharp focus as soon as you are pregnant, and you may be wondering just how much of your everyday life needs to be reevaluated. This section looks at some common concerns and will, I hope, help you weed out the myths and scares from the more sensible pregnancy precautions.

Of course it is impossible to eliminate risk from life, and pregnancy is no exception. In case you are feeling overwhelmed by worry, let's begin by getting things into perspective. About 4 in every 100 babies have an abnormality at birth (congenital abnormality). Most are due to genetic causes (see pp.144–45 and pp.416–19), with only a small proportion resulting from factors such as drugs, infections, and environmental hazards. Even if you are concerned that you have been exposed to something that may have harmed your baby (known as a teratogen), remember that you are more likely to be run down by the proverbial bus than to have a baby affected by one of these hazards in pregnancy.

...it is impossible to eliminate risk from life, and pregnancy is no exception.

An unexpected event

If you are pregnant unexpectedly or not as prepared as you hoped, you may be feeling a mixture of shock and disbelief coupled with anxiety. Instead of dwelling on what you were meaning to do or might have done to prepare for pregnancy, focus on adopting a healthier lifestyle now. Promise yourself that you will eat a well-balanced daily diet (see pp.43–49). Cut out alcohol and keep caffeine intake to a minimum. If you are a cigarette smoker, stop today.

If you are still in the first trimester, start taking folic acid supplements (see p.51) immediately, because these will help protect your baby from neural tube defects such as spina bifida (see p.146 and p.419). Although women are advised to begin taking the supplement while they are trying to conceive, starting now and continuing until the 13th week is definitely worthwhile—so don't think that it's too late.

Another possible concern is that this unexpected baby may have been exposed to all sorts of potential damage when you did not know that you were pregnant. You might be worried, for example, that you had too much to drink at a party, or that you were taking medicines before you knew that you were pregnant. In practice, the most commonly prescribed drugs are antibiotics and fortunately there are very few that harm a developing embryo (see p.35). There is also the possibility that you were using contraceptives around the time of conception and you may now be wondering if this has the potential to cause problems. I've addressed these concerns below, but let me begin with some reassurance: the reality is that most tiny embryos destined to make it through pregnancy and become a live, take-home baby are very resilient.

Smoking

If you are still smoking now that you are pregnant, you need to be aware of the problems that this can cause for your growing baby. During the first three months of pregnancy, smoking can directly reduce the ability of the developing placenta to invade the wall of the uterus and grow. If you continue

IF YOUR CONTRACEPTION FAILED...

If you conceived your baby while you were using contraception, you are probably wondering if this is going to cause problems for your pregnancy. In most cases, there is nothing to worry about.

▶ **If you were taking hormonal contraception**, such as the pill, simply stop taking them because they are no longer useful. Oral contraceptives contain estrogen to inhibit ovulation, and progesterone, which makes the cervical mucus less penetrable by sperm and cells in the lining of the uterus less receptive to implantation by the tiny embryo. Progesterone-only contraceptives and three-monthly injections, such as Depo-Provera, have similar effects on the cervical mucus and lining of the uterus. Now that you have become pregnant these effects are no longer a consideration. There is no evidence that the hormones in

modern-day preparations cause problems for the developing embryo and fetus.

▶ **Barrier methods of contraception** that involve spermicides are not harmful to a developing embryo, so there is no need to worry about them.

▶ **If you used postcoital contraception** such as the morning-after pill and are nonetheless pregnant, you may be distressed that the method has failed, but again, no harm will have come to the developing baby.

▶ **When pregnancy occurs with an intrauterine contraceptive device (IUD) in place** there is an increased risk of miscarriage because of the presence of a foreign body in the uterus; the inflammatory response that it causes; and the increasing risk of infection ascending along the IUD strings

in the vagina. If the strings or the device are visible on vaginal examination, it is best to remove the IUD. This does not further increase the risk of miscarriage but it does reduce the risk of miscarriage at a later stage in the pregnancy due to infection.

However, if neither strings nor IUD are visible, then it is best left where it is. It is unlikely to cause problems in a pregnancy that progresses to term since the baby develops within the fluid-filled amniotic sac. The IUD will be outside the sac and is usually delivered with the placenta.

▶ **If you have been sterilized** (tubal ligation) and find that you are pregnant, you should consult your doctor promptly. Your tubes have been damaged mechanically by the procedure, putting you at risk of an ectopic (tubal) pregnancy (see p.81 and p.423).

to smoke in later pregnancy, you will be reducing the supply of oxygen and nutrients to your baby and increasing your risk of premature delivery, placental abruption (see p.428), fetal growth restriction (see p.429), and stillbirth. Quit now; your obstetrician can help you. Be aware too, that if you are exposed regularly to cigarette smoke you become a passive smoker with similar risks to your baby's health. If your partner smokes, encourage him to quit. Recent bans on smoking in some bars and restaurants have helped, but avoid smoky places such as the areas outside train stations.

You may be offered carbon monoxide testing at your booking appointment and at subsequent visits. This is a noninvasive method to help assess whether someone smokes—some women find it difficult to admit they smoke, given that the pressure not to smoke in pregnancy is so high and that makes it difficult to ensure that they are offered appropriate support.

Alcohol

Any pregnant woman who has a heavy and regular alcohol intake is at increased risk of experiencing pregnancy complications. During the first few months of pregnancy, large quantities of alcohol can cause a distinctive pattern of abnormalities in the baby, known as the fetal alcohol syndrome (see p.435): these include failure to thrive after birth, damage to the nervous system, and poor childhood growth. Regular heavy consumption throughout pregnancy can have more toxic effects on the fetus. Because of this, you should not drink alcohol at all when you are pregnant.

If there were one or two occasions when you drank a bit too much in the weeks before you knew that you were pregnant, try not to worry unduly—but please make sure that you stop drinking now.

Recreational drugs

I have no intention of preaching about the rights and wrongs of taking recreational drugs but I do want to pass on some facts: drugs such as cocaine, heroin, and ecstasy have the potential to be a serious problem for a developing fetus. They all cross the placenta and enter the baby's bloodstream, increasing the risk of miscarriage (see p.431), placental abruption and premature delivery of a baby that is usually growth restricted (see p.429). After delivery, the baby may suffer serious withdrawal symptoms and possible brain damage, and will therefore inevitably be kept in the hospital under close observation for several weeks as well as social service intervention. If you want to have a trouble-free pregnancy and a healthy baby, avoid all recreational drugs.

> *…the reality is that most tiny embryos destined to become a live, take-home baby are very resilient.*

ENVIRONMENTAL HAZARDS

Potential hazards in the environment are of particular concern in early pregnancy when your baby's major organs and body systems are developing rapidly. Much of what you hear on this topic will be anecdotal, so I've included below evidence-based information on some of the most common concerns.

Although many environmental factors are blamed anecdotally for causing miscarriages and fetal abnormalities, most of them are not based on specific evidence. A brief summary of those that cause most anxiety is included here.

CONTACT WITH CHEMICALS

It is almost impossible to avoid all contact with chemicals in daily life, but you can try to minimize your exposure to them.

▶ **In your home:** Avoid inhaling vapors from gasoline, glue, cleaning fluids, volatile paints, household aerosols, and oven cleaners. Read the labels on any chemicals that you are thinking of using and if you are unsure about safety, don't use them. If you are stripping old paint and redecorating your house, keep the rooms well ventilated. If you think that the paint you are removing is so old that it may contain lead, delegate the decorating to someone else.

▶ **In the workplace:** A wide variety of solvents used in manufacturing industries can be the cause of problems to pregnant women who are overly exposed to them during their work. Fat-soluble organic solvents, found in paints, pesticides, adhesives, lacquers, and cleaning agents, can cross the placenta and inhaling these compounds may lead to complications.

Women at risk are those working in factories, dry cleaners, pharmacies, laboratories, garages, funeral homes, carpentry workshops, and artists' studios, to name but a few.

However, a recent study concluded that potential damage can be avoided if employers provide well-ventilated working premises and pregnant women are vigilant about wearing protective clothing and avoiding fume-filled areas.

One last point about chemicals: if your partner uses any of the above chemicals and/or vinyl chloride (found in plaster) in his work, avoid handling his clothing while you are pregnant. The same applies to clothes contaminated by pesticides.

X-RAYS

Large doses of ionizing radiation are known to cause problems in the fetus, and as a result, doctors are rightly concerned about the potential problems of X-rays during pregnancy.

However, it is important to know that modern X-ray machines emit much less radiation than they used to and focus much more accurately on the part of the body being investigated.

The only risk of fetal abnormality would occur if you had undergone a series of abdominal or pelvic X-rays (at least eight) before week eight of pregnancy, and even then, the risk is only 0.1 percent (one in 1,000 cases). A single chest or abdominal X-ray will cause no harm, so even if you unwittingly had one when you were pregnant, rest assured that it will not have damaged your baby.

The other point to mention is that there are some problems in pregnancy that really do need to be investigated with X-rays. If this proves to be the case, it is highly unlikely that they will cause harm.

If you work in a hospital you will be required to wear a protective lead jacket whenever you are in contact with X-rays. Female radiographers are usually moved to other duties in their department while pregnant, although the risks to their baby are negligible because of the strict safety rules that apply at all times.

COMPUTERS

Even if your job demands that you work in front of a computer screen for long periods of time each day, your baby is not at risk. Similarly, equipment that produces ultraviolet and infrared radiation, such as laser printers and photocopiers (and the microwave in the kitchen), is safe to use in pregnancy. There is also nothing to support the allegation that miscarriage and pregnancy problems are more common in women who live near electrical substations, electromagnetic fields, radio stations, and telephone poles. Ignore scary stories about them.

CELL PHONES

Our use of cell-phone technology has exploded over recent years, giving rise to concerns that it may lead to various health problems. Fortunately, the fears that exposure to radiofrequency radiation in the part of the head closest to the phone may result in an increase in brain tumors have been unfounded. Similarly, the suggestion that the babies of mothers who are regular cell-phone users during their pregnancies have more behavioral problems and hyperactivity syndrome in infancy and more likely to develop childhood cancers is weak. The specific absorption rate (SAR) value of your phone is the amount of radiation that your body absorbs when you are using the phone. The amount of energy emitted by your phone depends on how strong the signal is. The more powerful the signal the less energy your phone needs to work and the lower the SAR value. So only using your phone when you have a strong signal is one way to reduce the level of radiation you are exposed to.

SAFETY OF ULTRASOUND

One of the questions I am asked most frequently by my patients is whether they should be concerned about ultrasound, particularly if they have early pregnancy problems that require repeated ultrasound examinations. Happily, I feel confident in stating that ultrasound causes no problems for mother or baby, thanks to several studies that have looked carefully at this issue.

Although it has been suggested that ultrasound waves might cause changes in the membranes of cells, theoretically affecting the development of the embryo and later growth of the fetus, there is no scientific evidence to support this claim. A Swedish study found no association between repeated pregnancy scans and childhood leukemia. Several other large studies followed up babies who had been scanned on many occasions during pregnancy and found no serious developmental abnormalities.

On a more functional level, many women worry that the vaginal ultrasound probes used to scan in early pregnancy may cause them to bleed, or aggravate a pregnancy that is threatening to miscarry. This is not the case, and avoiding a vaginal scan may mean missing out on vital information about your pregnancy.

ULTRASOUND Even if you need repeated scans, they will not harm your baby.

Illness during pregnancy

Staying well can never be entirely within your control, but while you are pregnant, you need to be even more cautious than usual about avoiding infections—especially in the first three months.

Of course, this is much easier said than done and as many as 1 in 20 pregnant women contract an infection during pregnancy. This may seem an alarming figure, but the majority of these infections are completely harmless. Only a small number are capable of causing damage to the fetus or the newborn baby.

Since the majority of viral infections are caught from other people, the only way to avoid them completely is to become a hermit—not a practical solution. Young children are the most common reservoirs of infection, so it is sensible to minimize your contact with any child who has a rash or unexplained fever. If you work with young children, the very least you can do is insist that any child with a fever is sent home promptly.

AVOIDING INFECTIONS
It is impossible to avoid all contact with infections, especially if you already have a child.

Childhood infections

The two serious viral infections for pregnant women are common childhood infections—chickenpox (see p.412) and rubella, also known as German measles (see p.412). Chickenpox can cause miscarriage in the first eight weeks of pregnancy and carries a 1–2 percent risk of congenital abnormalities affecting the limbs, eyes, skin, and brain, together with growth problems in later pregnancy if it is contracted between eight and 20 weeks. Pregnant women exposed to chickenpox infection may be given varicella-zoster immune globin (within four days of exposure) to reduce the severity of the disease. If you are infected by the rubella virus for the first time when you are in early pregnancy, you are at risk of miscarrying; if the pregnancy continues, there can be severe effects on the fetus, including deafness, blindness, heart defects, and being developmentally challanged. Fortunately, this is now rare, since almost all women of childbearing age, and men, have either had the infection or been immunized against it. If you are not immune, you will be advised to have a rubella vaccination as soon as you have delivered your baby. Meanwhile, you will need to be more careful to reduce your chances of becoming infected. Although rubella is a live vaccine and best administered to women before they become pregnant, I have met several patients who have become pregnant almost immediately after having the vaccination. Happily, I can tell you that no abnormalities have been detected in their babies. Mumps, measles, and polio are unlikely to be a problem for pregnant women thanks to the vaccination programs in the US.

Whooping cough (pertussis) is a serious bacterial infection, which usually begins as an ordinary cold with cough which then deteriorates into spasmodic bouts of severe coughing and wheezing, making it difficult to breathe. The characteristic "whoop" sound comes from the effort of trying to draw it over the swollen vocal cords. The highest rate of infection is among babies and teenagers, who are also at greatest risk of serious complications, which is why routine vaccination is recommended at two, three, and 24 months of life, followed by a booster in the teenage years. Postpartum women are offered a combination vaccine that will protect them and subsequently protect her baby until his or her vaccinatons can work.

RUBELLA Virus particles are visible as pink spots in this blood sample.

Colds, flu, and stomach bugs

Catching your work colleagues' cough, cold, and flu is inconvenient but it is highly unlikely that your baby will be affected unless you develop a very high fever, which is a recognized cause of miscarriage in early pregnancy. In the US, we are exposed to a flu epidemic every couple of years and there is no doubt that these epidemics are responsible for a large number of deaths, both fetal and adult. If you develop a fever with a cold or flu, your doctor will advise you about safe drugs to help you reduce the fever quickly. Practical steps such as sponging with lukewarm water, lukewarm baths, and cooling fans are also effective.

It is recommended that all women in pregnancy get vaccinated. Your risk of developing flu may be a serious consideration.

Although worrisome, an upset stomach or gastroenteritis during pregnancy is best treated with rest and plenty of clear fluids and is unlikely to affect the baby.

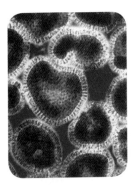

INFLUENZA The pink fringe around the core of each virus attaches to host cells.

Cytomegalovirus, toxoplasmosis, and brucellosis

Cytomegalovirus (CMV) is the most common congenital infection (i.e., affecting the unborn fetus or newborn infant). CMV affects approximately 2 percent of women in pregnancy and usually presents with nonspecific signs such as a general feeling of being unwell, tiredness, or lethargy. The time of highest risk of transmitting the infection to the fetus is the first trimester; it can cause serious neurological complications and hearing loss. The most common way of transmitting this virus is through bodily fluids such as saliva and urine, so it is advisable to thoroughly wash your hands after changing a

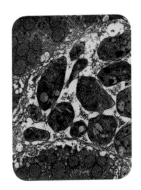

TOXOPLASMOSIS The single-cell, green parasites invade liver tissue (pink).

diaper and avoid kissing young children on the mouth or sharing drinks or utensils with them.

Most people catch toxoplasmosis because they have inhaled the infection from cat feces or have eaten unwashed vegetables, salads, or poorly cooked, infected meat (see p.50). Even though there is a good chance that you are immune to the parasite (particularly if you have a cat), always wash your hands carefully after handling a cat and avoid contact with strays. Have your cats wormed regularly and avoid contact with the litterbox. If you have to clean it yourself, wear gloves and wash the gloves and your hands afterward. You should also wear gloves to protect yourself from contaminated soil when gardening.

If you work on a farm or for a veterinarian, make sure to take precautions to reduce your risk of catching the bacterial infection brucellosis, which can cause a miscarriage. Avoid any work that involves lambing/calving or milking animals that have recently given birth, and be sure to wash your hands frequently.

Good hygiene around pets and other animals is especially important during the first trimester of pregnancy.

Other medical conditions

If you have an existing medical or mental health condition (see pp.408–411), such as a heart problem, diabetes, or bipolar disease you will need special care during your pregnancy and should see your doctor as soon as you know that you are pregnant (ideally, you should do this before you start trying to conceive). Do not, on any account, take it upon yourself to discontinue taking any prescribed medication—see your doctor.

If you have already had surgery before you discovered that you were pregnant, you may be concerned about the effect of the anesthesia or the procedure itself on your baby. In the very early weeks the risk of miscarriage is quite high, particularly if you have a laparoscopy or similar procedure that involves placing instruments into the uterus cavity or abdomen. This is why fertility specialists are always particularly careful to make sure that the woman is not pregnant when they perform invasive tests or diagnostic procedures. An emergency appendectomy also increases the risk of miscarriage. But that said, I have taken care of many women who discovered they were pregnant after having surgery and suffered no ill effects. If your pregnancy survived the surgery, rest assured that your baby will not have been harmed by the anesthesia.

TAKING MEDICINES IN PREGNANCY

The information below can be used as a guide for treating a minor problem, but keep all medicines to a minimum and see your doctor if a problem does not respond to treatment promptly.

ANTIEMETICS If you need to take an antiemetic due to severe morning sickness, your doctor will advise on the safest types available.

ANTIHISTAMINES Some prescribed and common over-the-counter types are contraindicated in pregnancy. If you suffer from hay fever, skin itching or another form of allergy, consult your doctor before taking antihistamines. Most over-the-counter preparations are safe, as are nasal steroid sprays.

ANALGESICS Acetaminophen is probably the best option during pregnancy. Avoid ibuprofen (or other nonsteroidal anti-inflammatory drugs), and ergotamine (migraine medicine).

ANTIDEPRESSANTS Do not stop taking antidepressants before talking to your doctor, who may advise you that you should continue to take them during pregnancy to avoid a relapse.

ANTIBIOTICS If you are prescribed antibiotics for an infection, the penicillin cephalosporin, ciprofloxacin, and Metronidazole family of drugs will not endanger your baby. If you are allergic to penicillin, erythromycin is a safe

alternative. The antibiotics below should be avoided because they can cause problems in early pregnancy.

▶ **Tetracyclines** can cause discoloration and deformity of a baby's teeth and bones.

▶ **Trimethoprim** is dangerous in early pregnancy because it blocks the action of folate and may cause poor growth in later pregnancy.

▶ **Chloramphenicol** is used to treat typhoid fever, and can cause abnormal blood reactions in the baby. The chloramphenicol contained in eye drops or ointment will not harm your pregnancy.

▶ **Streptomycin** can cause hearing loss in the fetus.

▶ **Sulfonamides** are broad-spectrum antibiotics that can cause jaundice in newborns and severe allergic reactions in mothers.

ANTIHYPERTENSIVES

If you are taking medication for high blood pressure, ask your doctor whether you should change your drugs, ideally before you become preganant. Never stop your medication without seeking advice.

ASPIRIN You may be asked to take a low-dose aspirin once a day if you are at an increased risk of preeclampsia; ideally, this is started from 12 weeks gestation.

LAXATIVES Constipation can be treated by adding fiber to your diet and by drinking plenty of water (see p.187). If you need to take a laxative, opt for the bulk-forming cellulose types such as Metamucil. Avoid senna-based laxatives: they irritate

the gut, which has the potential to trigger uterine contractions.

ANTACIDS Most antacids are effective and safe to use to treat heartburn and indigestion (see p.187). If you need iron tablets (see p.48), take them separately because antacids reduce their absorption.

DIURETICS Some fluid retention is to be expected in pregnancy and you should not take diuretics, including "natural" herbal types, in an attempt to deal with it. If your legs, feet, or fingers become very swollen this can be a sign of preclampsia (see p.426); see your doctor right away.

COLD AND FLU REMEDIES

Read the labels carefully because most contain antihistamines and caffeine, which should be avoided in pregnancy. Acetaminophen plus a hot drink are usually just as effective.

STEROIDS Creams containing steroids for skin disorders should be used sparingly but are unlikely to cause problems. Steroid inhalers for asthma (see p.409) are trouble-free. If you are taking oral steroids for a disorder such as Crohn's disease (see p.409), see your doctor for advice. Anabolic (body-building) steroids should never be used in pregnancy since they can have a masculinizing effect on a female fetus.

COMPLEMENTARY THERAPIES Always make sure you see a qualified practitioner (see pp.437–8).

TROUBLE-FREE TRAVEL

Judging by the number of questions I am asked about the safety of traveling when pregnant, this is an issue that is at the top of most women's list of concerns. However, there is no evidence to suggest that travel increases your own or your baby's risk of potential complications.

Whenever and however you travel, it makes sense to plan your trips. Long car trips can be tiring, you should arrange regular stops to stretch your legs and get some fresh air.

Driving should not be a problem as long as you are well and feel comfortable behind the wheel.

Do not listen to any scary stories about seat belts pressing on your uterus and unborn baby: you are both much safer behind a seat belt than not. However, three-point seat belts with the straps placed above and below your belly are best.

Your trip to and from work can sometimes turn out to be the most stressful of all, especially if you commute. If you have to stand during a long bus or train trip, ask for a seat. Even in early pregnancy, you may feel tired and nauseous and just as desperate to sit down as you will be later on in pregnancy.

AIR TRAVEL

Freak accidents and terrorist attacks aside, I do not think that air travel needs to be avoided in normal pregnancies. However, long-haul flights do increase the risk of deep vein thrombosis. Compression stockings are effective at reducing this risk. You may have heard that the reduced cabin pressure on long-haul flights can adversely affect a pregnancy, but this notion is difficult to explain scientifically. Your baby is surrounded by a thick muscular uterine wall and a pool of amniotic fluid, which protect it from any physical damage. Furthermore, the changes in your blood circulation and respiratory systems will ensure that your baby receives enough oxygen and nutrients, even if the oxygen supply in the outside world is a little lower than usual.

BUYING FRUIT Peel fruit or wash it thoroughly with bottled water before eating.

However, if you have had previous miscarriages, there are considerations about travel in general, and particularly air travel, to think about. Traveling by air will not cause you to miscarry, but you may not want to expose yourself to a situation where you would worry that you might be to blame if something goes wrong. Miscarrying in the middle of a flight, or in a country where you do not speak the language, is something that needs to be avoided. Most airlines do not accept pregnant women after the 34th week of pregnancy but not, as is often assumed, because the mother's lower oxygen levels are likely to induce labor. It is because some 10 percent of women deliver prematurely, and airlines want to avoid the risk of dealing with this in midflight. (For more advice on travel in the third trimester, see p.251.)

FOREIGN TRIPS

If you are planning a trip abroad, make sure you have valid medical insurance and find out where to seek medical help while away. Also, check with the Centers for Disease Control (see p.437) or ask your doctor whether or not you are at risk of endemic infections.

If at all possible, avoid traveling to countries where malaria is rife because the high fever that accompanies the infection increases the risk of miscarriage. During pregnancy, you are more likely to catch malaria, which will often be more severe and unpredictable than normal, because your immune response alters in pregnancy.

If your trip to a malarial zone is unavoidable, make sure that you take prescription antimalarial medicine and continue taking them after you leave the country and for as long as is recommended after your trip.

New antimalarial drugs are continually coming on the market so talk to your doctor about which drugs are safe in pregnancy. As a general rule of thumb:

▶ Chloroquine will not harm your baby, nor will proguanil if you take folate supplements at the same time.
▶ Mefloquin and pyrimethamine should be avoided in the first 12 weeks of pregnancy but if you have taken these drugs already, do not panic. A malarial infection is far more likely to cause problems in pregnancy than the drugs you have

IMMUNIZATION Your doctor will advise which travel vaccines are safe.

taken to prevent infection. For the latest advice on antimalarial medicine and immunization, contact the Food and Drug Administration (see p.437).

TRAVEL IMMUNIZATION

You also need to check which, if any, immunizations are recommended for your destination and whether any of them should be avoided during pregnancy. Your altered immune response in pregnancy makes immunization more unpredictable; as a rough rule of thumb, vaccines prepared with live viruses are best avoided. Cholera, polio, rabies, and tetanus vaccinations are considered safe for pregnant women, but the safety of yellow fever and typhoid vaccinations is unclear.

TIPS FOR STAYING WELL WHILE AWAY

▶ Be meticulous about washing your hands before you eat.
▶ Drink plenty of bottled water, particularly in hot climates.
▶ Avoid unpasteurized foods and shellfish.
▶ Resist the temptation to buy delicacies from street vendors (they may have been recycled

or reheated and may contain harmful bacteria).
▶ Peel fruit and avoid watermelons, which are often spiked and then immersed in water to make them larger and juicier.
▶ Refuse ice in drinks—it may be made from contaminated water.

Diet and exercise

LIKE MANY WOMEN, YOU PROBABLY STARTED TO IMPROVE your diet and fitness levels while you were trying to conceive—but if that is not the case, this is the time to start. Both you and your baby will reap the benefits. Most of the information you need is included here, but there are also additional features on diet and exercise specific to your particular stage of pregnancy in the sections covering each trimester.

From the moment you conceive, your body provides all your baby's nutrients. From now until the birth, everything you eat will be broken down into molecules that pass from your own bloodstream into your baby's via the placenta. Until after the birth, you will breathe and eat for your baby so it is important that you try to do the very best that you can. This is a major responsibility and I am well aware that the subject of what you should be eating during your pregnancy can be a source of anxiety. Without becoming obsessive about amounts and portions, you need to know what to eat to meet all those basic requirements as well as which foods are better avoided. You should find most of the answers that you need in this section.

Being pregnant also raises a host of new questions about fitness. Whether you have an established exercise regimen or whether you are not quite as committed to going to the gym or walking to work as you hoped that you might be, you are likely to be wondering what kind of physical activity you should be doing now. All sorts of myths suddenly crawl out of the woodwork: don't get too hot—it harms the baby; don't jump around—you'll miscarry; don't do sit-ups—you'll cramp your baby's growth. Ideas such as these, based on a mixture of anecdote and misconception, used to be passed down through the generations without much questioning of the wisdom behind them. However, we are fortunate that our knowledge and thinking on the subject of exercise in the first trimester has evolved considerably in recent decades. So while in your mother's day, the idea of playing a game of tennis during early pregnancy would have raised eyebrows, today, doctors advise the majority of pregnant women that moderate exercise throughout the nine months is beneficial. However, you should avoid high-impact sports, contact sports, and scuba diving.

...all sorts of myths about diet and exercise suddenly crawl out of the woodwork.

Your pregnancy weight

As with all pregnancy topics, well-meaning but unsought advice on diet will be given to you in bucketfuls—much of it conflicting and confusing. The one thing that is certain is that this is a time in your life when you really do need to eat sensibly and healthily.

The first thing to understand is that the old adage of "eating for two" is a recipe for problems. You really do not need to eat more than 2,000 calories a day (the same as any nonpregnant woman) in the first trimester. As soon as your body recognizes you are pregnant, your metabolism changes to make the most efficient use of food. Extra calories will not benefit your baby and will be deposited as fat, which may be hard to get rid of after your baby is born. In the last trimester you may need to increase your calorie intake, but by no more than 200–300 calories per day, which is the equivalent of a banana and a glass of milk.

At the other end of the scale, women who have spent most of their adult lives dieting may require a major shift in their psychology during pregnancy. The Western obsession with body image can make it difficult for some women to accept that they will be putting on about 29 pounds (14kg). Now that you are pregnant, fad diets are out, and real meals and healthy portions of food are definitely in. Studies of babies born to African women who had malnutrition during pregnancy and while breastfeeding found them to have an IQ significantly lower than normal, which should convince anyone that there is no place for vanity dieting in pregnancy. If you eat sensibly and exercise regularly, you can have your former body back very quickly after your baby is born.

If you are gluten intolerant, it is important to supplement your diet with calcium, iron, folic acid, and vitamin B_{12}. Seek support from your midwife or dietician if you have concerns about any aspects of your diet or weight gain.

> *Now that you are pregnant, fad diets are out, and real meals and healthy portions are definitely in.*

Your ideal starting weight

Ideally, you should be of normal weight when you are trying to conceive, because being significantly under- or overweight can affect fertility. If a woman's body mass index (see right) falls below 17, her periods usually become irregular or cease, which often leads to delays in conceiving. There are also good reasons for addressing a weight problem now that you are pregnant. An underweight pregnant woman is at risk of becoming anemic (see p.424), delivering prematurely or having a small for dates baby (see p.429). So if your BMI is less than 18.5 please consult your doctor who will advise you on how to reduce these risks (see p.42).

Obesity

In the developed world the problem of obesity has now reached epidemic proportions and is having a major adverse impact on our general health and well-being. In the US, the increased prevalence of obesity has resulted in more than 50 percent of women of reproductive age becoming overweight (BMI 25–29.9) or obese (BMI 30 or greater) There is no doubt that obesity increases the risk of immediate obstetric and fetal complications at every stage of pregnancy, delivery and the postpartum period. However, obesity also leads to an increased risk of long term health problems for both the mother and her baby.

If you are significantly overweight or obese, you may have experienced conception delays and will definitely be at greater risk of miscarriage, fetal abnormalities and less accurate prenatal assessments of your baby's growth and well-being, for example on an ultrasound scan. Later complications such as high blood pressure and preeclampsia (see p.426), gestational diabetes (see p.427) and thrombosis (blood clots in your legs) are also more likely and as pregnancy progresses you will become increasingly uncomfortable and unnecessarily tired. All of these prenatal problems, together with the fact that your baby is more likely to be larger than average, may lead to an increased risk of complications at the time of delivery. Obese women have higher rates of induction, failed induction, poor progress in labor and are more likely to experience anesthesia complications, require a cesarean section or instrumental vaginal delivery, or suffer an extensive perineal tear when compared to lean women.

After the birth, obese women are at greater risk of hemorrhage, endometritis, wound infections and the breakdown of the wound, prolonged hospital stays and deep vein thrombosis (see p.424). Their babies are very likely to be overweight, which increases the risk of shoulder dystocia and other birth injuries, low apgar scores and the need for admission to a neonatal unit. In addition these infants are at greater risk of stillbirth and neonatal death.

Obesity results in a disordered metabolic state. Body fat is not inert padding—it is an organ that is actively producing many chemicals and breakdown products which cause all kinds of short and long term damage to our bodies. The more overweight you are the greater the risks, which is why any interventions that reduce an overweight woman's prepregnancy BMI and limit her weight gain during pregnancy should be actively encouraged. So if you are overweight, being actively careful with your diet can really help to improve the situation and will have no adverse effects on your baby's nutritional intake.

YOUR BODY MASS INDEX

The best way of knowing if you are either underweight or overweight at the beginning of your pregnancy is to figure out your body mass index (BMI).

▶ **This is calculated using a simple equation:** Divide your weight in pounds by height in inches squared then multiplying by a conversion factor of 703.

▶ **Most doctors use the following BMI ranges:**
Underweight: below 18.5
Normal: 18.5–24.9
Overweight: 25–29.9
Obese: over 30
Dangerously obese: over 40.

▶ **Here is the calculation** for a woman who is 5'5" (65 inches) tall, weighing 150lb: 150 divided by 65x65 then multiply by 703 = 24.96. This woman's BMI has a BMI of 24.96—just within the normal range.

YOUR WEIGHT GAIN OVER 40 WEEKS

Although every woman's pregnancy is different, these guidelines for the amount and rate of weight gain over a normal pregnancy will help you determine whether you are putting on weight too slowly or too fast in each trimester.

If you are of average height and weight, you can be expected to gain about 25–35lb (11–16kg) during the 40 weeks of a normal pregnancy. As you can see from the chart, there will be times in your pregnancy when you put on weight faster than others. During the first trimester, your weight should change very little, and most of your weight gain will take place from the second trimester onward. Of course, the chart can only be used as a guide: allowances need to be made for individual variations between women and between pregnancies. For example, if you are expecting twins, you are likely to gain in the region of 35–40lb (16–18kg).

WHAT MAKES UP THE WEIGHT?

Your overall weight gain can be roughly divided into two different categories:
▶ **the weight of your baby**, the placenta, and amniotic fluid
▶ **your own increased body weight** to support the pregnancy. This includes the increasing weight of your uterus and breasts, the increase in your blood volume, and your increased fat reserves, combined with a variable amount of water retention.

EXCESS WEIGHT

Most of the weight gain mentioned is governed (for the most part) by the natural needs of the pregnancy. However, the amount of fat that a pregnant woman deposits on her body depends on what and how much fat and carbohydrate she eats. Gaining 6lb (3kg) in body fat is to be expected—90 percent of this will occur during the first 30 weeks and most of it will be shed during breast-feeding. However, if you lay down excess fat during pregnancy, breast-feeding will not reduce it— it will continue to sit there.

AVERAGE WEIGHT GAIN

Baby	7–9lb/3–4kg
Placenta	1½lb/0.7kg
Amniotic fluid	2lb/1kg
Maternal fat	5½lb/2.5kg
Increased blood + fluid	3lb/1.5kg
Water retention	5½lb/2.5kg
Breasts	1lb/0.5kg
Uterus	2lb/1kg
Total	**25–35lb/11.5–16kg**

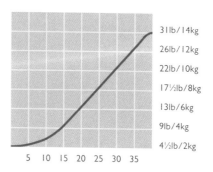

Weight gain over 40 weeks In a normal pregnancy, there is little gain in the first trimester. Most women then gain about 1½–2lb (0.7–1kg) a week until the last 1 or 2 weeks, when little further weight gain occurs.

The perfect pregnancy diet

The first thing to make clear is that there are two types of pregnancy diets—the textbook diet and the realistic diet that reflects your everyday life. This section is all about the second kind.

A healthy diet in pregnancy contains the right balance of carbohydrates, protein, fats, vitamins, and minerals. In this section I have listed the foods that are the best source of these essentials and explained why they are important in building your baby's body and helping keep yours healthy. Having said that, I know that there will be times when there is some discrepancy between the ideal diet and what you actually eat, particularly in the early months when nausea and vomiting can make well-planned eating nearly impossible. Pregnant women are sometimes made to feel guilty and obsessive about diet and it is not my intention to fuel those sentiments. Women are led to believe that they will be harming the health and intelligence of their unborn child if they eat too much chocolate or have a day when the most they can face is a cracker, a boiled egg, and a few pickled onions. I want to stress that if you try to eat as varied a diet as possible, you will be doing all that is necessary.

Pregnant women are sometimes made to feel guilty and obsessive about what they eat.

Protein

From the very early weeks of pregnancy, your protein requirement will increase by 15–20 percent. Proteins are the essential building blocks for your own muscles, bones, connective tissues, and internal body organs and have the same key role in the healthy growth of your baby. They are made up of 20 different amino acids, 12 of which are produced within the body and so are termed nonessential amino acids. The other eight amino acids are referred to as essential amino acids because they have to be supplied from food. These are found in the right proportions in first-class proteins—such as meat, poultry, fish, eggs, and dairy products. Amino acids are also found in nuts, grains and legumes, soybeans, and tofu; these are known as second-class proteins because they need to be eaten in combinations to supply the necessary amino acids. If you are vegan or lactose-intolerant you will need plenty of these.

Not all protein-rich foods have the same nutritional value: some are higher in fat than others and some have additional vitamins and minerals. So while red meat may be a first class source of protein, it contains much more fat than chicken. Fish is low in fat and high in vitamins, and oily fish, such as sardines, contains essential unsaturated fatty acids, which are particularly beneficial for your developing baby's brain. However, due to the danger posed by mercury to the developing fetus, the FDA advises women who are or might become pregnant to restrict their consumption to 12oz (375g) per week of cooked fish, selecting a variety of smaller fish rather than the long-lived, larger ones such as swordfish and marlin. You need 2–3 servings of protein-rich foods each day, a typical serving being 3oz (85g) of red meat or poultry; 5oz (150g) of fish; 1–2oz (30–60g) of hard cheese; or 4oz (125g) of legumes, grains, or cereals.

Carbohydrates

Pregnancy is not the time to be on a low-carbohydrate-based diet because starchy foods are a useful source of energy. There are two types of carbohydrates, simple and complex. Simple carbohydrates, as found in cookies and sodas, are high in sugar and are of little nutritional value. They give you a quick burst of energy because they are absorbed quickly, but only have short-term benefits. The exception is fructose in fruit. Since fruits are a good source of vitamins, minerals, and fiber, try to eat about five portions a day. Complex carbohydrates are found in starchy foods such as pasta, whole-grain bread, brown rice, potatoes, and legumes. They are the mainstay of a healthy diet, providing a slow and steady release of energy because their starch has to be broken down into simple carbohydrates before it can be absorbed into your bloodstream. Unrefined whole-grain flour, rice, and pasta are good sources because they retain valuable vitamins and minerals and also have a high fiber content to help prevent constipation. Try to include 4–6 servings a day of any of the following: one slice of whole-grain bread; 2–4oz (60–125g) of whole-wheat pasta, brown rice, and potatoes; 2oz (60g) of cereal.

RECOMMENDED PREGNANCY FOODS: HOW MUCH TO EAT EACH DAY

3–4 servings of vegetables such as broccoli and salad

4–6 servings of carbohydrates such as whole-grain bread

2–3 portions of protein—fish, chicken, red meat, and legumes

Fats

Although you should try to limit your fat intake when you are pregnant, don't cut out fat completely. Fats have nutritional value in that they help build cell walls in the body and supply important vitamins needed by your growing baby. Generally speaking, dietary fats divide into less healthy saturated fats, which come from animal sources, and healthier unsaturated fats, sourced from vegetable oils and fish, which are important for the development of your baby's nervous system. Typically, fried foods and fatty meats and meat products such as burgers and pastries are loaded with unhealthy saturated fats. Eat too many of these and you will pile on extra pounds as well as encourage the buildup of fat deposits on the lining of your blood vessels, which increases your risk of heart disease later in life. For a healthy fat intake, trim the fat off meat, use butter sparingly and choose low-fat varieties of dairy products and low-fat hard cheeses. Wherever possible choose foods that are rich in unsaturated fats.

> *Some of the virtually no-fat varieties of yogurt are loaded with sugar…*

Dairy products

Eating dairy products provides you with a balanced mixture of protein, fats, calcium and vitamins A, B, and D. Milk is a great standby in pregnancy so if you enjoy milk-based drinks include plenty of the low-fat variety, which contains the same amount of calcium and vitamins as whole milk. Low-fat dairy products are preferable to the full-fat varieties, but some of the virtually no-fat varieties of yogurt are loaded with sugar and as a result are high in calories. Try for 2–4 servings of dairy foods each day; typical portions are 1–2oz (30–60g) of hard cheese or 1/3 pint (200ml) of low-fat milk.

2–4 servings of low-fat dairy products such as milk

1–2 servings of iron-rich foods such as eggs and fortified cereals

5 portions of fruit to provide fiber and vitamins

ESSENTIAL VITAMINS AND MINERALS

Your own health and that of your developing baby relies on a regular supply of vitamins and minerals, most of which have to be obtained from the food you eat. The chart below is a guide to the best sources. Vitamins and minerals tend to break down during cooking and processing so choose food that is as fresh as possible.

	BEST SOURCES	BENEFITS
VITAMIN A	orange fruit and vegetables—peaches, melons, mangoes, apricots, carrots, and peppers; green vegetables; egg yolk; oily fish such as herring	antioxidants; important for eyes, hair, skin, and bones; help fight infections; but can be toxic in excess (do not take take supplements or eat liver or any organ meat)
B VITAMINS	poultry, pork, beef, and lamb; cod; dairy products; eggs; brewer's yeast; green vegetables, such as Brussels sprouts and cabbage; nuts, especially pecans, peanuts, and walnuts; fortified cereals; whole-grain bread and pasta; oranges; mangos; bananas; avocados; figs; sesame seeds	helps energy production and protein release from food; maintains healthy skin, hair, and nails; essential for nervous system and brain function; assists in the production of antibodies to fight infection and of oxygen-carrying hemoglobin in red blood cells (B_{12} supplements may be needed if you don't eat meat or dairy food)
FOLIC ACID	green vegetables such as broccoli, spinach, and green beans; fortified cereals, legumes (peas and chick peas); and yeast extract such as Marmite	helps prevent neural tube defects in the fetus; aids red blood cell formation and protein break down in the body (400mcg supplement recommended daily)
VITAMIN C	kiwifruit, citrus fruit, sweet peppers, black currants, potatoes (especially the skins), and tomatoes	assists growth and repair of tissues (skin, teeth and bones); aids iron absorption; antioxidant properties

Folic acid Early in pregnancy, eat foods rich in folic acid such as green beans and legumes.

Fetus Folic acid helps prevent neural defects at a very early stage of fetal development.

Food sources Vitamins and minerals are absorbed most effectively from food with little danger of overdose—any excess is eliminated naturally.

	BEST SOURCES	BENEFITS
VITAMIN D	oily fish such as herring, salmon, and sardines; butter; margarine; cheese; also provided through exposure to natural light	enhances calcium absorption; increases rate of mineral deposits in bone; can be toxic if taken in excess (do not take supplements)
VITAMIN E	eggs; nuts, such as hazelnuts, pine nuts, and almonds; sunflower seeds; green vegetables such as broccoli and spinach; avocado; vegetable oils	maintains healthy skin, nerves, muscles, red blood cells, and heart; important antioxidant—protects against free radicals, which can damage body tissues
IRON	red meat; eggs; apricots, raisins, and prunes; canned sardines, crab, and tuna in oil; fortified cereals; sesame seeds (liver and kidneys are rich in iron but should be avoided in pregnancy)	essential for oxygen-carrying hemoglobin production in red blood cells of mother and fetus; builds and maintains muscles
CALCIUM	dairy products; eggs; small bony fish such as sardines; soy products; most nuts; fortified cereals; leafy green vegetables, especially broccoli	essential for healthy bones, teeth, and muscles in mother and fetus; also helps with conduction of nerve impulses
ZINC	beef; seafood; nuts; onions; corn; bananas; whole-grain foods (iron-rich foods block absorption)	necessary for growth and energy; aids healing of wounds; supports the immune system

Calcium Boost your intake of calcium-rich foods such as cheese in pregnancy.

Fetal skeleton Calcium builds fetal bones and teeth and preserves the health of your own.

Vitamins

Vitamins are vital for good health, both yours and your baby's. There are five of them in total—vitamins A, B, C, D, and E and, with the exception of vitamin D, all have to be obtained from the food we eat. We need about 40 minutes of light (not necessarily sunlight) per day to produce sufficient vitamin D. Vitamins A, C, and E are antioxidants, which have an important role in protecting the body from the damaging effects of free radicals—chemicals produced from the waste products of the oxygen that we breathe. Antioxidants help mop up the chemicals and stop them from injuring our body cells.

Some vitamins such as the B and C vitamins are not stored by the body, so you need to make a particular effort to ensure that you have an adequate daily intake when you are pregnant. Furthermore, some vitamins, such as vitamin C, break down quickly when exposed to air and heat, which is why raw fruits or vegetables are preferable to the cooked variety in many instances. Similarly, frozen vegetables contain more vitamins than the canned variety, which lose most of their vitamins during processing.

Minerals

> When it comes to minerals, your baby will behave like a parasite, taking everything it needs...

Your diet also needs to include sufficient amounts of minerals and trace elements, the most important of which are iron, calcium, and zinc. These chemicals make an important contribution to the way our bodies function, but like vitamins, cannot be synthesized by the body, so have to be supplied from food. High levels of iron and calcium are particularly important during pregnancy because they help support your baby's development. When it comes to minerals, your baby will behave like a parasite, taking everything it needs from your body's reserves, so don't leave yourself feeling tired and unwell because they have been depleted by the demands of your developing baby.

Iron is essential for oxygen-carrying hemoglobin in red blood cells and also helps maintain healthy muscles. It clears from your body very quickly, so you need to eat iron-rich foods (see table p.47) on a daily basis. Although requirements vary from woman to woman (and the best way to boost iron supplies is a matter of debate), what is absolutely certain is that your blood volume will double during pregnancy, so you will need more than your usual level of iron to protect yourself from anemia (see p.424).

Animal sources of iron are better absorbed than the iron found in fruit and vegetables, although iron-rich food such as apricots and prunes have the advantage that they are also a good source of fiber, which helps prevent constipation. Pregnant women used to be advised to eat large amounts of liver because it is rich in iron, but it is now known that the high levels of vitamin A may cause birth defects. So avoid liver and liver products such as sausage and pâté.

An important point is that iron absorption is improved when iron is taken with acidic drinks that contain vitamin C, such as orange juice. On the other hand both milk and antacid drugs reduce the absorption of iron, so have a glass of milk between meals and increase your intake of iron-rich foods if you need to take antacids for indigestion.

You do not need to take over-the-counter or prescribed iron supplements unless you are anemic when you start your pregnancy or you develop iron deficiency during it. There is no need to expose yourself to the common side effects of iron—constipation or stomach upsets—unless a blood test establishes that you have a low blood count.

Zinc is a mineral that helps encourage growth and is also important in the immune system, wound healing, and digestion. Zinc absorption can be blocked by iron, especially supplements, so if you have to take these, try not to take them at the same time as zinc-rich foods (see table).

Calcium supplies are important for your bones and teeth and while you are pregnant, your baby will be drawing all its calcium from you. This process starts very early on—your baby's bones begin to develop from the fourth to sixth week—so ideally you should make sure that your calcium intake is high before pregnancy and remains so throughout. The old saying that "for every pregnancy, a mother loses a tooth" has its origins in the fact that low levels of calcium led to poor teeth in later years. Even if you think you are getting enough calcium, see your dentist. All dairy products are good sources of calcium, as are nuts and leafy vegetables, particularly broccoli (see table, p.47). An increasing variety of cereals and juices are now fortified with calcium and you might consider switching to these while you are pregnant.

Salt is used in high quantities in many processed and prepared foods, since it acts as a natural flavor enhancer and preservative. You can add a little salt when you cook, but remember that too much can be harmful as it encourages fluid retention, which in turn may lead to high blood pressure.

VEGETARIANS & VEGANS

If you are vegetarian, vegan, or allergic to dairy products, you need to discuss with your doctor the best way for you to obtain adequate levels of iron, calcium and vitamin B_{12}, in particular, as you may well be deficient in one or all of these elements. Vitamin B_{12} exists naturally only in animal products, although it is found in yeast extract spreads and fortified cereals. If you are a vegan, you may need to take supplements during pregnancy and while you are breast-feeding.

▶ **For a vegan diet** you can combine different plant proteins in order to receive your full complement of essential amino acids. For example, a handful of nuts or a portion of peas can be combined with a serving of rice or corn. To ensure you have enough iron, eat extra portions of green beans, cereals and dried fruits such as apricots, raisins, and prunes.

▶ **If you are vegetarian**, you should eat more dairy products and eggs to ensure you keep up your absorption of proteins, vitamin B_{12}, calcium, and iron.

AVOIDING FOOD POISONING

When cooking involves little more than a ready-made meal and a microwave, it is easy to become relaxed about hygiene. In pregnancy you are more vulnerable to infections and need to be more careful about what you eat and how you prepare it.

Much of what follows may seem excessively cautious and I want to emphasize that if you follow basic rules of hygiene, such as always checking sell-by dates on labels, and throwing away any suspect food, you would be very unlikely to be infected by any troublesome or potentially dangerous bacteria.

Having said that, some foods are known to carry bacteria that can be harmful to you, and sometimes to your baby as well. A severe episode of food poisoning can trigger a first-trimester miscarriage so it is worth being extra vigilant, especially during the first three months. As well as the cautions below, avoid eating raw shellfish.

SALMONELLA BACTERIA

These bacteria are found primarily in eggs and chicken. Infection with salmonella causes vomiting, nausea, diarrhea, and fever—effects that usually kick in within 12–48 hours of eating the affected food. Your baby will not be affected, since the bacteria do not cross the placenta, but you should see a doctor as soon as you suspect an infection.

Salmonella bacteria are killed by heat, so cook chicken thoroughly and avoid food containing undercooked or raw eggs such as mayonnaise, ice cream, cheesecake, and chocolate mousse. When you cook eggs make sure the whites and the yolks are both solid.

Note that free-range chicken and eggs are not free of salmonella but a smaller percentage of them carry the bacteria than caged hens.

LISTERIA INFECTION

Although listeria infection is rare, it can have fatal effects on your unborn baby (see p.413). It is found occasionally in pâtés, unpasteurized soft cheeses, such as brie and camembert, and blue cheeses. It can also be found in other chilled foods and prepared meals that have not been stored with very high standards. Avoid these foods in pregnancy and stick to hard cheeses and those made from pasteurized milk (the pasteurization process kills the bacteria). Cottage cheese and mozzarella are also safe. Drink only pasteurized and UHT milk. If you are in any doubt, boil milk before you drink it. Avoid unpasteurized goat's or sheep's milk and products.

E-COLI

This is another relatively rare bacteria that can be extremely dangerous to anyone who contracts it, since it can ultimately lead to kidney failure and death. It is primarily found in cooked meats and pâtés that have been stored at the wrong temperature. Again, I would advise you to avoid pâtés and to make sure that you buy cooked meats from hygienically proven sources. Always check sell-by dates and if you are in any doubt about the freshness of a particular food, throw it out.

TOXOPLASMOSIS

This is a relatively common infection, which produces only mild flulike symptoms but which can be highly dangerous for your unborn baby (see p.413). It is derived from a parasite that is present in animal feces, particularly cats' feces, and is also found in raw or undercooked meat. While you are pregnant, make sure that you cook your meat thoroughly and wash your hands after preparing it. You should also wash all vegetables and fruit thoroughly (see opposite).

Should I take vitamin supplements?

The only vitamin supplement that you should be taking routinely during pregnancy is folic acid—one of the B vitamins that is particularly important during the first trimester of pregnancy because it reduces the risk of the fetus developing neural tube defects, such as spina bifida (see p.146 and p.419). There is some evidence that folic acid may also reduce the risk of other types of congenital abnormality and birth defects.

Start taking a 1g tablet of folic acid (available from your pharmacist) as soon as you know you are pregnant; if your pregnancy was planned, you may have been taking this dose for three months or more. A normal diet of folate-rich foods (see table, p.46) will provide you with another 200mcg of folate, bringing your total daily intake up. If you have had a previous pregnancy affected by a neural tube defect, are taking epileptic drugs, or have a BMI greater than 30, you should be advised to take a higher dose of 5mg while trying to conceive and for the first 12 weeks of your pregnancy. This dose is only available by prescription.

Two other supplements you might take (only on your doctor's advice) are vitamin D and iron. Do not rush off and buy other supplements "just to make sure;" if you eat a healthy diet you get all you need. All pregnant women are given a prescription for prenatal vitamins. I should add that your body absorbs vitamins and minerals more effectively from food—folic acid being the only exception. By taking supplements you may overdose on certain vitamins, such as vitamin A, which can be harmful to the fetus in high doses.

> *If you eat a healthy diet you will be getting all the vitamins and minerals that you need.*

FOOD HYGIENE BASICS

▶ **Wash your hands** before and after handling food; take extra care with raw meat and poultry.
▶ **Make sure that raw food**, especially raw meat, is stored separately from prepared food to avoid contamination.
▶ **Use separate chopping boards** and knives for raw meats and wash both with very hot water and detergent afterwards.
▶ **Wash fruit carefully before eating**. Most fruit is treated with pesticides and ethylene oxide,

which is used to ripen fruit, has been shown to cause miscarriages.
▶ **Thoroughly wash** all vegetables and salads, peeling and topping carrots to remove all traces of soil. Wash your hands after picking fruit or gathering vegetables.
▶ **Take care to defrost frozen food**. In the microwave, turn the food around a few times so that it is fully defrosted.
▶ **When you are reheating** previously cooked food in the

microwave ensure that it is piping hot everywhere. Do not reheat a frozen dish a second time.

What to drink

Pregnant women need to drink eight large glasses of water—the equivalent of 2 quarts (1 liter)—every day. If you cannot bear the thought of drinking so much water, try herbal teas instead. Fruit juice and milk are also good but not as effective as water. The better hydrated you are, the less tired you will be, since dehydration causes muscle fatigue, which in turn leads to a general feeling of fatigue. You are also less likely to suffer from constipation. Think of your kidneys as a waterfall—the more water you flush down them the better.

HERBAL TEAS These teas do not contain caffeine, which is best avoided in pregnancy, and many women find that they prefer the taste of them.

Caffeine in tea, coffee and soft drinks is now on the list of substances "to be avoided in pregnancy," even though studies linking a moderate intake of caffeine to problems during pregnancy are, at best, inconclusive. However, a recent Italian study reported that more than six cups of coffee a day was accompanied by a higher rate of miscarriage. That said, so many women stop drinking coffee during their first trimester, they rarely need any encouragement to cut down their intake. The problem with caffeine is that it acts as a diuretic, draining you of much-needed fluid. It also interferes with the absorption of iron, calcium, and vitamin C, thus negating all your healthy eating habits. Chocolate also contains caffeine, and however much you tell yourself that it also contains high levels of brain-boosting magnesium, it also contains too much sugar and fats to make it a food you should eat in anything other than small quantities.

Is alcohol safe?

Many of us drink alcohol—albeit in small amounts—and one of the most common questions I am asked is whether it is safe to drink in pregnancy and if so, how much? There is no known safe level of alcohol use in pregnancy. So I would advise you to avoid alcohol consumption completely throughout your entire pregnancy.

There is no doubt that drinking large quantities of alcohol during this period (especially binge drinking) can cause fetal abnormalities. Alcoholic mothers put their babies at risk of fetal alcohol syndrome (see p.434), the effects of which are numerous and severe: the main ones include intrauterine growth restriction (see p.429) and facial anomalies, followed by learning and behavioral difficulties. When the child is a little older, other neurological and behavioral problems come to light. This syndrome is more common than people think and serves to illustrate the point that alcohol and pregnancy (particularly in the first trimester) do not mix. Not surprisingly, nature has an unerring way of pointing us in the right direction in early pregnancy, and many women cannot face the taste or smell of alcohol.

Exercise in pregnancy

There are particular reasons why being fit during pregnancy is a good idea. The next nine months are going to be a testing time for you physically so if you can boost your general fitness now, you are likely to cope much better with pregnancy, labor, and birth.

Why do we now advise women to keep fit during pregnancy and in what way do you and your developing baby benefit? It used to be thought that during exercise, blood flow to the uterus was reduced and this put the baby at risk, but many studies have shown that even strenuous exercise is of no risk to the baby, especially in the first half of pregnancy. Sufficient uterine blood flow is directed to the placenta during exercise and the uterine circulation is also able to increase the amount of oxygen it extracts from the mother's blood to compensate for the exercise requirements.

Female professional athletes who do endurance training at high levels during their pregnancy tend to have poor weight gain and low-birth-weight babies. They are usually advised to reduce their training schedules in later pregnancy. Studies that involve monitoring the fetal heart rate show that usually there is a short period after exercise when the fetal heart rate and temperature rise, but that in neither instance is this a danger to the baby. The conclusion is that in pregnancy you can exercise at up to 70 percent of your capacity (see box) without putting your baby's growth at risk.

Many women I meet are worried that if they continue to exercise in early pregnancy, they may precipitate a miscarriage. This concern has deep-rooted origins in the Western world: in the past it was generally assumed that strenuous exercise might disturb the implantation of a young embryo. The reality is that a pregnancy that is at risk of miscarrying will miscarry, even if the mother wraps herself in blankets or spends the day in bed. Similarly, a healthy pregnancy that is destined to be successful will continue to be so and it will take a lot more than a few bursts of physical exertion to dislodge it. So do remember that exercise is very unlikely to cause problems in the development of early pregnancy. If you doubt these words, then

A SAFE HEART RATE

▶ **To calculate your personal safe heart rate** when exercising, subtract your age from 220 and then figure out 70 percent of that figure. This will give you the rate of heartbeats per minute that you should be aiming for during exercise.
▶ **So, for example, if you are 30**, you will calculate 70 percent of 190 (220–30), giving you an exercising heart rate of 133 beats per minute.
▶ **To find your heart rate during exercise**, take your pulse for 20 seconds, then multiply by three. Alternatively, consider buying a heart rate monitor, available from most sporting goods stores.

just reflect on the fact that, worldwide, most pregnant women are exposed to much more physical exertion in their day-to-day lives than the majority of Western women. Furthermore, the population explosion continues in countries where pregnant women do hard physical labor in early pregnancy.

Worldwide, pregnant women are exposed to much more physical exertion in their day-to-day lives than the majority of Western women.

Benefits of exercise

As for the benefits of exercise in pregnancy, they are unquestionable. Doing any form of aerobic exercise, such as swimming, cycling, or brisk walking, makes the heart pump faster than normal and so increases stamina. This means that your heart muscle is able to pump blood round your body more efficiently and work less hard at times of physical stress. This will be of particular value in late pregnancy when you find yourself climbing stairs with a great deal of extra weight around your middle and also during labor.

Anaerobic activity, such as pilates, yoga, isometrics, or working with weights, is more resistance-based and is designed to exercise the muscles as well as improve flexibility. If you exercise in a gym, you probably do a mixture of aerobic and anaerobic exercises. The golden rule is that any exercise you are already doing on a regular basis is safe to continue throughout the pregnancy, as long as there are no pregnancy complications such as pain or bleeding. You should try to exercise regularly and make sure you warm up and cool down gradually and stop exercising if you experience any pain or discomfort. You are the best person to know how hard you can push your body and will instinctively know what your limits are. If you are feeling exhausted, it is your body's way of telling you to slow down.

What kind of exercise?

If you normally come out in a rash at the very thought of doing any exercise but feel that for once in your life you are prepared to give it a try, then do not be overly ambitious. If you stick to activities you will a) enjoy b) not become bored with after two months and c) be able to fit into your weekly schedule, you cannot go wrong. Be realistic and you will be much more likely to succeed. Obviously, you would not be advised to take up a strenuous or difficult sport at this stage, so consider some of the activities below, all of which are safe to pursue throughout pregnancy. Whether you exercise regularly or not, you should think about including one of them in your fitness schedule.

Yoga is excellent for flexibility and general well-being (see opposite page). Choose a class that focuses on prenatal exercises rather than a general class. You will find yoga is a wonderful form of exercise in the latter stages of pregnancy when the delivery begins to loom closer and you are in need of

YOGA IN PREGNANCY

Many of my patients extol the virtues of yoga for
its all-around effects on strength and flexibility. Before
you start, make sure that the yoga class you choose is
appropriate for pregnant women. Relaxation techniques
are very safe to do.

One of the core components of all forms of prenatal exercise is relaxation. Learning a variety of comfortable positions and breathing exercises will enable you to recover your equilibrium when you feel overwhelmed by your pregnant state and will give you some strategies to help you deal with the more testing times during labor and birth.

The key elements of any form of relaxation exercise are these:

▶ Find a time and place where you will not be disturbed and lie down in a position in which you can remain comfortable for 5-10 minutes.

▶ Close your eyes, keep your head straight and your neck relaxed, and drop your lower jaw.

▶ Breathe evenly and effortlessly.

▶ Clear your mind of worrying thoughts or lists of things to do—this will take practice.

▶ Before you resume your activities, take a few deep breaths, roll on to your side and get up slowly.

BUTTERFLY POSE Put your feet together against the wall and drop your knees to each side without straining, as wide as feels comfortable.

BREATHING TOGETHER

The belief that different types of breathing can help women deal with the pain of labor is central to many childbirth philosophies (see pp.248–9).

Joined breathing (shown left) is a yoga exercise designed to foster harmony between you, your partner and your growing baby. Sit in a position where you can both place your hands comfortably over your baby and uterus. As you breathe together deeply and slowly, focus on the muscles of your uterus, being careful not to tense the muscles in the rest of your body.

GENTLE LIFTS Head and shoulder lifts can be a good alternative to sit-ups during pregnancy.

relaxation. Yoga has the added benefit that it teaches you to control your breathing—a useful technique for the big day itself.

Swimming is the perfect three-in-one exercise, enabling you to develop stamina, flexibility, and muscle tone all at once. Like most women, you are likely to find swimming wonderfully relaxing in pregnancy, especially in the later stages when all of your extra weight is supported by water. Exercise in water has the added bonus of eliminating the possibility of overstraining any part of your body. The hardest part is getting back onto dry land and feeling the full force of your pregnant weight again.

Walking is easy to fit in as part of your daily routine—walking an extra 10 minutes on your way to or from work, or walking to a nearby grocery store rather than automatically jumping into your car are good ways to start. You will find that your ability to walk longer distances reduces later on in your pregnancy.

Cycling is useful because your leg joints are not under excessive strain from your increased weight. It develops stamina and tones your lower body, so it is an activity that can be continued right through to the birth.

Adapting your exercise regimen

During the first trimester, it is safe to do sit-ups, especially if you are used to doing them on a regular basis, although you should not do any more than usual. If you are not used to doing them, try gentle head and shoulder lifts:

• Lie flat on your back with your legs bent at the knees and your feet flat on the ground, shoulder-width apart. Your arms should be alongside your body.

• Raise your arms toward your knees while at the same time lifting your head and shoulders about 6in (15cm) off the ground. Do this exercise 10 times. If you have toned abdominal muscles they will help to support your belly, taking pressure off your back muscles and spine. The most usual advice is that you should stop sit-ups in the fourth month of pregnancy when your waistline begins to expand and the exercise may feel uncomfortable.

When you become pregnant, the level of the hormone relaxin begins to rise. The purpose of this hormone is to relax the ligaments, especially those in the pelvic area, in preparation for childbirth. Consequently, all your ligaments become looser, but this leaves you more prone to injury, if you put too much strain on them. Unlike muscles, which do regain their shape after the birth, your ligaments do not recover if you stretch them excessively. Keep this in mind if you do any exercise that involves lifting weights. After the first trimester, you should reduce the weights you lift to protect your pelvis and your lower spine to avoid putting any strain on your abdominal area. This advice also applies to any form of heavy lifting, whether it be grocery bags or toddlers. Of course, at times this will be unavoidable, so make sure you adopt a safe way of doing it (see p.193).

As your pregnancy advances

There are certain activities, such as squash, skiing, and horseback riding, which are not particularly advisable for women to continue beyond the fourth month of pregnancy, largely because they carry the risk of high impact injuries. However, if you are used to these sports and do not have a history of miscarriage, there is no reason why you shouldn't continue to do them during the first three months but check with your doctor first.

Sports such as tennis and golf can be continued as far into the pregnancy as you feel comfortable —you may have to stop when your enlarging belly starts getting in the way but it won't be because of a potential risk to your baby.

Your ability to withstand the biggest physical upheaval that your body is ever likely to know— labor and childbirth—is certain to be enhanced if you are reasonably fit, as will your ability to return to your pre-pregnancy shape. I hope you will not rush to the gym and work out until you are red in the face and aching all over, but rather will make a vow to keep active and fit during the months ahead. Remember, too, that although moderation and regularity are key, so is your enjoyment.

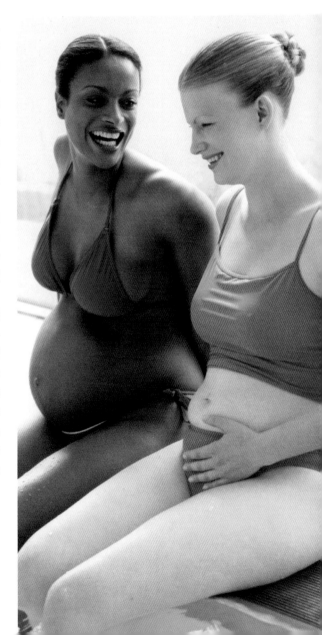

SWIMMING Whether you exercise in the water, swim lengths, or float and relax, your buoyancy supports your extra weight and prevents damage to your ligaments and muscles.

Work and maternity rights

BEFORE YOU INFORM YOUR EMPLOYER AND COLLEAGUES at work that you are pregnant, you need to give some careful thought to the issue of your maternity rights. Now is the time to find out what benefits are available to you and make plans for working during your pregnancy, timing your departure, and adapting your life after your baby is born.

Many women are justifiably worried that they will be perceived differently at work from the moment they announce they are pregnant, whether this is a first or a subsequent pregnancy. Some employers assume that you will be less committed to your career from this point onward and caught up in the exciting developments of your pregnancy. As a result, you may have serious concerns about your current job security and about your future career.

Although maternity rights and benefits in the US are not nearly as generous as the rest of the developed world, certain rights are protected by federal law, such as the Family Medical Leave Act (FMLA). This legislation requires employers with 50 or more employees to allow up to 12 weeks of unpaid leave for various family and medical reasons, including childbirth, and guarantees that you will return to the same position or one of equal seniority and salary when you return. Other legislation protects your right to working conditions that are safe both for you and your unborn child through the creation of the Occupational Safety and Health Administration (OSHA), and guarantees that your pregnancy must be treated as a medical condition for which you cannot be discriminated through the Pregnancy Discrimination Act (PDA). Further legislation has been enacted in twenty states, the District of Columbia, and Puerto Rico that extends the provisions of the FMLA by varying degrees.

Making plans

If you are aiming to return to your job after the birth, figure out some clear ideas about what you want to do and explain your plans to your employers or colleagues. It is entirely within your rights to say that you intend to return to work after a certain period of maternity leave and subsequently change your mind. However, try not to vacillate between the two options—it won't inspire your employer to have long-term confidence in you.

YOUR RIGHTS AND BENEFITS

Rights for parents have improved in recent years and are changing all the time—this is a summary of the current situation. However, your state, union, or employer may offer a better deal so find out what parenting packages are available.

HOLDING ON You can work right up until your due date, if you'd like. But if you are off sick with a pregnancy problem in the last four weeks, your boss can insist you start your leave.

FAMILY MEDICAL LEAVE ACT (FMLA)

If a company has 50 or more employees working within a 75-mile radius, and you have been employed by that company for the past year and have 1,250 hours, you can expect:
▶ 12 weeks of unpaid leave to take care of your new baby or newly adopted child
▶ continuation of any health insurance that you held over this 12-week period
▶ reinstatement into the same job or one that is equal in rank and salary.

EXPANSIONS OF FMLA PROVISIONS

Expansions of the federal law by local legislatures extend coverage to employees of smaller companies, increase the amount of leave, and, in California, Minnesota, New Jersey, New York, Rhode Island, and the District of Columbia, make some provision to pay mothers a portion of their regular salary through temporary disability insurance.

Alaska Applies to companies with more than 21 employees within 50 miles; Leave: 18 weeks over a one-year period, for birth or adoption.

California Leave: the period of medical disability (6–8 weeks up to a maximum of 4 months); Pay: 55% of weekly wage for 6 weeks, funded through employee payroll tax.

Connecticut For companies with more than 75 employees, leave is extended to 16 weeks over 2 years.

District of Columbia Applies to companies of 20 or more. Leave: up to 16 weeks of medical leave, plus up to 16 weeks of family leave. Pay: beginning in 2020, employees will receive a percentage of their weekly salary for 8 weeks.

Hawaii Covers all working women; fathers and adoptive parents not covered by the FMLA can take 4 weeks family medical leave per year.

Iowa Applies to companies with 4 or more employees. Leave: period of medical disability, up to 8 weeks.

Kentucky Extends FMLA coverage to all workers adopting a child under age 7. Leave: 6 weeks.

Louisiana Applies to companies with 25 or more employees. Leave: up to 6 weeks for a normal pregnancy; up to 4 months for a pregnancy with complications.

Maine Applies to companies with at least 15 employees. Leave: up to

10 weeks in 2 years for family and medical leave.

Massachusetts Applies to companies with 6 or more employees. Leave: 8 weeks.

Minnesota Applies to companies with 21 or more employees who have worked for at least for 12 months and at least part time. Leave: up to 6 weeks for the birth or adoption of a child. Pay: Low-income families who would otherwise qualify for child-care subsidies can take advantage of the At-Home Infant Care (AHIC) program to fund their own maternity leave by taking care of their babies themselves and keeping the subsidy.

New Hampshire Applies to companies with 6 or more employees. Leave: period of medical disability due to pregnancy and childbirth (usually 6–8 weeks).

FEDERAL LEGISLATION IN A NUTSHELL

▶ **Family and Medical Leave Act (FMLA)** Guarantees up to 12 weeks of unpaid leave for certain family and medical reasons.

▶ **Pregnancy Discrimination Act (PDA)** Pregnancy must be treated as a potentially disabling medical condition and cannot be used as a cause for discrimination.

▶ **Occupational Safety and Health Administration (OSHA)** Working conditions must be safe for both pregnant mother and fetus.

New Jersey Covers women who work for companies with at least 50 employees, who have worked more than 1,000 hours in the past year. Leave: 12 weeks over 2 years, not to exceed 6 weeks in a year. Pay: pregnancy disability insurance pays up to two-thirds of an employee's weekly salary for 6 weeks after the delivery date.

New York Leave: 12 weeks. Pay: A new program is increasing the percentage of pay in phases over the next few years. In 2018 employees receive 50% of the average weekly wage for 8 weeks; in 2019 they will receive 55% for 10 weeks; 2020 they will receive 60% for 10 weeks, and in 2021 employees will receive 67% of their weekly pay for 12 weeks.

Oregon Covers people who work for companies with at least 25 employees, who have worked more than 25 hours per week in the past 180 days. Leave: up to 12 weeks of family medical leave, 12 weeks for pregnancy disability, and 12 weeks for caring for a child with a serious illness at home.

Rhode Island Leave: 12 weeks. Pay: the Temporary Caregiver Insurance program provides up to 4 weeks of wage replacement benefits.

Tennessee Covers women employees. Leave: up to 4 months for maternity disability.

Vermont Applies to companies with 10 or more employees.

Washington Applies to companies with at least 8 employees. Leave: 12 weeks. Pay: In 2020 Washington will begin offering paid family leave benefits.

Wisconsin Leave: recipients get 6 weeks for the birth or adoption of a child.

PATERNITY LEAVE

Paternity leave is rarely paid in the US, although a few companies do offer new fathers paid time off, and a handful of states now mandate paid parental leave that extends to the father. However, an increasing number of fathers are using vacation or sick time to take care of their newborn children, or are taking advantage of the 12-week unpaid leave provided under the FMLA.

PARENTAL LEAVE

Both parents now have the right to take time off work to take care of their child, make arrangements for the child's welfare, or just to strike a better balance between work and family life. Parents who have worked for at least a year with one employer are each entitled to take up to a year unpaid parental leave per child, up until a child's fifth birthday.

REGULAR UPDATES

Rights and benefits are changing constantly. For updates, your best source may be the American Baby web site, at www.AmericanBaby.com.

MAKING PLANS
Exploring options for your future working life will make you feel more secure and give your colleagues confidence in you.

Think through how long you would like to take off (you can always change the exact dates later on) and what you would like to happen to your job while you are away. There may be an established procedure in your workplace for maternity cover but if there is not, you can help by suggesting how best to achieve this cover. Does an additional outside individual need to be hired? Can an existing colleague switch roles for a while? Don't undersell yourself by suggesting that someone can fill in for a couple of days a week or a few hours a day while you are away. First, this begs the question of what you have been doing for a full five days. Second, you do not want to return after your leave to find an overflowing inbox or a list of things that have not been done.

If you are self-employed, the ground rules are much the same as they are for an employed woman: be clear about your future plans; tell people how long you are taking off; and inform them of any maternity cover that will be arranged.

Part-time and flexible work

You may have already started to think about whether you want to work full time, part time, or flextime after your baby is born. Many employers will now consider a request to work part-time or flexibly to enhance the work/ life balance, but such policies do not give you an automatic right to do so. You may discover that a precedent for part-time or flexible work has already been established in your workplace, but even if your preferred working option has never been tried before, it does not mean that it is impractical. Present all the arguments in favor of your suggestion—most importantly, how it will benefit the employer as a whole—and draft a specific plan. Make sure that you have already thought out solutions for any problems that might be raised and be prepared to listen to any counter-offers that your employer might present. Also, keep in mind that both you and your partner are entitled to take a year's parental leave (unpaid) during your child's first five years if you've worked for your present employer for at least one year.

When work is a hazard

Although some women worry that their work is potentially harmful to their unborn child, hard evidence showing the link between certain occupations and a risk to an unborn child is scant. The few types of work that are potentially hazardous are strenuous jobs involving long hours; shift work; jobs that require a lot of lifting or standing for long periods of time; and work that exposes women to certain chemicals or harmful environmental factors (see pp. 30–31).

• If your job involves a lot of standing, look for ways to reduce this in the second half of the pregnancy. This is not because long periods of standing will harm your baby, but they will exacerbate common pregnancy problems such as fatigue, backaches, varicose veins, and swollen legs and ankles.

• If your work involves heavy lifting, make sure that you always lift correctly by bending your knees and keeping your back straight (see p.193). After the second trimester, try to reduce the weight of what you lift and the frequency, or request a switch to a less strenuous role until you stop work.

• Doing long hours or shift work can be very tiring and affect you and your baby. If you are working shifts, remember to eat well, take breaks, and get enough sleep. You need to conserve your energy, particularly in the third trimester, so if there is no alternative to your current job, consider starting maternity leave a little earlier. Although not specific to pregnancy, working long hours and doing shift work also increases the risk of diabetes and cardiovascular disease.

• Some jobs, such as those in dentistry, medicine, and certain industries, involve exposure to chemicals, X-rays, or other toxic substances. Your doctor or employer should be able to clarify whether your work is putting your baby at risk. If this is not possible, you have the right to be suspended at full pay.

LEAVING AND RETURNING TO WORK

▶ **How can I maximize my employee benefits?**
You should check with your employer to determine the company's maternity leave policies before you get pregnant. In some cases, you need to have worked for an employer for a set period before you become eligible under the Family Medical Leave and Pregnancy Discrimination Acts.

▶ **How should I negotiate the best maternity-leave package possible?**
You should discuss your maternity-leave plans with your employer approximately 12 weeks before your baby is due. Make sure that you check whether you are covered by any local laws or collective bargaining agreements that expand upon or limit the provisions of the Family Medical Leave Act.

▶ **How can I protect my rights under the Family Medical Leave Act (FMLA)?**
You should inform your employer in writing of your intentions at least 30 days before you go on maternity leave.

▶ **When should I finalize the details of my maternity leave with my employer?**
You should give your employer a letter summarizing your plans and what you have agreed upon, about 10 weeks before your baby is due.

▶ **How can I maintain good relations with my employer and hold my job until I return?**
About 6–8 weeks after the birth, you should tell your employer whether you expect to return to work on the date anticipated.

▶ **WEEKS 0–13**

The first trimester

During the first trimester, your baby will evolve from a cluster of cells into a recognizable fetus measuring about 3in (80mm). All the major organs, muscles, and bones will be formed. Until the placenta becomes mature enough to take over, your pregnancy is supported by maternal hormones, which also contribute to early symptoms such as nausea and fatigue. Although you may not look pregnant in the first trimester, you will almost certainly feel pregnant.

CONTENTS

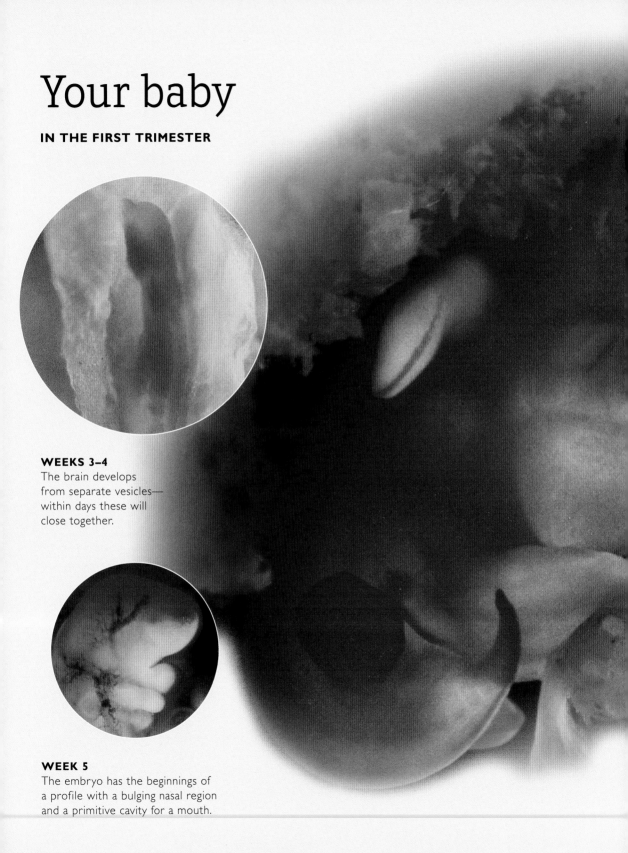

Your baby

IN THE FIRST TRIMESTER

WEEKS 3–4
The brain develops
from separate vesicles—
within days these will
close together.

WEEK 5
The embryo has the beginnings of
a profile with a bulging nasal region
and a primitive cavity for a mouth.

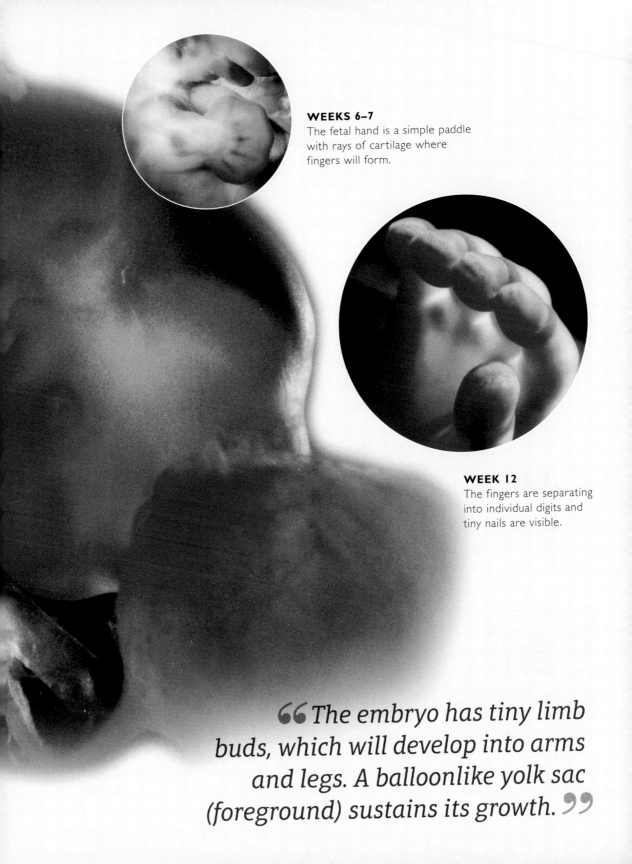

WEEKS 6–7
The fetal hand is a simple paddle with rays of cartilage where fingers will form.

WEEK 12
The fingers are separating into individual digits and tiny nails are visible.

"The embryo has tiny limb buds, which will develop into arms and legs. A balloonlike yolk sac (foreground) sustains its growth."

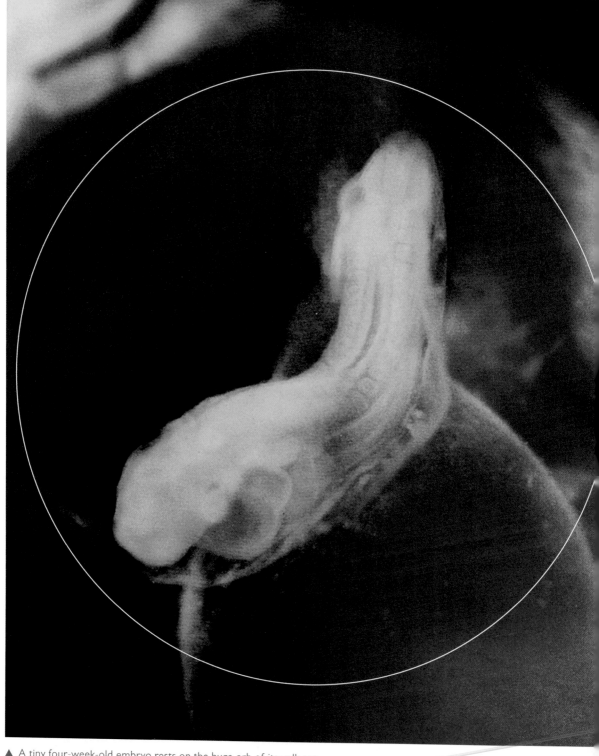

▲ A tiny four-week-old embryo rests on the huge orb of its yolk sac.

1	2	3	4	5	6	7	8	9	10	11	12	13	14	15	16	17	18	19	20

▶ **WEEKS 0–6**　　▶ **WEEKS 6–10**　　▶ **WEEKS 10–13**　　▶ **WEEKS 13–17**　　▶ **WEEKS 17–21**

▶ **FIRST TRIMESTER**　　　　　　　　　　　　▶ **SECOND TRIMESTER**

▶ **WEEKS 0–6**

The developing baby

THE FIRST SIX WEEKS OF PREGNANCY are an extremely creative time. Just three weeks after your last menstrual period, the newly fertilized egg divides repeatedly to form a cluster of cells called a blastocyst. It floats down into the uterus and embeds itself in the lining.

At this early stage when you do not even know that you are pregnant, many of the foundations of your pregnancy are being laid down. This tiny blastocyst produces chemical messengers that will send signals to your body to prevent your menstrual period from starting and prepare itself for the journey to come. At about the time of implantation, the clustered cells that will become your future baby have already become more specialized and have somehow acquired the capacity to know to which part of the body they have been assigned.

Three different layers of cells develop and each type will create different parts of your baby's body. The outer layer, or ectoderm, forms the skin, hair, nails, nipples, and tooth enamel together with the lenses of the eyes, the nervous system, and the brain. The middle layer, or mesoderm, will become the skeleton and muscles, the heart and blood vessels, together with the reproductive organs. The innermost layer, the endoderm, gives rise to the respiratory and digestive systems including the liver, pancreas, stomach, and bowel together with the urinary tract and bladder. Once a cell has been directed or programmed to have a specific function, it cannot change to become another type of cell.

By the beginning of week five, the cluster of cells is just recognizable as an embryo and is visible as a tiny nub of tissue on the ultrasound scan. Although little bigger than a nail head, all the building blocks for your baby's vital organs are already in place. The primitive heart begins to form and starts to circulate blood. At this early stage, it is a simple tubelike structure.

The position of the spinal cord has been decided and a row of dark cells appears down the back of the embryo. These cells then fold lengthwise and as they close together, they become the neural tube. At the top of the row, two

10 x life size

At four weeks, the embryo measures about 0.078in (2mm), roughly the size of this dash –. By the end of this six-week period it will double to 0.15in (4mm) in length.

large lobes of tissue become visible, which will become the brain. A digestive system is in place, although it will be many months before it is capable of functioning. A tube now extends from the mouth to the tail of the embryo and from this tube the stomach, liver, pancreas, and bowels will develop. All of the above organs and tissues are covered by a thin layer of translucent skin.

What does the embryo look like?

This is the question that most women ask and thanks to modern ultrasound scan techniques, it is now possible to see the tiny embryo on screen and describe it. By week six, the ball of cells has changed dramatically and now resembles a tadpole or a rather odd-looking shrimp. At the large head end, gill-like folds are visible, which will later become the face and jaw. The embryo's rudimentary heart bulges out of the mid portion of the body and by the sixth week it can be seen beating, or rather fluttering, on a vaginal ultrasound scan, although it is not always possible to see this if you have an abdominal ultrasound performed. Little budlike protuberances start to appear on each side of the embryo, and these will become legs and arms. Very soon these limb buds will develop nodules at their ends and these will become hands and feet.

The support system

As soon as the blastocyst starts to implant into your uterine wall, a support system for the embryo begins to develop. At this early stage, all its needs are met by the yolk sac, a balloonlike structure attached to the embryo by a stalk, which will continue to supply sustenance until the placenta is fully developed. The embryo floats in a fluid-filled bubble called the amniotic sac, which is covered by an outer protective sac called the chorion. The outer layer of the

EMBRYO AT SIX WEEKS ···

By week six, the beginnings of a nose can be seen on a head that nods over the bulging heart.

Two sets of limb buds will later develop into arms and legs. The body ends with a prominent tail.

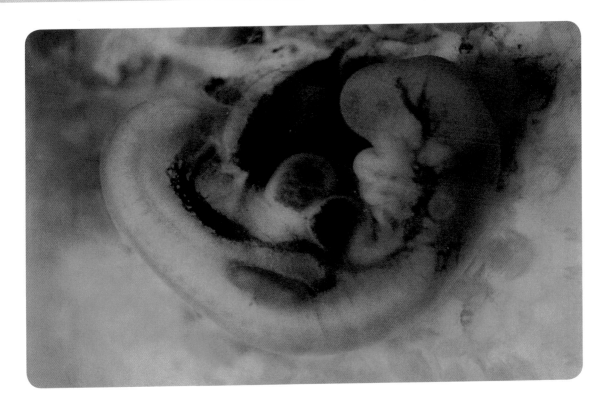

chorion will become the early placenta, and already little fingers of tissue called chorionic villi are starting to sprout and set up their future access to your circulation. Later on they will transfer nutrients and oxygen to your baby from your blood supply.

FIVE WEEKS
The embryo looks somewhat like an oddly shaped shrimp, floating in a fluid-filled bubble called the amniotic sac.

Your changing body

In the first six weeks of your pregnancy you will not look pregnant and you may not be feeling pregnant, yet your body is already undergoing major changes in response to a massive surge in pregnancy hormones that started immediately after conception.

Even when your period is not due for another seven days or so and you are unlikely to have realized that you are pregnant, several changes are taking place inside you. Your body is producing a flood of pregnancy hormones. In particular the levels of the estrogen hormone are higher than normal to help thicken the uterine lining and provide a rich environment for the tiny implanting embryo. The hormones human chorionic gonadotropin (HCG) and progesterone help to keep the embryo embedded, and progesterone also ensures that your cervical mucus thickens and forms a protective plug, which will seal the uterus off from vaginal infections for the duration of the pregnancy.

The overall size of your uterus is enlarging. When you are not pregnant, your uterus is roughly the size of a large plum, but during the course of your pregnancy it will expand to between 500 and 1,000 times its normal size. By the end of the sixth week of pregnancy, it will be the size of an apple, and although you will not feel any change, a doctor examining you internally would be able to feel the difference. However, it will not be possible to feel the uterus through your abdominal wall until the end of this trimester, when it rises above the pelvic brim and enters the abdominal cavity.

A boost in metabolism

Not surprisingly, these early pregnancy developments are accompanied by significant changes in the way your body functions. Virtually every organ system in your body needs to adapt to cope with the increasing demands that your pregnancy will make upon it. Your metabolic rate rises by as much as 10 to 25 percent during pregnancy to allow sufficient oxygen to reach the tissues of all the organs that are increasing in size and in their level of activity. To achieve this, the amount of blood being pumped through the heart every minute—the cardiac output—must increase by a total of 40 percent before 20 weeks, and this adjustment starts early in the first trimester. This boost in blood flow to almost every organ in your body is already underway. The blood supply to your uterus has doubled and the increased blood supply to the vagina, cervix, and vulva results in these tissues taking on a blue/purple coloration, distinctive of pregnancy. This change in color used to be one of the most common methods used by doctors to diagnose a pregnancy before more sensitive pregnancy tests were available. The blood flow to your uterus, kidneys, skin, and breasts will continue to increase until the very end of pregnancy.

To ensure that no area of the body is deprived of blood, the total volume has to increase from approximately 10½ pints (5 liters) before pregnancy to around 14¾ to 17pints (7–8 liters) at term. This process takes place gradually throughout pregnancy, but the volume of plasma, the watery component of the blood, starts to rise in the first six weeks to fill the newly formed blood vessels in the placenta and other growing organs. The volume of red blood cells also needs to rise to prevent the blood becoming too dilute and ensure that it has sufficient oxygen-carrying capacity, but this increase is slower and will not be noticeable until the beginning of the second trimester.

Some women are so sensitive to changes in their body, they know they are pregnant even before their period is due...

How you may feel physically

Some women are so sensitive to changes in their body that they know they are pregnant even before their period is due—and well before it is late or they have taken a pregnancy test. So you may know you are pregnant simply because you "feel" pregnant.

The sensation has been described as a strange and rather overwhelming sense of calm and fullness. For others it is a feeling of tenderness and tingling in the breasts that is much more pronounced than the usual symptoms that occur before their period starts. You will soon start to notice more changes in your breasts: they will feel heavier and look noticeably larger. Your nipples will continue to tingle, and you may see a change in the color of the areola around the nipple, and the appearance of visible veins on the surface of your breasts. These changes are due to the high levels of estrogen that are needed to provide a nurturing environment for the embryo.

You may notice that your bladder has gone wild and that you need to urinate more frequently both day and night—this symptom usually continues until the end of the first trimester. The reasons for this are two-fold. First, the blood supply to the kidneys increases by about 30 percent and results in more blood being filtered, which produces more urine. Secondly, the enlarging uterus presses on the bladder, which effectively reduces the amount of urine that can be stored because the bladder is prompted to empty itself at an earlier stage.

Fatigue and a tendency to feel overly emotional and tearful are common. These symptoms are entirely normal and merely reflect the fact that your body is dealing with a flood of pregnancy hormones to prepare it for the months ahead.

For many women a heightened awareness of smell is the first sign that suggests something has changed in their body and that they are pregnant.

A heightened sense of smell

Many women tell me that a heightened awareness of smell is the first sign that suggests to them that something has changed in their body and that they are pregnant. It is not just that smells are stronger—they are also different. Similarly, you may experience a strange metallic taste in your mouth, develop cravings for certain foods, and find yourself unable to stomach some others. I cannot explain these symptoms scientifically and can only hazard a guess that this is one way in which our bodies try to protect the tiny embryo from foods, drinks and other substances in our environments that are slightly suspect, since this change in our perception of certain smells is often linked to a distaste for alcohol and tobacco, coffee, tea, and fried foods.

No early signs

Although some women know that they are pregnant even before their period is due, many others do not experience any early signs of pregnancy. Furthermore, if their periods are very irregular, they may not realize that they are pregnant for weeks or even months. When the embryo embeds into the lining of the uterus more, between the eighth and tenth day after ovulation, there is sometimes a little bleeding; this can mislead a woman into thinking she is having a light period and is therefore not pregnant The same is true for those women who, for reasons that we do not fully understand, continue to have light periods throughout their pregnancy. I know that many women worry if they do not experience clearly recognizable symptoms of pregnancy in the first few weeks and think this means that their pregnancy is weaker or at risk. This is not the case. There is no right or wrong way to feel at this stage of pregnancy and no one symptom—or lack of—will have any bearing on your ability to carry a healthy baby to term. Signs and symptoms of pregnancy are very individual and, just as no two women will have an identical labor, no two women will have the same start to pregnancy or an identical journey through it.

PREGNANCY AFTER IVF TREATMENT

If you are undergoing invitro fertilization (IVF), the treatment cycle begins by stimulating your ovaries with hormones to produce multiple eggs. These eggs are collected on or around Day 13

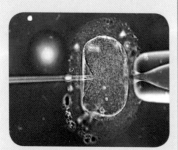

FERTILIZATION Intractyoplasmic sperm injection (ICSI) involves injecting a single sperm into an egg.

and mixed with sperm in the laboratory. If fertilization is successful during the next 48 hours, one or, occasionally, two embryos will be transferred back into your uterus on about Day 16.

A blood test on Day 27 will check for raised levels of human chorionic gonadotropin (hCG)—the first sign that the IVF treatment may have been successful. However, it is possible to have a positive test, only to discover, in a few days time, that the hCG levels have fallen because the embryo has not implanted successfully.

Your first ultrasound scan usually takes place 5–6 weeks after the start of the treatment.

If all is well, there will be a small pregnancy sac visible in the uterine cavity. If no sac is present there is a risk that you have an ectopic pregnancy (see p.81 and p.423). Sometimes the scan reveals several sacs, suggesting twins or triplets, but quite frequently the extra sac(s) disappears (vanishing twin syndrome). Multiple pregnancies are at greater risk of miscarriage, congenital abnormalities, and prematurity, which is why IVF clinics are concerned to keep the number of embryos they transfer to the minimum.

Another scan at 6–7 weeks should show a fetal pole and beating fetal heart. Your pregnancy is underway!

Your emotional response

Undoubtedly your feelings at the beginning of pregnancy are very much dictated by your personal circumstances but if, like many women, you are feeling emotionally unpredictable, your surging pregnancy hormones are almost certainly playing a part.

HIGH EMOTION
Some women are bursting to pass on their exciting news; others want to hug their secret to themselves for a little longer.

As I described in my chapter on conception, the mix of panic and elation that you may be feeling from one moment to the next may not simply be a reaction to the new future that is about to unfold. Most pregnancy books explain that fragile emotions and tearfulness gradually settle down and become less troublesome but I am increasingly of the view that they continue, since they are hormonally linked. Basically, we get used to them and, practical creatures that we are, we learn to ignore or accommodate them.

If your pregnancy was planned or has been difficult to achieve, you may feel on an almost permanent high in these early days and bursting to pass on your exciting news. However, for some women the knowledge that they are pregnant is something very private that they want to share only with their partner, or closest family and friends in the early weeks. Others, particularly those who have miscarried in the past, do not want to tempt fate by declaring to the world at large that they are pregnant, just in case something goes wrong, and prefer to wait until the pregnancy has progressed to the second trimester before announcing their news.

Deciding whether you want to tell other people about your pregnancy at this stage is a very individual issue and there is no right or wrong way to deal with it. As we all know, family dynamics are complex and only you will know the best way to break the news to your partner, mother, sister, in-laws, or friends. The only piece of reassurance I can offer is that, by and large, most family members and friends will be thrilled by your news and the fact that you want to share it with them. Indeed, the only significant problem you are likely to meet, in my experience, is a deluge of well-meaning advice and offers of help.

Many women in early pregnancy tell me that they are concerned they will upset friends and family members who are having fertility problems, have lost a baby, or experienced a pregnancy complication in the past. I think it is practically impossible to protect everyone you know from their own raw emotions and sad memories. In any event, they will have to come to terms with the situation sooner or later. However, I must add here that I am repeatedly impressed by patients in my office by how generous they can be toward other women who announce they are pregnant, even when the news must bring them personal heartache. In general, I think it is best to be open about the news that you are pregnant. I suspect that you will feel pleasantly surprised by the warmth of the response that you receive.

How your partner may feel

Whatever your decision is about when and how to reveal your pregnancy, if you are part of a couple, your partner is likely to be the first to share your news. Remember that it is not only women who have a mass of conflicting emotions at this time—many men feel much the same. Although most will be pleased at the thought of becoming a father, I think there is a fundamental difference in how men and women feel about pregnancy in the early stages. For a start, men have nothing tangible to relate to in the first few weeks. Until a baby can be visualized on an ultrasound scan or can be felt and seen moving inside you, your partner may find it hard to feel as involved as you might want him to be. Women on the other hand are already conscious of the fact that a live baby is growing inside their body, they feel physically and emotionally different and soon they will start to look different, too.

YOUR PARTNER
Becoming a father may be the best news possible or something that he needs time to come to terms with.

I do feel it is important to talk to your partner about how you are feeling, but it is all too easy to make your pregnancy the sole topic of conversation. Try not to feel disappointed or resentful if you find he is less than fascinated by your early symptoms and prefers to settle down with a crime novel rather than the pregnancy book that you are finding irresistible.

Just like you, he may need time to come to terms with the news and how it will impact on your lives. Remember that he may be feeling very apprehensive about the responsibility that looms ahead, especially if he will be financially supporting you for a while. Although there are likely to be few noticeable changes in the next few months, he will be acutely aware that he is moving into another phase of life. Calm, relaxed discussions during which he can talk openly about his feelings will help to prevent any misunderstandings from building up. So much of the focus during pregnancy is on the mother-to-be that your partner's opinions can end up being overlooked. For example, it is assumed nowadays that fathers should be, and will want to be, present at the birth. For some men, nothing could be further from the truth. Similarly, partners are often actively encouraged to attend childbirth classes. Great, if he wants to be included and is eager to join you, but pressuring him to do things a certain way, or making him feel that he is letting you down if he does not immediately adopt the role of ideal prospective father, may become the source of future conflicts. Having said that, don't worry if at this stage he says that he is not coming to any classes or into the delivery room—most men do change their mind. It may be worth finding out if there are any father-only classes, which he may feel more comfortable attending.

What your partner may be thinking

The thoughts going through your partner's head may include:
- Will our relationship stay the same?
- Will I still be able to go over to my friends' place/watch football?
- What happens if one of us has to quit working?
- How much do I want to be involved in the pregnancy?
- Is she more fragile than she was before?
- What happens if something goes wrong?
- Do I want to be at the birth and what will I be expected to do?
- Will she give the baby all the attention?
- Will I be a good father?

Having a baby on your own

Inevitably, much of the discussion in this book assumes women have a partner and that partner is a man, but I'm conscious that society is more complex than this. Many readers will be embarking on pregnancy alone and it is not my intention to make them feel excluded. If you have chosen to be a single parent, you have probably given plenty of thought to the way you will handle the next nine months and life with a baby. If lone parenting has been forced on you by circumstances, you may be feeling overwhelmed by the the prospect as well as by practical and financial considerations. Try to put a support network in place. Ask a relative or friend to share your pregnancy and to be with you for key events, such as the first scan, and for the birth. If this is difficult, find out about local childbirth classes for single mothers and single-parent support groups (see Useful Contacts, pp.437–8). Having close friends who share similar experiences can make all the difference.

> *Until a baby can be visualized on an ultrasound scan or can be felt and seen moving inside you, your partner may find it hard to feel as involved as you might want him to be.*

Your prenatal care

As soon as you know that you are pregnant, make an appointment with your gynecologist or obstetrician. It's a good idea to get to know your doctor at the beginning of your pregnancy, especially if, like most healthy women, you rarely see him or her normally.

The first thing the doctor will do is ask you for the date of your last menstrual period so that your estimated date of delivery (EDD) can be calculated. An average pregnancy lasts from 37 to 40 weeks from the first day of your last period, so your doctor will add 40 weeks on to that date using a chart or a pregnancy wheel. Much confusion and potential distress can be avoided if you always use weeks, rather than months, to measure the stage that your pregnancy has reached. The accuracy of the EDD calculated in this way will depend on whether you have a regular 28-day cycle. If yours is shorter, longer or irregular, the doctor will aim to adjust the EDD accordingly and may suggest that you do not rely on this date until you have had your first scan, when the age of your pregnancy and your future EDD can be decided very precisely.

Your doctor may also carry out a few basic tests such as testing your urine for the presence of sugar or protein and measuring your blood pressure. More detailed tests will be performed at your prenatal visit at the end of the first trimester or the beginning of the second (usually at 10-12 weeks).

Arranging your first prenatal visit

This first contact with your doctor will also give you the opportunity to discuss the options for care during your pregnancy. If you have had a baby before or have strongly held views on childbirth, you may know exactly what type of prenatal care you want and where you want to receive it. However, if this is your first baby, you will probably welcome a detailed explanation of what is available and some guidance as to which option is best suited to your individual needs. For this reason, I have included a comprehensive section on prenatal care and birth choices at the end of this section (see pp.84–91).

Your first formal prenatal visit will happen toward the end of the first trimester. (This may take place at the doctor's office of your choice or alternatively in your own home if you are going to be cared for by a certified-nurse midwife.) Even if you are hoping for a home birth, make sure that you go to visit a hospital or birthing center. You will be able to discuss your pregnancy with midwives and doctors, who will be able to offer you valuable

Even if you are hoping for a home birth, make sure that you go to a hospital to see what it looks like around the unit.

YOUR ESTIMATED DELIVERY DATE

Look on the chart for the month and then the first day of your last menstrual period (printed in bold type). Directly below it is the date that your baby is due—your estimated delivery date.

Month	1	2	3	4	5	6	7	8	9	10	11	12	13	14	15	16	17	18	19	20	21	22	23	24	25	26	27	28	29	30	31
January	1	2	3	4	5	6	7	8	9	10	11	12	13	14	15	16	17	18	19	20	21	22	23	24	25	26	27	28	29	30	31
Oct/Nov	8	9	10	11	12	13	14	15	16	17	18	19	20	21	22	23	24	25	26	27	28	29	30	31	1	2	3	4	5	6	7
February	1	2	3	4	5	6	7	8	9	10	11	12	13	14	15	16	17	18	19	20	21	22	23	24	25	26	27	28			
Nov/Dec	8	9	10	11	12	13	14	15	16	17	18	19	20	21	22	23	24	25	26	27	28	29	30	1	2	3	4	5			
March	1	2	3	4	5	6	7	8	9	10	11	12	13	14	15	16	17	18	19	20	21	22	23	24	25	26	27	28	29	30	31
Dec/Jan	6	7	8	9	10	11	12	13	14	15	16	17	18	19	20	21	22	23	24	25	26	27	28	29	30	31	1	2	3	4	5
April	1	2	3	4	5	6	7	8	9	10	11	12	13	14	15	16	17	18	19	20	21	22	23	24	25	26	27	28	29	30	
Jan/Feb	6	7	8	9	10	11	12	13	14	15	16	17	18	19	20	21	22	23	24	25	26	27	28	29	30	31	1	2	3	4	
May	1	2	3	4	5	6	7	8	9	10	11	12	13	14	15	16	17	18	19	20	21	22	23	24	25	26	27	28	29	30	31
Feb/Mar	5	6	7	8	9	10	11	12	13	14	15	16	17	18	19	20	21	22	23	24	25	26	27	28	1	2	3	4	5	6	7
June	1	2	3	4	5	6	7	8	9	10	11	12	13	14	15	16	17	18	19	20	21	22	23	24	25	26	27	28	29	30	
Mar/Apr	8	9	10	11	12	13	14	15	16	17	18	19	20	21	22	23	24	25	26	27	28	29	30	31	1	2	3	4	5	6	
July	1	2	3	4	5	6	7	8	9	10	11	12	13	14	15	16	17	18	19	20	21	22	23	24	25	26	27	28	29	30	31
Apr/May	7	8	9	10	11	12	13	14	15	16	17	18	19	20	21	22	23	24	25	26	27	28	29	30	1	2	3	4	5	6	7
August	1	2	3	4	5	6	7	8	9	10	11	12	13	14	15	16	17	18	19	20	21	22	23	24	25	26	27	28	29	30	31
May/Jun	8	9	10	11	12	13	14	15	16	17	18	19	20	21	22	23	24	25	26	27	28	29	30	31	1	2	3	4	5	6	7
September	1	2	3	4	5	6	7	8	9	10	11	12	13	14	15	16	17	18	19	20	21	22	23	24	25	26	27	28	29	30	
Jun/Jul	8	9	10	11	12	13	14	15	16	17	18	19	20	21	22	23	24	25	26	27	28	29	30	1	2	3	4	5	6	7	
October	1	2	3	4	5	6	7	8	9	10	11	12	13	14	15	16	17	18	19	20	21	22	23	24	25	26	27	28	29	30	31
Jul/Aug	8	9	10	11	12	13	14	15	16	17	18	19	20	21	22	23	24	25	26	27	28	29	30	31	1	2	3	4	5	6	7
November	1	2	3	4	5	6	7	8	9	10	11	12	13	14	15	16	17	18	19	20	21	22	23	24	25	26	27	28	29	30	
Aug/Sep	8	9	10	11	12	13	14	15	16	17	18	19	20	21	22	23	24	25	26	27	28	29	30	31	1	2	3	4	5	6	
December	1	2	3	4	5	6	7	8	9	10	11	12	13	14	15	16	17	18	19	20	21	22	23	24	25	26	27	28	29	30	31
Sep/Oct	7	8	9	10	11	12	13	14	15	16	17	18	19	20	21	22	23	24	25	26	27	28	29	30	1	2	3	4	5	6	7

advice as to how you can best achieve your aim. It is often presumed that obstetricians and other doctors are against the idea of home delivery. This is indeed generally the case. Birthing centers that are located close to hospitals are homelike and minimize the risk should complications arise.

Unless you have a specific problem that requires urgent attention, you will probably not have an appointment for several weeks. Indeed, some women do not visit their obstetrician until they are about 5 weeks pregnant. The exceptions are women who have been unlucky enough to experience a previous pregnancy problem, such as a miscarriage or a late pregnancy complication. If this is the case for you, your doctor will probably arrange an immediate prenatal visit and possibly an early scan. If you have an existing medical condition, such as diabetes, your prenatal care is also likely to start sooner rather than later.

Many women used to feel disappointed that no one seemed to be very interested in their pregnancy until they go to their first doctor visit. They also felt confused because there was no one they could talk to about what to do and where to go to find out more about being pregnant. As one first-time mother told me: "I am very grateful that I am just a normal pregnant woman with no medical problems, but this is all new to me. This is a special time for me and I am out of my depth."

It is also one of the reasons why it is very important for pregnant women to be able to access as much information as possible both to reassure them and help them understand what is going to happen next. In an ideal world with limitless health-care resources, we would be able to offer women instant access to the prenatal care they choose at the moment that their pregnancy test is positive. In the real world, the best solution I can offer is an informative book.

Common concerns

You may already have a few questions about the early stages of your pregnancy and about your general health. These are definitely worth discussing with your doctor at your first consultation.

I've included some common early pregnancy concerns below but if you cannot find your particular problem here, check the other examples in the next two sections of the first trimester: weeks 6–10 and weeks 10–13. Do not hesitate to bring up any others you may have with your doctor; it is always better to have advice on them sooner rather than later. It is important that you have all the reassurance you need to make the next few weeks as enjoyable as possible.

Previous pregnancy problems

If you had previous pregnancy problems, such as a miscarriage, an ectopic pregnancy (see opposite and p.423), or complications in later pregnancy such as preeclampsia (see p.426), then your doctor will probably arrange for you to have an early ultrasound scan. For those women who have experienced miscarriages in the past, a scan is often very helpful in allaying fears that history is repeating itself. But do remember that all you need to see at this stage is a healthy sac in the uterine cavity. While most scanners may detect a tiny fetal heart beat at 5–6 weeks, every pregnancy is unique and as long as a pregnancy sac can be seen in the uterus, the absence of a fetal pole (a tiny

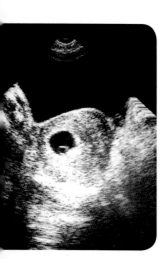

FETAL POLE The tiny embryo is just visible as a white speck on the yolk sac, which is floating in the dark circle of the amniotic sac.

rectangular blob within the sac) or a heartbeat at this early stage, is common. The most likely explanation is that your embryo implanted a few days later than you had calculated or expected, so a little more time is needed to see these landmarks in your pregnancy. This may be the case if your menstrual cycle is irregular, or you are not sure exactly when you conceived.

Early ultrasound scans are also particularly valuable for women who have had a previous ectopic pregnancy, in which the embryo develops outside the uterus. The most common site is in the fallopian tube, although ectopic pregnancies are also found on an ovary or in the abdominal cavity. A scan establishes whether there is an early pregnancy sac in the right place inside the uterus. If there is no evidence of this, you will probably undergo a series of blood tests to establish your blood levels of hCG hormone. If these are raised and yet there is still no sign of an intrauterine sac on the scan, you may need to undergo laparoscopy or drug treatment to stop the ectopic pregnancy from continuing and prevent more damage to the fallopian tube.

pregnancy sac with embryo
fallopian tube
uterus
ovary
cervix

ECTOPIC PREGNANCY
The sac and embryo have developed in the fallopian tube where there is insufficient room for them to grow.

DIET AND EXERCISE

I am all too aware that nausea and tiredness in the first trimester can be a setback to your best-laid plans to eat the healthiest diet possible and stay fit and supple during pregnancy.

In the first trimester, what you eat is particularly important because, during these weeks, your baby's vital organs will be formed. The early development of the heart, liver, brain, and nervous system, for example, takes place in this period. Having said that, I know from personal experience that nausea and sickness in the early months of pregnancy can make it impossible to follow your ideal diet.

▶ **Eat little and often**. Small meals and snacks are easier on the stomach than three large meals per day. Try nibbling on a piece of bread or some dry biscuits early in the day.

▶ **Keep healthy snack foods** such as fruit, nuts, and pieces of cheese easily available, so that your blood sugar level never gets too low.

▶ **However little solid food** you can cope with, try to remember to drink plenty of fluids regularly. If you feel happier skipping your daily exercise now that you are pregnant, that's fine, but do remember that exercise is very unlikely to cause any problems in early pregnancy and, if you maintain your exercise program, it will help you keep fit and healthy. I can guarantee you that, once the first three months are over, you are likely to feel a renewed vigor that will propel you back, if not to the badminton court, at least to the pool for some wonderfully relaxing and beneficial swimming sessions.

Urinary tract infections

Although it is normal for pregnant women to experience urinary frequency in the early weeks of pregnancy, the possibility of a urinary tract infection should not be overlooked. If you feel tingling when you urinate, the urgent need to urinate, lower abdominal pain or discomfort, or notice blood in the urine, you may have an infection and need prompt treatment with antibiotics.

Urinary tract infections are very common in pregnancy because progesterone makes the urinary tract more relaxed. This makes it easier for bacteria to enter the urethra and reach the bladder, where they cause inflammation or cystitis. Because the bladder is also relaxed in pregnancy, the infection can spread easily up the ureters and infect the kidneys, a condition known as pyelonephritis. Symptoms usually develop suddenly, and include a high temperature and shivering, pain over the bladder and kidneys, with severe discomfort in the abdominal region, which may radiate down to the groin. Antibiotics will resolve the infection promptly. Left untreated, the infection can cause permanent scarring and damage to the kidney. The bacteria are present in 2–5 percent of pregnant women without causing symptoms, but screening and treatment reduces the risk of pyelonephritis.

An existing condition

If you have a general medical problem, however minor, you should discuss this with your doctor as soon as you know that you are pregnant. Your doctor will help you to decide on the best type of care for your needs, and if you are taking any medication, you will be advised on whether the dosage needs to be altered. On no account should you stop taking prescribed drugs without first seeking advice.

On pages 408–10 you will find details about existing medical conditions such as diabetes, epilepsy, high blood pressure, thyroid disease, renal disease, heart disease, and inflammatory bowel disorders, which need specialists' care. If you suffer from one of these (or a disorder that is not included in this list), your doctor may suggest prenatal care with a particular specialist for your condition. You will be seen quickly, particularly if you are taking medications that may need to be altered. Throughout your pregnancy, you will be monitored by your obstetrician and, possibly, a physician with specialized knowledge of your particular condition. Your part of the bargain is to try to stay healthy and to follow their advice rigorously.

❝ *I can guarantee you that, once the first three months of pregnancy are over, you are likely to feel a renewed vigor...* ❞

OLDER MOMS-TO-BE

Over recent years there have been very significant societal changes in our reproductive patterns. The average age of a woman giving birth for the first time has continued to rise steadily and the number of babies born to mothers aged 35 years or older has increased dramatically.

One in five of all births are currently to women over 35 and the number of babies born to women over the age of 40 is at an all time high, having doubled in the space of ten years. If you are over 40 years of age and this is your first pregnancy, you may have heard someone describe you as an "elderly primigravida" or seen in your medical records that your age is considered to be a risk factor. Although many women who are able to become pregnant at this age will go on to have a successful outcome, there is no doubt that women over 40 years of age are more likely to experience problems conceiving a baby, staying pregnant, and suffering from later pregnancy complications than younger women.

That said, many of these complications can be identified or predicted to minimize the consequences. So there is no reason why your pregnancy should not have a successful outcome as long as you have regular prenatal care.

A woman's fertility starts to decline rapidly during her thirties because the quality of her eggs deteriorates with advancing age. Not only are her older eggs less likely to be fertilized but they have a greater chance of carrying a genetic or chromosomal abnormality. This may prevent her from becoming pregnant or lead to an early miscarriage, because the resulting embryo is so abnormal that it cannot develop any further: for example Trisomy 16, the most common abnormality found in miscarriage tissues which is incompatible with life.

Importantly, some genetic abnormalities are not always fatal and the best known of these is Down syndrome or Trisomy 21 (see p.147), the incidence of which increases sharply after the age of 35 years. For this reason, older pregnant women are routinely given prenatal tests to diagnose possible fetal genetic and physical abnormalities (see pp134–43). Recent evidence suggests that advanced paternal age also increases the risk of chromosomal abnormalities and miscarriage.

Many important prenatal complications such as high blood pressure, preeclampsia, gestational diabetes, premature labor, placenta pravia, and poor fetal growth are more common in older mothers. And the older the woman, the more likely she is to have preexisting medical problems such as high blood pressure, diabetes, heart disease, or obesity to further complicate the picture. Lastly, women giving birth over the age of 35 have a higher chance of requiring medical interventions during labor (particularly if this is a first birth) and of developing complications in the postpartum period: for example, they are more likely to be induced, to require a cesarean section, or to have an episode of postpartum depression. I hope you will use this information to take the best possible care of yourself, in the knowledge that good prenatal care and support from family and friends will minimize the chances of these things happening to you.

As an older mother of twins myself, I am conscious of the fact that much of what I have said above may sound very negative, but my own experience was consistent with the increased risks of becoming pregnant in my late thirties. My twin daughters were born early at 33 weeks, delivered by emergency cesarean section, needed help with breathing and feeding, and spent the first four weeks of their lives in a neonatal intensive care unit. They were vulnerable premature babies at birth—but are now completely healthy young adults.

PRENATAL CARE AND BIRTH CHOICES

Where you decide to have your prenatal care depends both on the type of birth you would like and on where you live and what facilities are available to you. Choosing the right kind of care goes a long way toward making your pregnancy a pleasurable time in your life.

An open-minded approach

When I was pregnant and looking for a book to read that would help me understand my special condition, I was struck by how polarized most

GOOD ADVICE Your prenatal care team take care of your health in pregnancy and offer advice and information.

books are on the particular issue of where I should have my baby. They divided roughly into two camps: those written by obstetricians who seemed to believe that the only place to have a baby is in a specialized hospital unit, and those written by vehement supporters of natural childbirth and home births who suggest that hospital units are designed to make women feel lost and vulnerable by taking away control of their own bodies during childbirth. Like most readers, I was left with the feeling that I would have failed as a mother, missed out, or been cheated of something special if medical interventions were required. I concluded that both types of books were unrealistic and unhelpful because they did nothing to make me feel calm and confident about the potentially unpredictable journey I had just embarked upon. On reflection, that is when I decided another type of book was needed and that I should sit down and write it.

Before you decide

The main goal of prenatal care is to maintain your health and well-being during pregnancy and to help you produce a healthy baby, and this involves detecting any condition that may adversely affect either of you as early as possible. During your prenatal care, you will also be given information

TYPES OF PRENATAL CARE

OBSTETRICIANS

A physician who specializes in pregnancy and childbirth is an obstetrician. Your choice of obstetrician may be based on the advice of family and friends, or you may choose an obstetrician on the list of providers for your particular medical insurance. Your obstetrician may be the same physician who has been providing your gynecological care. Your obstetrician may provide all of your prenatal care or, alternatively, may just be on call in case of complications. In urban centers, he or she may be affiliated with more than one hospital, so you may have a choice about where you give birth, but in rural settings, the choice is likely to be more limited.

FAMILY PRACTITIONERS

These physicians provide basic care for most conditions, including normal (uncomplicated) pregnancies and childbirths. However, if you have a difficult pregnancy, or previous pregnancies or deliveries were not straightforward, your medical care is more likely to be supervised by an obstetrician.

MIDWIVES

While physicians focus on the pathological, midwives focus on the normal process and are interested not only in your uterus and how it functions but also in your general well-being. Midwives may deliver babies at home, in the hospital, or in a birthing center but they must always have a physician on call in case of an unforeseen emergency. Midwives fall into several categories based on their education and training.

▶ **Certified nurse-midwives**
These are designated by the initials CNM after their names. CNMs are graduates of midwifery programs that are accredited by the American College of Nurse-Midwives and they are both nationally certified and licensed in their particular state according to the relevant state law. They may be part of a practice with other midwives and obstetricians or, alternatively, they may work in a hospital or birthing center. CNMs have authority to write prescriptions in 48 states and the District of Columbia. Their fees must be reimbursed by Medicaid in all 50 states; in addition, 33 states mandate reimbursement of midwifery fees by private medical insurers.

▶ **Certified midwives (CM)**
Starting in 2000, an additional level of certification was created for those who started out in a health-related field other than nursing before switching to midwifery. They go through the same educational and licensing system as certified nurse-midwives. This category is not yet established across the US. CMs and CNMs assisted 312,129 births—more than 11 percent of all vaginal births, or 7.8 percent of total US births—during 2010.

▶ **Direct-entry midwives**
Midwives who enter the field through a combination of self-study and apprenticeship with other midwives, or through a private midwifery program, are known as direct-entry midwives. Their fees are covered by Medicaid in only 10 states, and using their services is actually illegal in 15 other states. Be sure to check your insurance coverage carefully to make sure that your midwife's fees will be reimbursed.

NURSING STAFF

You may be helped by a variety of nursing staff from the time you enter the hospital or birthing center, depending on the size of the center and the particular needs of you and your baby, because each nurse has a different role, particularly in larger medical centers. These include specialists in labor and delivery who help both mother and newborn child, postpartum nurses specifically for you, and neonatal nurses for your child.

❝ *Midwives focus on the normal process... but also on your general well-being.* ❞

A CHANGE IN APPROACH

The issue of choice in childbirth has been a matter of fierce debate over the years, much of it prompted by the dissatisfaction many pregnant women began to feel with the more traditional hospital-based patterns of care.

A "medicalized" approach to childbirth developed over the last 50 years as a by-product of advances in the way labor and birth could be managed. As a result women began to protest that they were being treated as part of a baby production machine, rather than informed participants. These widespread feelings prompted Doris Haire to write her ground-breaking *The Cultural Warping of Childbirth* in 1972. Comparing birthing practices in developed countries around the world, she illustrated how childbirth had become over-medicalized in the US, and prompted women to educate themselves.

▶ It was no longer justified on the grounds of safety to encourage all women to give birth in the hospital.

▶ Many women wished to have continuity of care throughout pregnancy and childbirth, and midwives were likely to be best placed to provide this.

▶ Greater choice in the type of care for pregnant women was needed.

▶ Provision for home birth or birth in a small maternity unit was largely unavailable in the US, despite increasing demand.

▶ Some traditional interventions during labor and delivery such as continuous fetal monitoring, epidurals, chemical stimulation of labor, and episiotomy were unnecessary or not evidence-based.

▶ The hospital environment was leaving some women feeling that they had lost control of their bodies and disappointed by their labor and birthing experience.

▶ Within the hospital, women should be able to exercise choice in the personnel who care for them.

▶ The relationship between the woman and her caregiver was of fundamental importance and needed to be recognized..

Doris Haire's report did much to raise public awareness of these issues. Two Congressional hearings have been held to discuss obstetric care. A law enacted in New York State as a result of a consumer legislative effort required all New York hospitals to publish rates of cesarean sections, VBACs, forceps use, induced labor, aumentation of labor, and episiotomy, enabling women to compare hospitals.

Hospital maternity units have become friendlier, more comfortable, and less clinical places in which to give birth. Although the number of home deliveries has not increased dramatically, there has been a very significant change in the way that hospital units provide care for women at the time of delivery. The emphasis is now on flexibility, one-to-one midwifery care, and minimizing medical interventions wherever possible.

I believe that in the majority of cases, obstetricians should work in partnership with the midwife and only become the principal provider of care when a woman is at risk of medical and obstetric complications. The key to this is finding better ways of identifying the women who are at greatest risk and require medical intervention.

❝ …hospital maternity units have become friendlier, more comfortable, and less clinical places in which to give birth. ❞

and health education that will prepare you for labor and motherhood. These two major principles will be followed through wherever care is provided.

Even if you feel it is too early at this stage for you to know whether you would prefer to deliver your baby in the hospital, in a birthing center, or at home, your choice of prenatal care is a more pressing issue because your first appointment will take place before the end of the first trimester. So take some time to read through the various options on page 85 to see what might suit you best.

What is available in terms of prenatal care varies from one region to the next: some women have many different options available to them, while others have only a limited choice. Your physician or midwife will be able to tell you what is available and which services are provided by your physician's or obstetrician's office and nearby hospitals. Make sure that you understand the details so that you can consider all of the options on your own time. Explore other sources of information before coming to a decision—talk to friends and neighbors, pick up any information pamphlets in your physicians' offices, and contact your health insurer as well as the American College of Obstetricians and Gynecologists and the American College of Nurse-Midwives (see p.437). There is a bottomless pit of information on the internet, but remember that much of this is not backed by professional expertise. Lastly, remember that you can always change your mind at a later date.

Childbirth in the US

Today most babies in the US are born in the hospital—about 92 percent according to the most recent statistics—with about eight percent born either at home or in birthing centers. In 1972, in *The Cultural Warping of Childbirth*, Doris Haire noted that childbirth practices were relying on the so-called common sense of doctors rather than on scientific fact. Haire criticized a number of obstetrical practices that were common in the early 1970s, including routine induction of labor; depriving women of support during labor and delivery; and requiring them to give birth on their backs. Hospital birth wasn't actually safer but infant mortality rates had fallen dramatically in recent years, making hospital birth appear to be safer. Undoubtedly, the major reason for this fall was an overall improvement in prenatal care and general living standards.

Now that women are so well informed about pregnancy and their general health, and living standards have improved, there is the opportunity to offer greater choice in their prenatal care and place of birth. Although there will always be women who need high-tech medical care, the majority are likely to have straightforward pregnancies with a normal vaginal delivery. In view of this, many women and their maternal caregivers now feel that the option of a home delivery should at least be considered for low-risk pregnancies.

Choosing a birthplace

Your choice of where you want to have your baby is likely to come down to one of the following options: in a hospital in a maternity unit, or in your own home. The two important factors to take into account when making a decision are personal preference and the safety of both you and your baby. Sometimes problems that arise in the pregnancy or previous complications make these factors incompatible, but it is usually possible to reach a compromise, as long as time is spent discussing the practicalities rather than either you or your medical caregivers sticking to a previously held bottom line.

In the hospital

If this is your first pregnancy, if you have current medical problems, or have experienced complications in a previous pregnancy, your obstetrician will probably recommend that you deliver your baby in the hospital. However, it is likely that you will have prenatal care with your doctor or one of the practice's doctors. There may be more than one hospital in your area and if you can choose which to go to, try to find out about the facilities and prenatal care options each one offers.

For example, if you want a midwife to deliver the baby, she will usually be based at one particular hospital or practice, and this may influence your decision. In addition, many hospitals now have a limited number of birthing rooms, which are designed to be much less clinical; in fact, much more similar to your own home than the standard delivery room. Most of these birthing rooms have soft lighting, music, comfortable chairs and large floor cushions, birthing balls and stools, and some units also provide a birthing pool.

In a birthing center

Family practitioner and midwifery units are more common in rural areas and are located either within the main hospital maternity unit or as a separate stand-alone unit within the hospital. Staffed by doctors and midwives who can offer continuity of care throughout your pregnancy, labor, and delivery, birthing centers are designed to deal with normal pregnancies and deliveries in a low-tech and informal environment. Those hospital maternity units that include a birthing center are probably offering pregnant women the best of both worlds in that the surroundings are less formal, continuity of care can be maintained, but expert medical help is literally just around the next corner if any problems requiring specialist expertise arise during the birth, such as difficult presentations that may require a cesarean delivery.

At home

If you are considering having a home birth, the first thing to do is talk to your obstetrician and midwife. If this is your first baby, most will have concerns about the safety of a home delivery. No matter how well your pregnancy goes, no one can predict what will actually happen during labor.

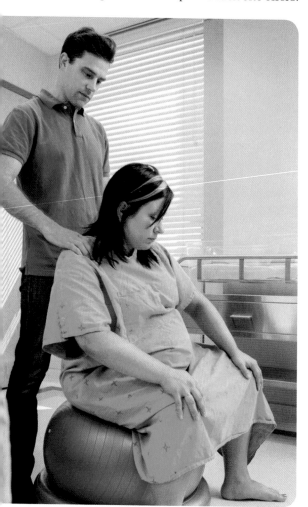

HOSPITAL BIRTH It is possible to have an active labor in the hospital, in a relatively nonclinical setting.

They will have similar concerns if you have had medical problems in the past or previous pregnancy complications.

You may be a potential candidate for a home birth if you have had one or more previous pregnancies that were free of complications and that ended in straightforward vaginal deliveries. Even then, no two pregnancies can be guaranteed to follow the same pattern, so be prepared to change your plans if this pregnancy turns out differently.

It is within your right to have your baby at home, but it will be your responsibility to identify the people who will take care of you, so find out whether your own obstetrician is prepared to care for you during pregnancy and home delivery. If he or she cannot help, there may be another local obstetrician that can. Your health insurer can give you the names of providers in your area with a special interest in pregnancy and childbirth and you can arrange to transfer your care to them.

Alternatively, you can choose to have a certified nurse–midwife give you all or most of your prenatal care and attend a home delivery. Or, you may opt to see a certified nurse-midwife who works within an obstetric practice. The American College of Nurse-Midwives (see Useful contacts, p.437) can put you in touch with a certified nurse-midwife. Also, some obstetric practices employ midwives.

Finding out more

Before you reach a final decision about the place of birth, ask other parents, midwives, and doctors about the reputation of the hospitals available to you. Then arrange a visit to one or two local units so that you can see for yourself and get a feel for whether you will be comfortable and relaxed there. Remember that feeling comfortable in labor is not just about physical issues such as the

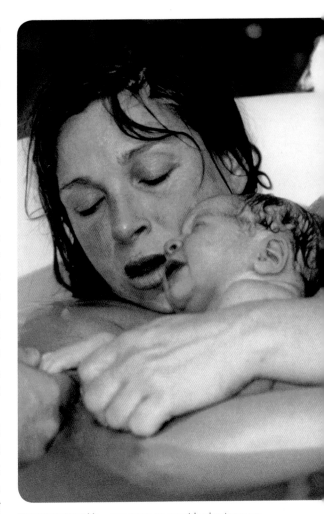

HOME BIRTH You may want to consider having your baby at home in the heart of your family if you have had a previous labor and delivery that was straightforward.

facilities and decoration of the rooms, it is also about the friendliness of the staff and the attitude they have toward birthing. I know that many women feel nervous about going into hospitals because they associate them with illness or possibly sad or unpleasant memories. But labor wards have a totally different atmosphere than any other department in a hospital. The patients in the labor wards are fit and healthy, in fact they are positively blooming, and the staff participates

QUESTIONS TO ASK ON A VISIT TO A MATERNITY UNIT

The very best way to find out what your local maternity hospital can offer is to arrange a visit. The following questions should help you decide on your preferred type of maternity care and delivery.

GENERAL ISSUES

▶ Does the maternity department have particular interests or offer any types of specialized services?
▶ What is the hospital's policy on different types of prenatal and delivery care? What is, for example, their policy on midwives assisting in the delivery?
▶ Is there 24-hour anesthesia service available?
▶ Is there a neonatal-intensive-care unit (NICU)?
▶ Does the hospital offer childbirth

classes and tours of the labor ward and postpartum facilities?
▶ What is their policy on birth plans?

LABOR AND DELIVERY ISSUES

▶ Are the labor nurses flexible about special requests and different types of delivery; for example, are they willing to encourage women to give birth in any position that feels comfortable to them, such as lying down, standing up, or squatting?
▶ How long are the labor nurses' shifts? A 12-hour-shift system gives you a better chance of being cared for by the same labor nurse throughout your delivery.
▶ What is the hospital policy on induction, rupturing the membranes, pain relief, and routine electronic monitoring during labor?
▶ Would they be worried if labor was slower than normal or than expected?
▶ Are partners, friends, and family welcome in the delivery room? Are the numbers limited?
▶ Is there a 24-hour epidural service?
▶ Do they make use of breathing techniques to help relieve pain?
▶ Is there a birthing pool or can you bring a rented one with you to the hospital?
▶ What are the forceps, vacuum, and cesarean section rates? Remember that a specialist or teaching hospital will have a higher

number of these than a small general hospital because they will be caring for women who are more likely to experience complications in labor.
▶ What is their rate of vaginal births after cesareans (VBACs)?
▶ Is there a hospital policy on episiotomy and repair of vaginal tears? Some hospital units have trained their labor nurses to suture but some rely on on-call doctors.

AFTER THE BIRTH

▶ Are single, private rooms available? If so, how many and what do they cost? Do they have private bathrooms? Are they reserved for women who have had difficult births? What size are the general maternity units and how many beds per room?
▶ What is the usual length of stay after the birth? (This will probably be longer after a first baby than for subsequent births.)
▶ Will your baby be with you at all times or is there a separate nursery?
▶ Are there specialist breast-feeding counselors available in the maternity ward? This can be enormously helpful when you are trying to establish breastfeeding.
▶ What are the visiting hours?
▶ Are special diets, such as kosher or vegetarian, available?
▶ Will you need to bring items such as pillows, towels, and diapers?

in helping you have a happy and successful ending to your pregnancy.

The who's who of maternity care

You will meet a variety of health-care professionals during your pregnancy, labor, and birth and also during the postpartum period. The following is a brief description of the role each person plays.

Your Family Practitioner will sometimes arrange for you to be seen in by an obstetrician if you are deemed potentially high risk. Many family practitioners participate in shared prenatal care with the hospital clinics and obstetricians particularly in isolated parts of the country.ome family practitioners may be involved in your delivery, either in the hospital or at home. All family practitioners contribute to postpartum care.

Midwives are nurses or other health-care workers who have had additional specialized training in childbirth. They are qualified to take responsibility for you and your baby before, during, and after a normal birth and if complications arise, will seek advice from an obstetrician. Certified nurse-midwives are licensed to practice in every state, and certified midwives in many states, and the number of live births they have assisted increases every year. Some hospital-based midwives have developed specialized skills in the management and care of pregnant women with specific problems such as diabetes, high blood pressure, infection, and other medical complications of pregnancy. Whichever type of midwifery care you receive, you will find that she (occasionally he) is an invaluable source of information, advice, and comfort.

Direct-entry midwives are much more limited in their training and, consequently, in the places they are allowed to practice; make sure that they are legal in your state and covered by your medical insurance before using one for your prenatal care and delivery, whether in the hospital or at home.

Obstetricians are doctors who are specialists in the care of pregnant women. If you do not already have your own obstetrician, you may get a recommendation from your gynecologist. The obstetrician is usually part of a practice of other obstetricians and is occasionally linked to a team of midwives. You may meet the midwife only if your pregnancy runs smoothly. If it is more complicated, you will not see a midwife during your prenatal care in an obstetric practice.

Pediatricians are doctors who are specially trained in the health of babies and young children. Every maternity unit works hand in hand with pediatricians to ensure that the babies they deliver are healthy and receive any medical help they need. A pediatrician is always present at the delivery of twins and higher multiple births, most instrumental deliveries (such as forceps), and cesarean sections. Every baby is checked by a pediatrician at birth and 5 minutes after birth, using the Apgar score, which assesses five key elements of a newborn baby's health: heartbeat, breathing, muscle tone, reflexes, and skin color.

Neonatologists are pediatricians with specialized skills in the care of newborn babies with problems. They run the neonatal intensive care unit (NICU). If your baby is born prematurely or is found to have a problem, a neonatologist will be involved in your baby's care.

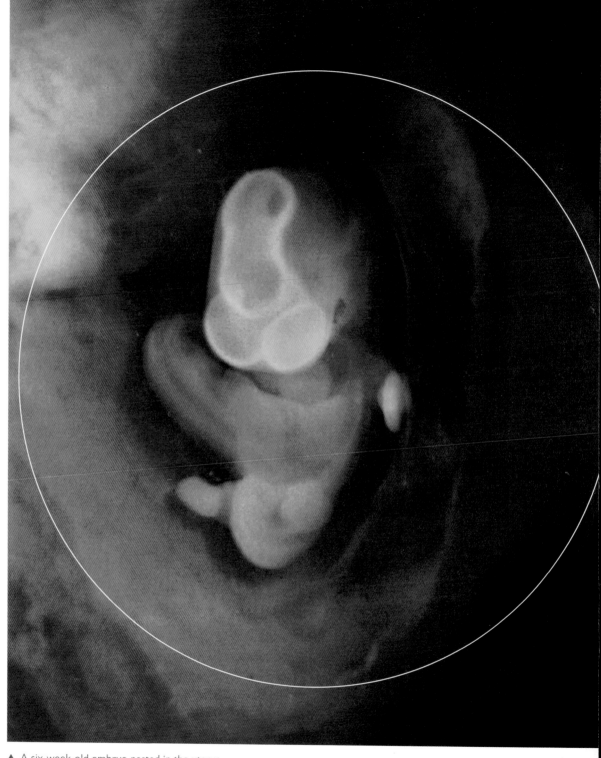

▲ A six-week-old embryo nested in the uterus.

| 1 | 2 | 3 | 4 | 5 | **6** | **7** | **8** | **9** | **10** | 11 | 12 | 13 | 14 | 15 | 16 | 17 | 18 | 19 | 20 |

▶ **WEEKS 0–6**　　▶ **WEEKS 6–10**　　▶ **WEEKS 10–13**　▶ **WEEKS 13–17**　　▶ **WEEKS 17–21**

▶ **FIRST TRIMESTER**　　　　　　　　　　　　　▶ **SECOND TRIMESTER**

▶ WEEKS 6–10
The developing baby

DURING THE NEXT FOUR WEEKS, your developing baby will quadruple in size and undergo dramatic changes in appearance. By week 10 the embryo has become a fetus and is starting to resemble a human being.

Several facial features can now be recognized on ultrasound as the body is straightening and the limbs developing. The head continues to grow more rapidly than any other part of the body in order to accommodate the developing brain. The back of the head grows at a faster rate than the front; as a result the embryo is curled over the front of the body and appears to be nodding at its bulging heart. However, the body has started to lose its earlier commalike shape. A neck is appearing, the back is straighter, and the tail is disappearing.

The head now has a high forehead and, as the primitive facial bones develop and fuse together, eyes, nose, ears, and a mouth become recognizable. The primitive eyes and ears, which were mere swellings on the head at six weeks, are developing rapidly. By the end of the eighth week, the eyes have grown in size and already contain some pigment. By the 10th week, they are easily recognizable but will continue to be hidden behind sealed lids and cannot function until later in the second trimester, when the nervous system is fully formed. At each side of the head the depressions that will become ear canals have deepened and the inner ear starts to form. By the eighth week, the middle ear, which will become responsible for both balance and hearing, is formed and by 10 weeks, the external part of the ear (the pinna) has started to grow low down on the fetal head. The nostrils and upper lip can now be seen and inside the mouth there is a tiny tongue, which already has taste buds. Tooth buds for the future milk teeth are in place within the developing jawbones.

Future limbs

More miraculous changes are happening as your baby's limbs take shape. The folds of skin making up the limb buds condense and form cartilage, which will later develop into hard bones. These cartilaginous limb buds grow rapidly and

2 x life size

By the end of the 6th week, the embryo is 4mm in length and weighs less than 0.03oz (1g). By week 10, the fetus will measure 1¼in (30mm) from the top of its head to its bottom (crown to rump) and weigh 0.1oz (3–5g).

wrists and paddle-shaped hands can soon be identified. The arms lengthen and by eight weeks, shoulders and elbows are present, making the upper limbs project forward. The webbed hands now develop separate fingers and by week 10, touch pads have appeared at the end of the stubby fingertips. The lower limb buds start to go through the same process, but the distinction between thighs, knees, calves, ankles, and toes progresses more slowly. Most of the muscles are now in place, and small, jerky movements can be seen on an ultrasound scan.

Inside the body

Inside the body, the central neural tube is now differentiating into the brain and spinal cord. The nerve cells multiply rapidly, aided by support cells called glial cells, and migrate along pathways into the brain, where they connect with each other and become active. This is the beginning of the neural network that will later transmit messages from the brain to the body. The fetus has also developed some very basic sensory perceptions. It can respond to touch although it is still too early for you to be able to feel movements.

By week 10, the embryonic heart has developed into the definitive four-chambered heart. The two atria receive blood from the fetal circulation, while the ventricles pump blood out to the lungs and rest of the baby's body. Valves develop at the exit of all four chambers to ensure that the blood is always pumped in one direction and cannot leak back into the heart. The heart is beating at a rate of 180 beats per minute—twice the speed of your own heart.

Although the digestive system is developing rapidly, it will be some time before it is functioning properly. The stomach, liver, and spleen are all in place

FETUS AT 10 WEEKS ···

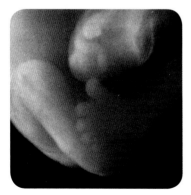

The fetal intestines still protrude from the abdominal wall.

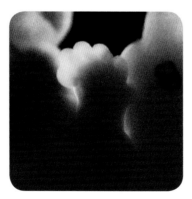

Shoulders and elbows develop and the arms project forward.

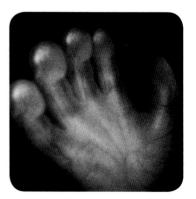

The webbed hands now develop fingers with touchpads.

THE AMNIOTIC SAC

The fetus continues to float within the amniotic sac, which is surrounded by an inner layer called the amnion and an outer layer called the chorion. These two layers are separated by a space (the extracoelomic cavity), which contains the yolk sac.

The small fingers of tissue called chorionic villi that sprout from the chorion are becoming concentrated in one circular area on the wall of the uterus; this will soon develop into the placenta. At this site, the villi are developing blood vessels and burrowing into the lining of the uterus, setting up their future access to your circulation.

Elsewhere, the chorionic villi are disappearing and the smooth chorion (known as the chorion laeve) is forming; this will fuse with the wall of the uterus in the second trimester, when the growing fetus has distended the uterine cavity more. The umbilical cord is now formed and blood is circulating through it, although the fetus is still receiving most nourishment from the yolk sac.

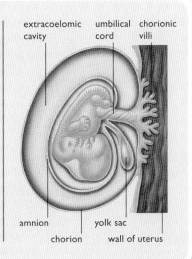

extracoelomic cavity
umbilical cord
chorionic villi
amnion
chorion
yolk sac
wall of uterus

and the intestines grow so fast that loops are formed and for a time, some of these actually protrude through the baby's abdominal wall.

By the end of the embryonic period, the new fetus has all its major organs and body systems, although the brain and spinal cord will continue to develop throughout your pregnancy. During this critical period of structural development, the fetus is highly susceptible to the damaging effects of a variety of drugs, viruses, and environmental factors (see p.30). It is very rare for congenital fetal abnormalities to develop after this time.

Your changing body

During the next few weeks, your uterus will grow significantly in size. By eight weeks it is the size of a medium orange and by 10 weeks the size of a small grapefruit. However, it cannot be felt through your abdominal wall because it is still behind your pubic bone.

This growth in the uterus can only be achieved by increasing the blood flow to it. In the nonpregnant state the uterus receives about two percent of the total amount of blood that is pumped through the heart per minute (the cardiac output). Very early in pregnancy, this percentage increases dramatically and by the end of this trimester 25 percent of your cardiac output will be directed to the uterus to deal with the demands of the placenta and baby. This increase in cardiac output is mainly due to the volume of blood pumped with each heartbeat (the stroke volume) since your heart rate (the number of times the heart beats per minute) increases only slightly during pregnancy. The thick muscular walls of the heart are relaxed by pregnancy hormones and this allows the heart to increase the volume of blood it contains each time it fills (diastole) without having to increase the force with which it pumps the blood during contractions (systole). To ensure that your blood pressure does not become too high as a result of the rise in cardiac output and blood volume, the blood vessels throughout your body also develop an increased capacity to hold larger blood volumes—once again due to the increase in pregnancy hormones, especially progesterone. This is why your systolic blood pressure falls only slightly during pregnancy, but your diastolic blood pressure is markedly reduced, a change that occurs early in the first trimester and only returns to normal nonpregnant levels near to the time of delivery.

Noticeable effects

As a result of these dramatic changes in your circulatory system, you will start to become aware of differences in the way your body is functioning. You already have noticed that you need to urinate more frequently as a result of your kidneys working much harder to filter your blood more efficiently. If your breasts had not begun to change earlier, they will almost certainly be larger, heavier, and more tender now, because the milk ducts are already beginning to swell in preparation for lactation and the areola surrounding your nipples will be larger and darker in color. The sweat glands (called Montgomery's tubercles) in the areola, which look like little pimples around the nipples, have also enlarged and start to secrete a fluid to lubricate the nipples. This is one of the most reliable signs of a first pregnancy, but since they do not shrink completely after pregnancy this cannot be depended on as a diagnostic sign for subsequent pregnancies. An outer ring of lighter-colored tissue called the secondary areola starts to appear on the breasts, together with more visible veins, as a result of the increased blood flow.

Changes in your skin

One of the first things you may notice is that your skin is either more acne prone or drier than usual, due to your high progesterone levels. Many women also develop spidery red lines called spider nevi on their legs and across their upper chest. These are small blood vessels in the skin that have expanded due to the increased production of estrogen. They usually fade away after

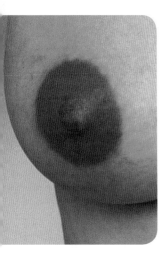

DARKENING AREOLA
The area around the nipples becomes larger and darker in color.

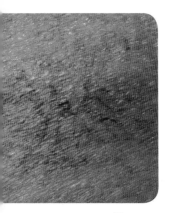

SPIDER NAEVI These tiny, spidery red lines on the skin are due to high levels of estrogen.

pregnancy and are nothing to be worried about. The blood supply to your skin has increased and because the veins are now much more dilated, you are better able to eliminate heat from your body surface. However annoying you may find your sudden intolerance of only moderately warm temperatures, this adaptation is essential because you need to be able to dispose of the rising heat that you generate from your increased metabolic rate and blood flow.

The skin of your genital area will start to darken and you will probably have noticed that your vaginal discharge has increased. This is due to the secretion of a watery substance which mixes with cells that are being shed from the vaginal walls. Normally this discharge is mucuslike and is usually clear or sometimes milky colored. It may stain your underwear but it should not be the source of any discomfort. If the discharge becomes yellow, develops an offensive odor, or causes you symptoms of itching or soreness, see your physician (see p.215).

...the reality is that many women are not physically sick, but they do feel very nauseous.

How you may feel physically

Some women sail through the first trimester feeling neither tired nor sick—indeed some will not even realize that they are pregnant. For the majority of women, however, the first three months are usually dominated by unpleasant symptoms—nausea, vomiting, and exhaustion.

No one can predict how you will feel during these early weeks of pregnancy because symptoms vary considerably from one woman to another and in each pregnancy that a woman has. Nor is there an accepted time for these common symptoms to start or finish. Some women feel exhausted from the time of the positive pregnancy test until they reach the second trimester, whereas others will only be troubled for a short time. Similarly, nausea may hit you hard and then disappear or be troublesome for many weeks.

Nausea and vomiting in pregnancy

Nausea and vomiting are, without doubt, the best known and most talked about side effect of early pregnancy—as many as 70 to 80 percent of all pregnant women will experience some degree of it. I have always thought "morning sickness" a poor description of the problem, since the reality is that many women are not physically sick, but they do feel nauseous. What's more, this feeling is rarely confined to the mornings—it can continue throughout the day or only be a problem in the evening. However, I do want to stress that it is also perfectly normal not to feel any nausea and if you happen to be one of these fortunate women, then count your lucky stars. Many women I meet in their early pregnancy worry that this is a warning sign that their pregnancy may be less robust or at greater risk of miscarriage. I promise you that you do not have to experience the misery of throwing up like clockwork every day in order to have a successful pregnancy.

No one has come up with the definitive answer as to why nausea and vomiting occur but there are several plausible theories. Like most baffling medical problems, nausea is probably the result of a combination of a number of different factors. One suggested cause is the high levels of the hormone human chorionic gonadotropin (HCG), which are present during the first trimester and then taper off at about 13 weeks. This would explain why nausea usually resolves spontaneously between the 16th and 20th weeks of pregnancy, although some women continue to feel sick for much longer.

Another theory is that nausea is linked to low blood sugar levels, since it often occurs first thing in the morning after many hours without food, or at the end of the day when you may be tired and in need of rest and nourishment.

Another possible explanation is that the flood of progesterone in pregnancy relaxes the smooth muscles in the digestive tract and slows down the passage of food through it. As a result, the food you eat and the digestive acids that are produced to process it remain in your stomach for longer periods of time. This is why you may feel nauseous and may occasionally vomit as well.

...there is a wide spectrum of foods that women either crave or develop a violent aversion to in early pregnancy.

HOW TO RELIEVE NAUSEA AND VOMITING

There is no magic cure, just a variety of remedies to try in different combinations to see which work for you. I often ask my patients about the individual remedies that helped them deal with their nausea and I have included some of their recommendations here.

▶ **Eat small, easily digestible meals** at regular intervals rather than just one or two large meals during the day. Dry toast, plain cookies, rice cakes, and savory crackers are good standbys when you

cannot stomach anything else. Cut down on the snacks when you start eating regular meals again or you will soon find that you are putting on unwanted pounds.

▶ **Stay clear of fatty food** because it can be particularly troublesome.

▶ **Bland foods** such as dry cornflakes and other cereals are a well-tolerated favorite and have the advantage that they are fortified with iron and vitamins. They are a good substitute for a meal when you can't manage anything else.

Whatever the cause, I know from personal experience how distressing and uncomfortable, not to mention inconvenient, nausea can be while it lasts. In addition to this, you may start to worry that your inability to keep any food or fluids down may be putting your baby at risk. I must reassure you that this is not the case. However little you are eating or drinking, your baby will getting the best part, including everything it needs to develop normally. You may be feeling terrible, but your baby is fine.

More serious vomiting

Occasionally, women vomit regularly and for such long periods of time (by which I mean weeks, not days) that they become dehydrated and weak because they cannot keep down fluids or food of any kind. This condition is called hyperemesis gravidarum and fortunately occurs only in about 1 in 200–500 pregnancies. However, if you do develop this problem, you will probably need to be admitted to the hospital briefly in order to have an intravenous line give you the necessary fluids, glucose, and minerals to help you rehydrate and stop you from feeling so weak and ill. If you do need to be hospitalized you will probably be advised to take some anti-vomiting medication (called antiemetics), either in pill form or via your IV line. These drugs are known to be safe in early pregnancy and will not have any harmful effect on your baby. Ever since the thalidomide disaster during the 1950s and '60s, doctors are extremely careful about what they prescribe to pregnant women to combat nausea. The antiemetic drugs that are now used have an excellent safety record, so do take them if you are advised to, because they will help you through a difficult period.

PEPPERMINT TEA
The refreshing taste of peppermint tea seems to combat the metallic taste that often accompanies feelings of nausea.

▶ **If you feel particularly nauseous** when you wake up in the morning, try nibbling on plain crackers before you get out of bed.

▶ **Some women swear by acupressure wristbands** (normally used to prevent motion sickness). These work by pressing on the acupuncture point known as P6.

▶ **Herbal teas** are also regularly recommended by my patients, especially peppermint tea, which has a refreshing taste that helps combat the horrible metallic taste in the mouth that frequently develops when you feel nauseous. For the same reason, brushing your teeth at regular intervals during the day can also be a source of relief.

▶ **Try taking small amounts of ginger** either as ginger ale, ginger tea, ginger capsules, crystalized or root ginger, or gingersnaps and other ginger cookies.

CATNAPPING No one knows why women feel incredibly tired in the early weeks of pregnancy, but while you feel this way, take a nap whenever the chance presents itself.

Food aversions and cravings

These often go hand in hand with nausea and vomiting, although they can also occur on their own. Again, we don't know why they happen, nor why there is such a wide spectrum of foods that women either crave or develop a violent aversion to in early pregnancy. I remember how puzzled I felt when I was suddenly unable to drink my morning cup of coffee, and when orange juice seemed so heavy that I found it undrinkable. The smell of roasted meat of any color was repellent to me and, although a cheese lover since childhood, I could not look the smallest morsel in the eye without feeling seriously nauseous.

I worried that the only sustenance I was offering my babies was grapefruit juice diluted with club soda, together with the occasional piece of bread and butter accompanied by an apple or a canned asparagus spear. But, compared to some of the stories about I have heard about the food cravings of pregnant mothers, the oddities I've just described seem quite mild.

Cravings for salty food such as pickles at strange times of the day or night are also common. Perhaps this is our bodies' way of telling us that we need salt, but no one really knows. Similarly, we have all heard about pregnant women who develop a pica: a desire to eat an unusual substance such as chalk, coal, or grass or to smell substances such as mothballs. I have no practical experience of these, but I can reassure you that I know of no data to suggest that they have ever caused any harm to a pregnancy. There are only a few foods in pregnancy—such as liver and unpasteurized cheeses for example—that are potentially dangerous and you will find more information about these in the diet section of this book (see p.50).

Feeling tired

I suspect that the time-honored phrase "fatigue is the female condition" was first coined to describe a woman in early pregnancy. The feelings of exhaustion in the first few months can be quite overwhelming. I can remember finishing an ordinary day at work, reaching home, and being just about capable of putting my key in the lock before collapsing in a heap at the bottom of the stairs. There was nothing wrong; I simply couldn't overcome my exhaustion and climb to the top. No one has so far been able to give a good scientific explanation for this fatigue, although theories abound. Some doctors believe that it is caused by the soporific effects of the high levels of progesterone, while others attribute it to the huge physiological changes that are taking place—the raised cardiac output, blood volume and oxygen consumption. The speed at which the tiny embryo is growing is another explanation, but you may find it difficult to understand how a baby that is small enough to fit into the palm of your closed hand can bring about such a dramatic change in your energy levels.

Like every experience in pregnancy, fatigue passes, but I am mentioning it here because it often raises concerns in partners and other family members. They see a woman who is usually full of energy appearing limp and exhausted and, since fatigue is so often equated with illness, they may be worried. Rest assured that after a couple of months of needing extra sleep and catnaps, fatigue lifts, so for the time being just go along with what your body is telling you to do.

> *The feelings of exhaustion that many women experience in the first few months can be quite overwhelming.*

Your emotional response

If you are suffering from wild fluctuations in mood, these are undoubtedly the result of major hormonal changes occurring in early pregnancy. One minute you may be talking excitedly about the future, and a few minutes later you find yourself weeping like an overflowing bathtub about some trivial issue.

You may also find yourself verbally lashing out at a harmless comment your partner makes and accusing him of not understanding how you are feeling. Since you yourself may not know how you are feeling, or why you are so emotionally fragile, you can recognize how difficult it is for him at this time. The only practical thing to do is talk to him about how confused you feel and reassure him (and yourself) that you have not undergone a permanent character change. Although your mood swings can be very intense, leaving you feeling helpless and out of control, keep in mind that they are temporary and just one of the many side effects of a completely normal pregnancy.

You may also be feeling anxious about the future, the birth and your ability to be a good parent. However well you are adjusting to the important changes that this new baby will bring to your life, when you are feeling tired and nauseous you may still be daunted by the prospect of it.

Common concerns

The most common concern at this stage of pregnancy is about miscarriage, the majority of which occur in the early weeks. But not every worrying symptom means miscarriage is inevitable and with each week that passes your pregnancy is becoming more secure.

As many as one in three women have some sort of bleeding during the first trimester ranging from brown to bright red spotting to large blood clots. In the majority of cases, it settles down and does not mean that there is any serious problem, since most women go on to have healthy babies. Having said that, I do understand how alarming bleeding can be.

As a safety measure you may be offered an early ultrasound scan, during which it may be possible to identify the pregnancy sac in the uterine cavity and the fetal pole and yolk sac developing. This will be very reassuring. Some women are frightened that a scan may increase the bleeding (it will not) or confirm their fears that the pregnancy has been lost. These feelings are understandable, but it is always the best plan to establish what is happening at the earliest opportunity.

I will never forget how distressed I felt when I experienced a heavy bleed at eight weeks into my pregnancy. I was sitting quietly in a meeting of a large number of medical colleagues and suddenly, without warning, I realized that

DECLINING RISK OF MISCARRIAGE

▶ **Miscarriage is the most common complication** of pregnancy and by definition can occur at any gestational age up until 20 weeks (see p.431). However, the vast majority of miscarriages occur very early on, even before the pregnancy can be recognized on an ultrasound scan.

▶ **If you are six weeks from your last menstrual period**, the risk of miscarriage has fallen to about 15 percent or 1 in 6

pregnancies. At this stage it is usually possible to see the yolk sac in your uterus and the fetal pole inside it on an ultrasound scan.

▶ **By eight weeks, the risk is much smaller** and, if a fetal heartbeat can be seen on the scan at this stage, your risk of miscarriage has fallen to five percent. Looked at more positively, this means that 95 percent of pregnant women with a fetal

heartbeat at eight weeks can expect their pregnancy to continue and to take home a baby at the end of it.

▶ **After 12 weeks**, the risk of miscarriage is no more than one percent. So the message here is that, as pregnancy progresses, the risk of miscarriage falls dramatically and, by the time you reach the end of this trimester you are very unlikely to experience this distressing event.

❝ *Bleeding in early pregnancy is always worth investigating and, even if it is heavy, does not necessarily mean that the pregnancy is over.* ❞

my seat was warm and wet and that I was bleeding. There was no pain and no warning—it just happened. I immediately assumed that I was miscarrying and after leaving the meeting as discreetly as possible for a woman with blood all over her clothes, I went home and cried. Happily, the scan showed two little embryos who appeared quite unruffled by the bleeding of the day before. Bleeding in early pregnancy is always worth investigating and, even if it is heavy, does not necessarily mean that the pregnancy is over.

Abdominal pain

Most pregnant women experience some abdominal aches and pains in early pregnancy. They are always a source of worry but try to remember that most of the time they simply reflect that enormous changes are occurring in your pelvic organs, particularly your growing uterus. All this growth occurs at the end of the same ligaments and muscles that were attached to your uterus in its nonpregnant state. So it is hardly surprising that the inevitable stretching of these ligaments results in some twinges and discomfort.

On the other hand, if your abdominal pains become constant or very severe, tell your doctor immediately, because they may be a sign that you have an ectopic pregnancy, which needs to be investigated and treated as a matter of urgency. Most ectopic pregnancies announce themselves during this stage of pregnancy and if you develop severe abdominal pain, your doctor will arrange for you to have an ultrasound scan to see if the pregnancy sac is in the uterine cavity. If there is no sac in the uterus, you will need more investigative tests, which may include a laparoscopic examination under general anesthesia.

Feeling dizzy

Dizziness and feeling faint or light-headed are also very common symptoms in early pregnancy. Most of the time these symptoms are harmless, but if they keep occurring they can become a cause for concern. If you feel faint or light-headed while sitting down, one of the most likely explanations is that your blood sugar levels are low. This is quite common in the first trimester when many women find it very difficult to eat properly; you can solve it by making sure that you keep a supply of small carbohydrate snacks with you and eating them regularly. If you find yourself feeling faint or dizzy when you stand up suddenly or have been standing up for a long period, it is because there is insufficient blood reaching your brain at that moment. Your blood supply has increased, but when you are upright, it pools in your legs and feet. When you stand up suddenly, the blood rushes into your legs and the supply to your brain is reduced.

Things to consider

At this stage of pregnancy there are rarely any pressing concerns —just a period of gentle adjustment to the idea that you are carrying a new life and a few strategies that will help lay a good foundation for the months ahead.

Visiting the dentist

There are several good reasons why you should visit a dentist regularly over the next 40 weeks. In pregnancy your gums become softened by pregnancy hormones and are more likely to bleed and become infected. Thorough toothbrushing and flossing, together with cleaning and plaque removal by your dentist, will help you limit your chances of tooth decay and gum disease during your pregnancy.

Your dentist will always try to avoid performing any tooth or jaw X-rays when you are pregnant, but if your dental problems are serious and causing you pain, let me reassure you that your mouth is a long way (in X-ray terms) from your tiny embryo, and that there are lots of gadgets to help ensure that the rays do not spread anywhere else. Local anesthesia is also perfectly safe, so you do not have to endure dental procedures without pain relief.

A pregnancy bra

Your breasts have begun to increase in size and, for some women, may have become very uncomfortable, even painful. Now is the time to invest in a couple of good pregnancy bras because sagging breasts will be the source of physical discomfort and backaches, not to mention distress at your own appearance. I remember thinking that there was no point in buying a new set of bras at this stage, because I wrongly assumed that I would grow out of them in a month or so and then need to invest in another set. The reality is that your breasts enlarge during the first three months of pregnancy and then do not change much until after the birth when you start breast-feeding at which point you need a completely different type of breast support. The best way to be sure that you buy the right type of bra is to find a department store that has

You will love this new baby every bit as passionately as you do your other child.

salespeople with particular expertise in fitting a pregnancy bra. A pregnancy bra has good support all the way around, including the underarm and back sections. Underwire bras are not a good option; the wiring will dig into your breasts and may harm the later development of breast milk ducts. If your breasts were full before you became pregnant, you may want to wear a bra to bed at night.

Women who have had cosmetic breast implants may be feeling especially tender now that their own breast tissue is growing. The skin over the breasts may feel taut and uncomfortable, too. You may be wondering if you will be able to breast-feed, and this largely depends on where the incisions were made to insert your implants. If cuts were made around the areola of your nipples, milk ducts and nerves that are essential to breast-feeding may have been severed. If incisions were made under your breasts, there is a good chance these are unaffected.

MATERNITY BRA
It is never too early to be fitted for a good supportive bra since your breast size increases right at the beginning but then will not change much more during the pregnancy.

Telling the rest of the family

If you have other children, you may well be anxious about how this new pregnancy will affect them and how they will react to their new sibling. There is no doubt that, while many older children are thrilled to learn that there will be a new baby arriving in the family, some younger children will not be so delighted. At this early stage I think it is probably best to wait a little while before telling your existing children that you are pregnant.

However, if you have experienced early pregnancy problems and needed to go into the hospital to resolve them, young toddlers may be extremely upset by your sudden disappearance and "illness." In the world of a young child, mommies are meant to be reliably available, rock-solid figures and disappearing at short notice can be disturbing. If you find yourself in this situation, my advice is to explain that you have not been well and be as honest as possible. The details you share with your children will depend on their ability to comprehend, but whatever you say to them, emphasize that you are going to get better quickly.

How can I love another child?

Some women worry that they will not love their new baby as much as their existing child and they can't imagine how this new addition will fit into their family set-up. I can assure you that such thoughts will seem ridiculous in a year's time. You will look back and not be able to imagine what life was like before this birth and love this baby every bit as passionately as you do your other child.

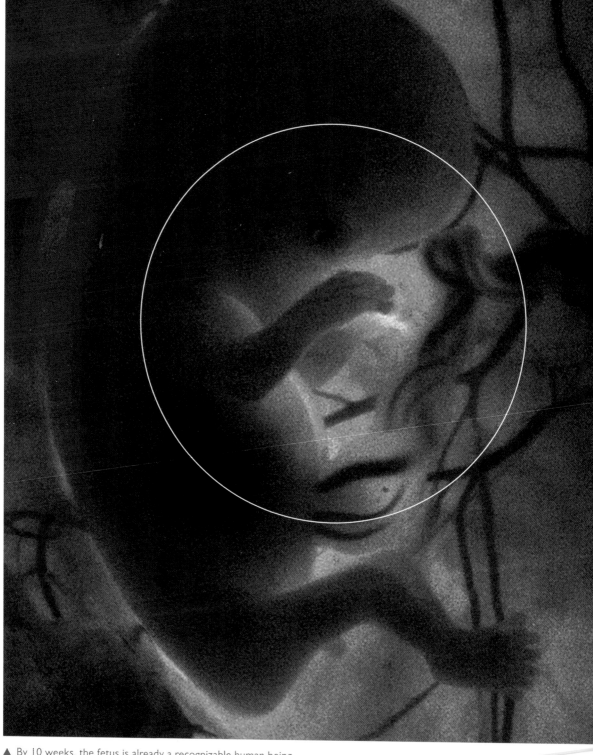

▲ By 10 weeks, the fetus is already a recognizable human being.

| 1 | 2 | 3 | 4 | 5 | 6 | 7 | 8 | 9 | 10 | 11 | 12 | 13 | 14 | 15 | 16 | 17 | 18 | 19 | 20 |

▶ WEEKS 0–6   ▶ WEEKS 6–10   ▶ WEEKS 10–13   ▶ WEEKS 13–17   ▶ WEEKS 17–21

▶ FIRST TRIMESTER   ▶ SECOND TRIMESTER

▶ WEEKS 10–13
The developing baby

**THE FETAL STAGE HAS BEGUN AND ALL OF YOUR BABY'S VITAL BODY ORGANS
ARE NOW IN PLACE. FROM THIS TIME ONWARD, YOUR BABY'S DEVELOPMENT
WILL BE CONCERNED ENTIRELY WITH THE GROWTH AND MATURATION OF THESE
MAJOR BODY SYSTEMS.**

During the next few weeks the fetus will grow rapidly and steadily, at a rate of
about ½in (10mm) per week and its weight will increase five-fold. If you have
an ultrasound scan at this stage, you will be amazed by how easy it is to
recognize the various parts of your baby's body and that it is already starting
to look more and more like a tiny human.

The fetal head is still relatively large, accounting for approximately one-
third of its length from the crown of its head to its bottom (the crown rump
length or CRL), but the growth of the rest of the body is starting to catch up.
The head is now supported by a recognizable neck and the features on the face
are better defined, since all of the facial bones are completely formed. The
forehead is still high, but there is now an obvious jaw line and chin and the
nose is more pronounced; 32 tooth buds are in place. The eyes are fully
developed and although still quite widely spaced, now appear closer to the
front of the face. The eyelids are still developing and remain tightly closed.
The external ears (pinna) become more clearly visible as they enlarge and
assume their adult shape. They have now moved from the base of the skull to
a higher position on the sides of the fetal head. The inner ear and middle ear
are completely developed. The fetal skin is still thin, transparent and permeable
to the amniotic fluid and a layer of fine hairs now covers most of the body.

life size

At 10 weeks, the fetus
measures 1¼in (30mm)
and weighs 0.1oz (3–5g).
By the 13th week, it is
3in (80mm) long and
weighs about 1oz (25g).

Your baby's limbs
The fetal body appears much straighter than it was just a few weeks ago. The
limbs are growing rapidly, and shoulders, elbows, wrists, and fingers can be
clearly seen. The lower limbs are developing too, but their growth will be at a
slower pace for some time. The fingers and toes are separating into individual

> ❝ *The fetus is now moving around quite vigorously inside the amniotic sac, producing small jerky movements …* ❞

digits and tiny nails are now present. At about 12 weeks, hard bone centers develop in the cartilage of the fetal bones, a process called ossification. As calcium continues to be deposited in these centers, the skeleton will gradually calcify and harden. This formation of hard bone continues long after your baby is born and will not be complete until adolescence. The fetus is now moving around quite vigorously inside the amniotic sac, producing small jerky movements of its body and upper limbs rather than simply free-floating. However, you are still unaware of its movements. The muscles of the chest wall are starting to develop and practice breathing movements along with the occasional hiccup and swallowing movements can be seen on the ultrasound. Even more excitingly, your baby is starting to make reflex responses to external stimuli. For example, if your abdomen is prodded, the baby will try to wiggle away from the intruding finger. If a hand or foot happens to brush against the baby's mouth, the lips purse and the forehead may wrinkle—this is the very first sign of the future sucking reflex. Similarly, if the eyelids are touched, an early blinking reflex can be seen. However, these are only reflex movements and it is generally accepted that the fetus does not have the ability to feel pain until about 24 weeks of gestation.

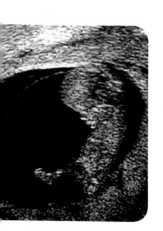

FETUS AT 10 WEEKS
A 2-D ultrasound reveals the fetus free-floating in the black circle of the amniotic sac. The faint white line is the sac's outer layer, called the chorion, which is still separate from the uterine wall.

Inside your baby's body

Inside your baby's body, the ovaries or testes have fully formed, and the external genitalia are developing from a small swelling between the fetal legs into a recognizable penis or clitoris. In theory, an experienced ultrasound scanner may be able to determine the sex of your baby at this early stage, but if you rely on the diagnosis now you might be in for a surprise at the birth.

The heart is now fully functional and pumping blood to all parts of the fetal body at a rate of 110–160 beats per minute. This is slower than it was a few weeks ago and will continue to slow down as the fetus becomes more mature. It may be possible to hear the heartbeat with a sonicaid machine placed over your lower abdomen, just above the pubic bone. This device uses Doppler ultrasound waves and is completely harmless to your baby. During the early weeks of pregnancy the embryo's blood cells were manufactured in the yolk sac, but by 12 to 13 weeks the sac is fast disappearing, and this essential task is taken over by the fetal liver. Later in the second trimester, the fetal bone marrow and spleen will make important contributions to the production of blood.

The chest and abdomen are gradually straightening out, and the intestines, which a few weeks ago were coiled around the umbilical cord in the amniotic cavity, are now firmly behind a closed abdominal wall. The fetal stomach is now linked to the mouth and the intestines—an important development because the fetus now begins to swallow small amounts of amniotic fluid. This will later be excreted as urine when the fetal kidneys start to function.

The volume of amniotic fluid at 12 weeks is about 30ml—less than 6 teaspoons. The amniotic fluid has many protective functions not the least of which is to provide a sterile swimming pool maintained at a constant temperature (slightly higher than your own) in which the baby moves freely. Later on, waste products excreted in the fetal urine will be absorbed from the amniotic fluid back into the maternal blood by passing through the placental membranes.

FETUS AT 13 WEEKS
The fetal arms have developed rapidly—elbows, wrists and hands with fingers are clearly visible on this three-dimensional ultrasound. The fetus makes reflex responses when its hand brushes its face.

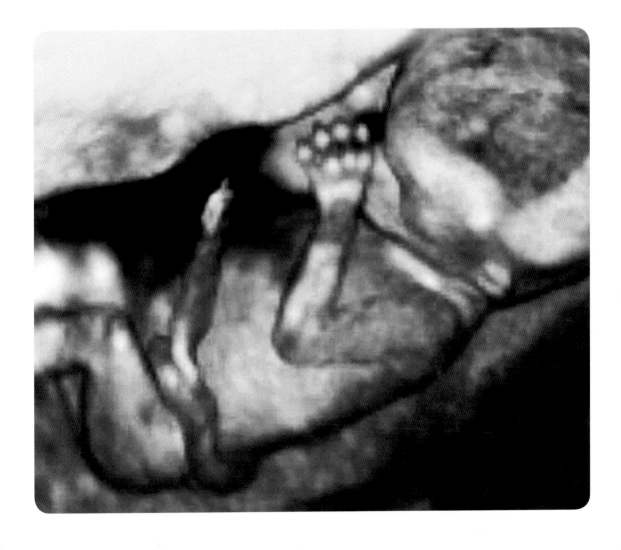

THE PLACENTA

Your fully-formed placenta is now your baby's life support system. It is a complex piece of biological engineering that supplies all your baby's needs via your bloodstream, yet also acts as a barrier against infections and harmful substances.

TThe placenta has been developing rapidly and by 12–13 weeks its structure is complete, although it will continue to grow in size throughout the remainder of pregnancy. By the end of the first trimester, it is fully established and performs a variety of essential functions for the rest of the pregnancy. In essence, the placenta is a sophisticated filtering system that allows your baby to breathe, eat and excrete. It also acts as a protective barrier that shields your baby from most infections and potentially harmful substances. In addition, the placenta is responsible for producing increasing quantities of hormones that help to maintain the pregnancy and prepare your body for birth and breastfeeding.

All of this placental activity uses up a considerable amount of energy, and the metabolic rate of the placenta is similar to that of the adult liver or kidney. Furthermore, the successful functioning of the placenta depends upon a good maternal blood supply to the spiral arteries in the uterine wall. This is why smoking and disorders such as high blood pressure and preeclampsia (see p.426) that reduce the blood flow to the placenta, have a dramatic effect on its function and on the growth of the fetus.

A NETWORK OF TREES

The placenta is best understood by imagining a network of branching trees made up of about 200 trunks. These divide into limbs, branches, and twigs, which are covered with an extensive network of chorionic villi, the majority of which float in a lake of maternal blood called the intervillous space. Some of the longest branches grow down into the decidual lining of the uterus. A few travel even farther, penetrating the deeper layers of the uterine wall to access the maternal blood vessels. These anchoring villi help to form the lower boundary of the intervillous space.

THE UMBILICAL CORD

The umbilical cord is now fully formed and consists of three blood vessels: a single large vein, which carries oxygen-rich blood and nutrients from the uterus to the fetus, via the placenta; and two small arteries, which transport waste products and oxygen-depleted blood from the fetus back to the mother for recharging. The three vessels are coiled like a spring to ensure that the growing baby can move around easily within the amniotic sac and are covered by a thick protective coating called Wharton's jelly.

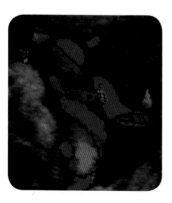

Doppler scan of the umbilical cord shows blood flow via a single large vein (red) and two arteries (blue).

THE LIFE SUPPORT SYSTEM

The extensive network of chorionic villi are bathed in maternal blood inside the intervillous space. Although nutrients and waste products can pass freely through the chorion (a thin membrane that surrounds the villi), the chorion acts as a barrier, too, protecting the fetus from infections and environmental poisons.

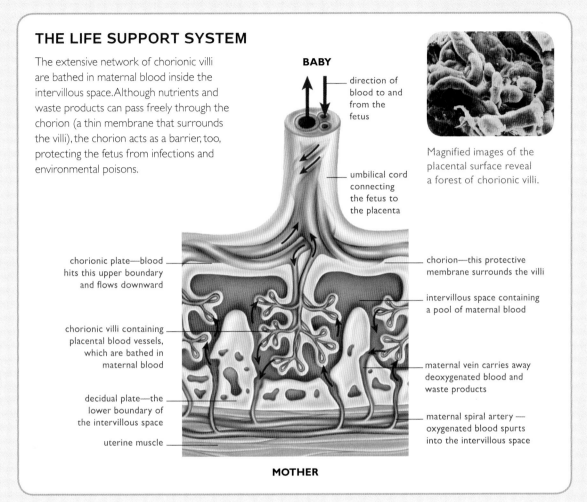

BABY

direction of blood to and from the fetus

umbilical cord connecting the fetus to the placenta

Magnified images of the placental surface reveal a forest of chorionic villi.

chorionic plate—blood hits this upper boundary and flows downward

chorionic villi containing placental blood vessels, which are bathed in maternal blood

decidual plate—the lower boundary of the intervillous space

uterine muscle

chorion—this protective membrane surrounds the villi

intervillous space containing a pool of maternal blood

maternal vein carries away deoxygenated blood and waste products

maternal spiral artery — oxygenated blood spurts into the intervillous space

MOTHER

OXYGEN AND NUTRIENTS

With every beat of your heart, blood from your arteries in the lower boundary of the intervillous space (the decidual plate) spurts into the intervillous space like a fountain. The blood hits the upper boundary (the chorionic plate) and then flows downward bathing the chorionic villi before seeping away through the veins in the decidual plate. The large number of blood vessels in the chorionic villi and the relatively sluggish flow of your blood through the intervillous space gives ample opportunity for oxygen and nutrients to pass to the fetal circulation. At the same time, carbon dioxide and other waste products from the fetus are transferred into the intervillous space and carried away in your blood.

SEPARATE CIRCULATIONS

Despite their very close proximity in the intervillous space, the maternal and fetal circulations remain entirely separate and never mix. They are separated by a thin membrane, in places as little as only one cell thick. This protects the developing fetus from infections, and other damaging substances such as pesticides, alcohol, and some drugs. Furthermore, any bleeding in pregnancy always originates from the pool of maternal blood, not from the fetus. The fetal circulation is protected, even if the placenta is damaged.

Your changing body

By the end of this trimester your waist will probably have thickened slightly and you will have put on a little weight. Your belly may be starting to swell too, but this is more likely to be due to mild bloating and sluggish bowels than to your growing baby.

At 10 weeks the size of your uterus will be equivalent to a large orange. By 12 weeks it will have reached the size of a large grapefruit and by 14 weeks it will be the size of a small melon. Sometime between 11 and 14 weeks, depending on your body weight and the size of your pelvis, your doctor will be able to feel the enlarging uterus through your abdominal wall just above the pubic bone. Of course, if you are expecting twins or triplets, the enlarging uterus rises above the pelvic brim at an earlier stage. Indeed, before the days of ultrasound, this was one of the first signs of a multiple pregnancy.

Your breasts will have continued to develop under the influence of progesterone and several other hormones, the production of which has been steadily increasing during this first trimester. Some previously small-chested women are alarmed to find themselves going up three or four bra cup sizes in the first 12 weeks of pregnancy. If this is the case with you, don't panic, you will probably find that your breasts stop filling out fairly soon and then will only increase a little more in the last month or so before the birth. If your breasts were full before you became pregnant, you may want to wear a bra to bed at night now for support. If they feel very tender, try using a soothing massage cream on the breast tissue and nipples.

By 12 weeks your uterus will be the size of a large grapefruit… by 14 weeks it will be the size of a small melon

Increasing your oxygen supply

Many women notice that they occasionally feel breathless by the end of the first trimester; a symptom that sometimes continues throughout pregnancy. The major changes that have occurred in your heart and blood vessels mean that almost every organ in your body is now working much harder, demanding a dramatic increase in your oxygen supplies. Your oxygen requirements rise by 15–20 percent during pregnancy, and half of this is used by the growing uterus, placenta, and baby. The other half is needed to fuel the work of the heart and kidneys, although some is directed to the respiratory muscles, breasts, and skin.

To achieve this extra supply, your lungs need to make a special adaptation to take in an increased volume of oxygen and expel an increased amount of waste carbon dioxide with every breath that you take. This is called the tidal volume of air and it increases by 40 percent during pregnancy. When you exercise or exert yourself, your tidal volume and oxygen consumption increase way above their pre-pregnancy levels. Although the exact mechanisms for achieving this are not understood, we do know that progesterone makes an important contribution. It effectively allows your lungs to hyperventilate or overbreathe and that is why you experience feelings of breathlessness.

How you may feel physically

By the end of the first trimester, you are likely to be getting back into your stride and feeling more like your old self. However, every pregnancy is different—there are no hard and fast rules governing how you should or should not feel.

For many women, the nausea and vomiting that often dominate the first 10–12 weeks of pregnancy has started to settle down, but for the less fortunate it may persist for a longer period of time. No one can predict a pregnancy experience exactly. However, as the nausea improves, you will be able to start eating normally once again, and if you have been worrying about how little nourishment you and your baby have been receiving, this is a very welcome change.

Your uterus has already enlarged considerably, and as the ligaments attaching it to the sides of your pelvis have been stretched, it is very normal to experience the occasional twinge or muscular ache. Of course, if the pain continues or becomes severe, seek advice from your doctor as soon as possible. Now that the uterus is moving up into the abdominal cavity, there should be less pressure on your bladder, giving you some respite from the constant calls to urinate.

You may still be experiencing bouts of severe fatigue, but in general, the complete exhaustion characteristic of the first 10 weeks of pregnancy is beginning to lift and you should be feeling more energetic. Undoubtedly, there will still be days when you feel awful but these are becoming less frequent now, and after a bad day you will be pleasantly surprised at how quickly you are able to bounce back. Indeed, some women feel physically exhilarated at this stage in pregnancy. Whatever the case is for you, take this period of transition at your own pace and aim to deal with life one day at a time.

...after a bad day you will be pleasantly surprised at how quickly you are able to bounce back

Your emotional response

If you have been troubled by mood swings, these are likely to be settling down now that you have had time to make some physical and mental adjustments to being pregnant. That said, there may still be times when you feel inexplicably anxious or irritable.

Just getting to the end of the first trimester eliminates an important source of worry because from now onward you are extremely unlikely to suffer a miscarriage or later pregnancy loss. The reality is that the vast majority of miscarriages have occurred well before 10 weeks of gestation. After 12 weeks, the risk of losing a pregnancy is no more than one percent.

Many women tell me this was the point in their pregnancy when their partner first started to come to terms with the idea that he was going to become a father, and that there was going to be a baby in the house in the not too distant future. Even if his reaction is not exactly what you had hoped for, it is a relief that someone else is at least sharing the realization of what being pregnant is all about. Up until now, there have probably been days when it has been difficult for both of you to believe that you are pregnant but from now on, there will be no doubt in your mind, particularly after your doctor visit and having enjoyed the excitement of seeing your tiny baby on an ultrasound scan for the first time (see p.124). Like many other important things in life, dealing with certainty is much easier than dealing with uncertainty.

Some women are able to relate to their unborn baby from a very early stage in the pregnancy, whereas others find it virtually impossible until much later in pregnancy, particularly when it is a first pregnancy. I want to say here that there is nothing odd about talking to your baby and generally including this new person in your everyday life, if that feels right for you. Nor is it odd if you cannot begin to imagine that the tiny fetus growing inside you is going to develop into a real human being. What you do and how you feel is in no way an indication of how good or bad a mother you will turn out to be; it is just another example of how individual we human beings are.

>
> *...this may be the point when your partner first starts to come to terms with the idea that he is going to be a father.*

Telling people your news

Now that you feel more secure about your pregnancy and no doubt convinced that everyone around you must have noticed that you are beginning to change shape, you will probably decide to tell people that you are pregnant. Sharing your news with friends and family is usually a cause for celebration although inevitably there will be at least one person for whom the subject of pregnancy is difficult. Only you can know who to tell and how to handle the situation but it is worth remembering that, a bit like deciding on a guest list for a wedding, there may always be someone who feels left out or upset by your news.

This is a good point to refer back to the section on work and maternity rights (see pp.58–63) because this is a subject that you should start thinking about sooner rather than later. You need to be clearly aware of your employment rights before you share your news with your boss or head of personnel.

EXCITING NEWS
Most women begin to feel more secure about their pregnancy as they come to the end of the first trimester—and are ready to share their news.

Your prenatal care

Your first detailed prenatal visit will usually take place during the next few weeks. The purpose of this visit is to identify potential problems and create a set of individual prenatal medical case notes that detail all your current and past medical and social histories.

For your first visit to the doctor, you will need to set aside a couple of hours during which you will be asked a large number of detailed questions about your general health and your previous medical and gynecological history. Your responses will form the basis of your medical notes. This visit provides an opportunity to discuss your prenatal care and make arrangements for any tests that you may want to have performed (see pp.134–43). Some obstetricians suggest that get copies of your records for the duration of your pregnancy so that you always have them at hand wherever and whenever you may need medical assistance. For the majority of pregnant women, these notes will record the fact that you are at low risk of any complications and can be reassured of a trouble-free pregnancy.

During your visit, it is very important that you discuss every aspect of your medical history, lifestyle, and social circumstances with your doctor. Sadly, 1 in 4 women suffer domestic violence during pregnancy, which is why your doctor will enquire sensitively when you are alone whether you need help or support.

Your previous pregnancies

Your past obstetric history (if any) is important because the outcome of previous pregnancies and any problems that you may have had will help your doctor assess whether you can be classified as low risk or high risk in your current pregnancy and the type of prenatal care you require. For every past pregnancy you will be asked about the gestation in weeks at the time of delivery, the weight of your baby, whether the labor was spontaneous or induced, the method of delivery and any complications that arose prenatally, during the delivery, and postpartum. If your previous pregnancy was complicated and you were at another hospital, your doctor may request that you get a copy of your records.

Some women feel sensitive about a previous termination of pregnancy and would prefer not to have this information recorded in their copy of any records, which theoretically could be picked up and read by anyone. You may have similar concerns if you had assisted fertility treatment to achieve this pregnancy

QUESTIONS AT YOUR FIRST PRENATAL VISIT

The following list is not exhaustive but should give you a feel for the sort of information that your doctor needs to know.

▶ **What is the date of your last period?** Your Estimated Date of Delivery (EDD) will be calculated according to this date (see p.79) so try to figure it out before your visit.

▶ **Have you had problems becoming pregnant and if so, how was this pregnancy achieved?** Assisted fertility treatments such as IVF increase the chances of a multiple pregnancy, which needs specialized care.

▶ **Have you had any problems in this pregnancy so far?** This includes major concerns such as bleeding and abdominal pain, but you should also mention minor problems such as vaginal discharge.

Your doctor can then arrange for you to have appropriate investigations and treatment.

▶ **Do you smoke cigarettes or use recreational drugs?** If you haven't managed to give these up yet, this is an opportunity to ask for help to do so.

▶ **Do you have a medical illness?** If you have an illness such as diabetes, asthma, high blood pressure, thrombosis (blood clots), kidney or heart disease, you may need to see a specialist during your pregnancy and the type and dosage of drugs that you are taking may need to be altered.

▶ **Are you taking any medications?** Make sure that you mention any medicines and preparations that you are taking whether they are prescription

drugs, over-the-counter medicines, or complementary remedies.

▶ **Do you suffer from any allergies?** It is important to record any allergies such as hay fever, asthma as well as any allergic response you may have to medications, foods, adhesive bandages, and iodine.

▶ **Have you ever suffered from a psychiatric illness?** You may feel that this is an intrusive question, but pregnancy can have a profound effect on some psychiatric disorders. It is important that you discuss any problems that you may have had in the past, so that your doctor can help minimize future problems. Postpartum depression, for example, is very likely to recur but it can be treated effectively if it is recognized quickly. The best way to (believing this to be a private matter between you and your partner). I can understand these feelings, but you do need to discuss every part of your history with the doctor to make sure potential future complications can be identified and prevented.

Your first physical examination

How detailed a physical examination you have at your first prenatal visit varies between doctor's offices. When I was a junior trainee in obstetrics, a thorough physical examination of the heart, lungs, abdomen, legs, skin, and breasts, together with a routine pelvic and vaginal examination and pap smear were performed routinely by a doctor. Now, the physical examination tends to be much less intrusive, partly because most pregnant women are generally fit and well, but also because we are conscious that prenatal care providers are not general physicians and specific medical problems are best dealt with by a referral to a specialist. Some doctors' practices include a routine

do this is to identify those women who are at risk before it develops.

▶ **Have you undergone abdominal or pelvic surgery?** Previous surgical procedures may determine how you should deliver your baby. An elective cesarean is sometimes the preferred option if you have had, for example, surgery to remove a fibroid from your uterus. On the other hand, a vaginal delivery may be a better option if you have lesions or scar tissue as a result of a gastrointestinal or bladder operation. Always mention previous surgery, however minor you think it may have been.

▶ **Have you ever had a blood transfusion?** A previous blood transfusion will alert your doctor that you may have developed atypical antibodies in your blood or

be at risk of a blood-borne infection such as hepatitis or HIV. These are extremely unlikely complications in the US because our transfusion service is carefully monitored and is considered one of the safest in the world. However, this may not be the case if you had a transfusion in some other countries.

▶ **Do you have a history of infection, particularly a sexually transmitted disease?** All doctors will screen you for immunity to rubella and possible infection with syphilis, gonorrhea, and chlamydia. You may also be offered screening for Hepatitis B, Hepatitis C, and HIV infection. An HIV test requires your permission. I strongly advise you to give full information about your possible exposure to infections and to have

any screening tests that your doctor offers. The implications of these tests are discussed later in this section. Ignorance is not bliss for pregnant women. Knowledge will provide you and your unborn baby with an opportunity to reduce the damage the infection can cause.

▶ **Do you have a family history of twins, diabetes, high blood pressure, thrombosis, tuberculosis, congenital abnormalities or blood disorders?** If there is one or more of these in your family history, this does not necessarily mean that you will suffer from them during your pregnancy but it does alert your doctor to watch for signs that a potential problem is developing.

examination of your heart and lungs, but these are not usually very informative when performed by doctors who are not specialists.

If you have never had a medical problem and this is your first pregnancy, the physical examination at your first prenatal visit may be confined to measuring your height, weight, and blood pressure, taking a urine sample, and examining your hands, legs, and abdomen. If you are high risk, your heart and lungs may be examined also.

Height

If you are less than 5ft (1.5m) tall, you may be concerned that your pelvis is also smaller than average and that this may lead to problems at the time of delivery. For the same reason, your shoe size used to be recorded because small feet can indicate a narrow pelvis. However, the reality is that your height and shoe size are not conclusive measures of the capacity of your pelvis; your ability to deliver a baby cannot be accurately assessed until you are in full-blown labor. So I think that worrying about your height at this stage in pregnancy is not helpful. I have seen plenty of very short women deliver big babies and, seen some small babies experience problems negotiating an ample pelvis in a tall woman.

PRENATAL VISIT Being candid about any previous pregnancies and your past medical history will help your doctor tailor your prenatal care.

Weight

Your weight at the first prenatal visit is a more useful measurement than either your height or your shoe size, because you are more likely to experience problems during pregnancy and at the time of delivery if you are either significantly under- or overweight (see p.41). Pregnant women are weighed at every prenatal visit. If you are significantly under- or overweight at the beginning of your pregnancy, or if you have diabetes or you develop gestational diabetes, your doctor will monitor your weight and eating habits throughout your pregnancy and may recommend that you follow a calorie-controlled eating program.

Legs and hands

The appearance of your legs and hands is another useful baseline measurement at your appointment and some doctors keep a regular check on this during

WHY YOU MAY NEED SPECIAL CARE

The following factors may mean that you need specialist prenatal care:
▶ Previous preterm delivery (before 37 weeks)
▶ Recurrent miscarriages
▶ Baby with a congenital abnormality
▶ Preeclampsia or high blood pressure in previous pregnancy
▶ Diabetes or gestational diabetes
▶ Previous thrombosis (blood clot)

▶ Previous birth of baby weighing more than 9lb (4kg) or less than 5½lb (2.5kg)
▶ Pregnancy with identical twins (see p.123)
These risk factors may mean that you need extra care during delivery:
▶ Previous cesarean section
▶ Previous long labor and instrumental delivery (forceps

or vacuum)
▶ Previous failed induction of labor
▶ Previous birth of baby weighing more than 9lb (4kg) or less than 5½lb (2.5kg)
▶ Excessive bleeding after a birth (postpartum hemorrhage)
▶ Problems with an anesthestic
▶ Urinary or bowel problems after a delivery
▶ Current twin pregnancy

pregnancy. The color and condition of your fingernails is a useful factor when assessing your general health since they can reflect your diet and whether you are anemic. Spider nevi (small broken veins with a spiderlike appearance) and reddening of the palms and soles are to be expected in pregnant women, but the sudden appearance of lots of broken veins or areas of bruising suggests that you need tests, including blood clotting tests.

The other sign that will be looked for is swelling or puffiness of your fingers, feet, ankles and lower legs, which may indicate problems with fluid retention. In later pregnancy, a degree of swelling in these areas is very common, especially at the end of a busy day, but any sudden swelling or progressive increase in swelling needs to be taken seriously since it suggests that you are at risk of developing preeclampsia (see p.426).

Abdomen

Your doctor will examine the size of your expanding uterus at and will also want to see if you have any scars from previous surgeries and exactly where these are positioned. Provide as much detail as you can about previous abdominal or pelvic surgery, because this can influence decisions about how your baby would be best delivered. For example, having your appendix out could have been a simple uncomplicated surgery leaving you with a small scar on your right side. But if your appendix burst and you developed peritonitis, you may have had major abdominal surgery as an emergency and have been left with a scar that extends all the way down your abdomen and dense adhesions inside your abdominal cavity. For similar reasons, whether the scar is smooth, puckered or tethered to underlying tissues is also useful information

...the reality is that your height and shoe size are not conclusive measures of the capacity of your pelvis...

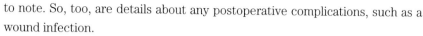

> " *Your blood pressure will be checked at every prenatal visit.* "

to note. So, too, are details about any postoperative complications, such as a wound infection.

Later in pregnancy, it is perfectly normal to notice stretch marks or stria appearing on your abdomen. However, if you suddenly develop livid stretch marks at this early stage in pregnancy it can be due to steroid medication or be a sign that you have an underlying hormonal problem. Your doctor will arrange for you to see a specialist promptly.

Vaginal and pelvic examinations

A routine vaginal examination is done at the first visit and sometimes at later prenatal visits are no longer considered necessary but if you have a discharge or have experienced some bleeding, your doctor may examine your cervix and take a swab to identify any infection.

We now recognize that trying to judge the capacity of your pelvis at this early stage does not really contribute very much to the planning of your prenatal care. However, there are situations when it may be helpful for you to be examined internally at this visit—for example, if you have had a previous cesarean section because you failed to progress in labor and the cause was attributed to the ischial spines of your pelvis being too prominent or your pubic arch being too narrow to let the baby through.

Breasts

Breast examination is may not performed at all visits but my opinion is that it should be. Fortunately, breast cancer is uncommon in women under the age of 40 years, but when it does develop in younger women, the tumor is usually estrogen dependent, which means that pregnancy can greatly accelerate both the local growth and distant spread of the abnormal cells. Obstetricians may not be the best clinicians to identify every suspicious breast lump, but it is probably better that they identify some rather than none at all, since early diagnosis and treatment may greatly improve prognosis. Advice and information about breast changes and self examination during pregnancy should be offered to all pregnant women.

Urine tests

You will be asked to produce a sample of urine at your all prenatal visits, which will be tested immediately with dipsticks specially treated to identify sugar, protein, and ketones (chemicals produced when fat is metabolized) in your urine. Normally, our kidneys filter out all of the sugar and protein from our urine. However, during pregnancy the increase in blood flow places more of a load on the kidneys, and as a result, the urine of pregnant women sometimes contains a small amount of sugar or protein. This always needs to be investigated further. Ketones are typically found in the urine of diabetic people, but in healthy pregnant women they are sometimes present if your metabolism is upset, such as when you have not been eating enough or have been vomiting. Your urine will be checked for ketones if you are unwell, unusually thirsty, or urinating more frequently than usual.

Glycosuria—sugar in the urine

It is common for pregnant women to have a small quantity of sugar in their urine (glycosuria) during the second or third trimester, but finding sugar in the urine this early in pregnancy is unusual. If it persists at subsequent prenatal visits it suggests that you may be developing gestational diabetes, a prenatal condition that affects around 5 percent of pregnant women (see p.427) and needs careful monitoring to minimize fetal complications. The US guidelines no longer recommend urine testing for glycosuria before the glucose tolerance test is done routinely at the end of the second trimester between 24 and 28 weeks (see p.212). However, your doctor will be checking your urine for protein at each visit, and if sugar is present you will be strongly advised to limit

MONITORING BLOOD PRESSURE

Checking and recording your blood pressure using the correct size of cuff is important because this first measurement will be used as the baseline reading against which all subsequent readings are compared during your pregnancy. Your blood pressure will then be measured at every prenatal visit regardless of where it takes place.

▶ **A reading of around 120/70mm Hg** is usual for most women. The first figure (120) refers to your systolic blood pressure, which means the pressure in the main blood vessels as your heart pumps blood around your body. The second figure (70) is the diastolic blood pressure— the pressure in your arteries when the heart is at rest. Both the systolic and diastolic readings are important, but if you have a diastolic blood pressure reading of 90 or more, your doctor will suggest that you be seen by a specialist team for advice.

▶ **A persistent increase of 15–20** to either or both of your baseline figures is usually a cause of concern since it suggests that you may be at risk of developing problems such as preeclampsia (see p.426). Of course, blood pressures vary a great deal between different women during pregnancy, so this figure should be regarded simply as a useful rule of thumb.

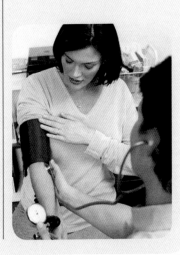

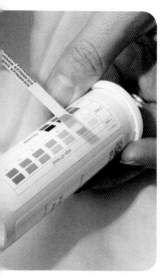

URINE TEST A dipstick is used to detect sugar in the urine. The chemically treated tip changes color according to the amount of glucose that is present.

your intake of sweet foods, particularly cakes, cookies, candy and chocolates, fruit juices and whole fruits with high sugar content (bananas, pineapple and melons). If the glycosuria persists, or your BMI is over 35, if you have a family history of diabetes or had gestational diabetes or a large baby in a previous pregnancy, or if you belong to a high risk ethnic group (including South or East Asian and Middle Eastern) you may be advised to have an early glucose tolerance test to see whether you have really developed gestational diabetes or have simply eaten too much sugar rich food just before your urine test.

Proteinuria—protein in the urine

There are several important causes of proteinuria in pregnancy so your doctor will want to investigate you thoroughly on any occasion that it is found. If you are found to have protein in your urine at your initial visit, you will be asked to produce a clean midstream sample of urine. To do this you will be given a special pack that contains a cleansing pad to wipe away organisms from your vulva and a sterile collection cup. After cleaning your vulva, you need to pass the first few drops of urine into the toilet and then collect a midstream sample of your urine into the cup to be sent off to the laboratory for testing.

The most common cause of proteinuria is an infection in your kidneys or urinary tract. You are more prone to these infections in pregnancy because the tubes that connect your kidneys to your bladder and your bladder to your urethra are relaxed under the influence of pregnancy hormones. This makes it much easier for infective organisms to gain access to your bladder and kidneys. Most importantly, the usual early sign of a urine infection—pain or discomfort when urinating, called cystitis—is often missing in pregnancy, which means that you can develop a full-blown infection of your kidneys (pyelonephritis) with very little warning. This is a potentially serious problem, because urine infections can cause the uterus to become irritable. If they are left untreated, they may lead to miscarriage or premature labor. Furthermore, repeated urine infections can cause permanent scarring of the kidneys. If an infection is diagnosed, you will be given appropriate antibiotic treatment and a further test will be performed approximately one week after completing the antibiotic course to ensure that the infection has cleared. Some doctors suggest that you have a midstream sample test as a routine part of your initial visit to avoid missing a silent urine infection that has not produced symptoms.

More rarely, protein in the urine at your initial visit can be a sign that you have underlying kidney disease, in which case you will probably be under the care of a specialist kidney (renal) doctor already. Occasionally, however, underlying renal disease is identified for the first time during regular urine

urine that is not caused by an infection or a previous history of renal problems, this will alert your doctors to the fact that you are a high-risk pregnancy and may develop preeclampsia or other complications at a later stage.

Your first ultrasound scan

Most, although not all, maternity units now offer pregnant women a dating ultrasound scan at around 10 to 12 weeks (see next page) to measure the baby's size. You may also be offered a Nuchal Fold Translucency Scan (see p.136) between 11 and 14 weeks for the early detection of Down syndrome.

If you are carrying twins, a scan at 12 weeks or earlier will identify which type of twins you are carrying (see p.125). This is an important detail to establish early on because it has implications for your prenatal care, and you will need a specialist's care. With twin pregnancies, there is a higher risk of miscarriage, intrauterine growth restriction, developing diabetes, as well as preeclampsia. You are more likely to have an earlier delivery date because of the slightly increased risk of stillbirth and complications occurring later in pregnancy.

All women are offered a scan at 18–20 weeks, which is usually referred to as the fetal anatomy scan (see pp.173–76). By this stage, it is possible to get a clear picture of the development of your baby's organs and body systems, and the majority of structural abnormalities can be detected during this scan. If a problem is found you will probably have more specialist scans but the majority of women will not need more scans.

ULTRASOUND SCANNING
An early scan is a valuable baseline because it can be used to establish the exact stage of your pregnancy.

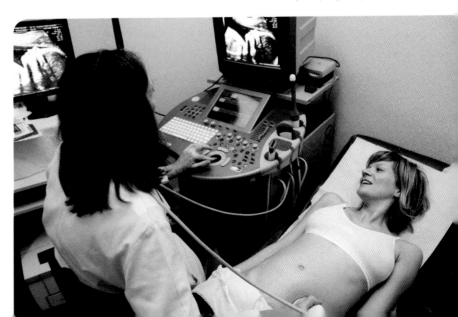

DATING ULTRASOUND SCAN

The dating ultrasound scan at 7–9 weeks measures the size of the fetus so that future prenatal care can be organized around your baby's gestational age— the earlier the scan, the more accurate the dating measurements will be.

HOW SCANS ARE USED

▶ **5–8 weeks** Pregnancy viability scan shows sac in uterus; after 6 weeks fetal pole and heartbeat detectable; dating using CRL measurement

▶ **7–9 weeks** Dating of pregnancy using CRL and BPD. Scan confirms growth, heartbeat and brain formation

▶ **11–14 weeks** Nuchal Fold Translucency Scan screens for Down syndrome; dates pregnancy

▶ **18–20 weeks** Detailed anatomy or anomaly scan examines baby for heart, kidney, bladder, spine, brain, and limb abnormalities; checks growth of head, body, and limbs; checks the position of the placenta

▶ **28 weeks plus** Detailed scans to detect placental problems, intrauterine growth restriction and volume of amniotic fluid

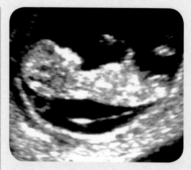

SCAN AT 10 WEEKS The fetal arms and hands are developing. A yolk sac is visible below the baby's head.

▶ **Ultrasound scans**, as their name implies, work by emitting high-frequency sound waves, which are sent through a pregnant woman's body using a handpiece called a transducer. These soundwaves are reflected back from the solid tissues of the developing baby and translated into images on a computer screen. There is no radiation involved in ultrasound, only sound waves.

▶ **During the hour before an abdominal ultrasound scan**, you will be asked to drink several glasses of water and to avoid emptying your bladder. You may find this a little uncomfortable but there is a good reason for it. When your bladder is

full, the ultrasound waves are reflected through this water-filled window lying immediately over the uterus and tiny baby, producing much clearer images.

You will be asked to lie down and lubricating gel is smeared onto the lower part of your abdomen to ensure good contact with the transducer. The sonographer, usually a doctor or technician, then moves the transducer smoothly forward and backward to produce ultrasound images on a computer screen which you can usually see also.

▶ **For a vaginal scan** a tubular probe is introduced into your vagina. You will probably need to empty your bladder because the probe is close enough to your uterus to produce clear pictures. Many women worry that a vaginal scan may be painful or will damage their pregnancy, but this is not the case. If you do have any vaginal bleeding afterward, it was going to happen anyway and was not caused by the vaginal probe.

▶ **The key measurements** taken at the 12-week dating scan are the crown rump length (CRL)—the distance between the top of your baby's head (crown) and bottom (rump)—and the biparietal diameter (BPD), which is the distance between the two parietal bones on each side of the baby's head. The size of the baby's limbs cannot be measured accurately while the baby is still in a curled position, so the length of the thigh bone or femur (FL) will not be used to assess fetal size until the middle of the second trimester.

Your baby's heartbeat will also be monitored—an extraordinary sight, as it beats fast and furiously.

If your dates don't tally with the measurements, it may be because your dates are wrong or there is a problem with the pregnancy. You will probably be asked to come back for another scan at a later date to make sure that all is progressing well.

▶ **Twin pregnancies** are often diagnosed during the 12-week scan, although they can be detected as early as a six-week scan when two pregnancy sacs are usually clearly visible in the uterus. At 12 weeks

the sonographer will be able to detect whether you are carrying identical (monochorionic) or nonidentical (dichorionic) twins by examining the thickness of the membranes that separate the two amniotic sacs in the uterus. If the junction between the placenta and the membranes of each twin resembles a T (the T sign), the twins are identical. However, if that junction looks like a lambda (λ) the twins are nonidentical.

DATING SCAN AT 12 WEEKS

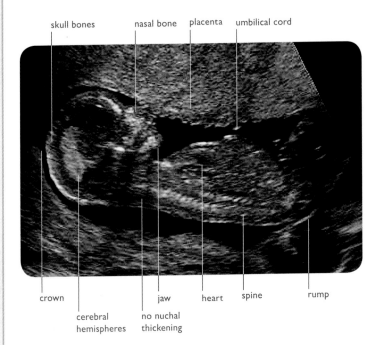

skull bones   nasal bone   placenta   umbilical cord

crown   cerebral hemispheres   jaw   no nuchal thickening   heart   spine   rump

INTERPRETING A SCAN Dense tissues such as bone appear white, while fluid-filled areas are dark. At 12 weeks this fetus has well-formed skull bones and a clearly defined spine. The heart is visible as a small dense area mid-chest that pulsates on screen. The placenta is visualized as a spongy mass connected to the blood-filled umbilical cord, which appears white because blood cells reflect sound waves.

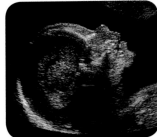

IN PROFILE The cerebral hemispheres are clearly visualized. The sharp profile shows that the nasal bone has formed.

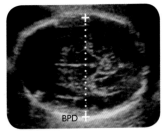

BPD

BIPARIETAL HEAD DIAMETER This is one of the baseline measurements plotted on a graph to monitor fetal growth.

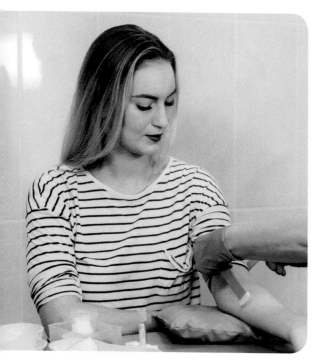

BLOOD TESTING
You are usually asked to provide blood samples for several blood tests at your first prenatal visit.

Blood tests

A variety of blood tests will be performed at your first prenatal visit. Your consent is needed for HIV testing. Some are routine for all pregnant women, but you may be offered additional tests depending on your medical and obstetric history.

Your blood group

This will be one of the four types found in humans: A, B, AB, or O. The most common is O, followed by A and B, and then the more rare AB. For each of these combinations, the individual is either Rhesus positive or negative, the most common type being Rhesus positive. So your blood type card reads: O Rhesus positive or negative, A Rhesus positive or negative, and so on for each blood group. Rhesus status is especially significant in pregnancy because an Rh-negative mother who carries an Rh-positive baby can develop damaging antibodies to her baby's blood (see p.128 and p.425).

Establishing your blood group early in pregnancy is very important. The journey through pregnancy is one of the few times in your life when you are at increased of risk of catastrophic bleeding requiring a blood transfusion, so it is essential that your exact blood group is known, included in your maternity records, and available at any time of the day or night. Until very recently, the most important cause of maternal death in the Western world was hemorrhage or blood loss, and in countries that do not have access to blood transfusion services, it remains so. We should never forget these facts or take them for granted. Of course, most women do not experience any bleeding in pregnancy and have no need of a blood transfusion. But for the few that do, knowledge of their blood type can save valuable time for the laboratory staff who are trying to cross-match stored samples of blood with the exact group of the pregnant woman. This is why you will have a blood sample taken if ever you are admitted to a maternity hospital with a problem that could result in a blood transfusion.

Hemoglobin level and blood count

The hemoglobin level in your blood is a measure of the oxygen-carrying pigment in your red blood cells. The normal level for women is 11.5 grams per liter of blood in your body. If the level is low, this means that you are anemic and you will be advised to eat foods with a high iron content (see p.47) and may be prescribed iron tablets as well. Anemia (see p.424) can cause you to feel very tired and may also lead to problems if you have excessive bleeding at the time of delivery.

The full blood count also analyzes numbers of red blood cells, white blood cells, and platelets to provide further information about your general health. For example, it may suggest that your anemia is not due to a lack of iron alone but other factors, such as vitamin deficiencies, which can be identified and treated.

Sexually transmitted diseases

Pregnant women are routinely screened for syphilis infection (see p.415) and offered prompt treatment with penicillin if they are found to be infected. After a long decline, syphilis is now on the rise in the US again, and so screenings for this disease are routine. It is important to remember that undetected infection with syphilis during pregnancy can be the cause of severe congenital and developmental problems in the baby. Since syphilis can be treated so swiftly and easily, I believe that we should continue to screen for it routinely during pregnancy. Sadly, the incidence of syphilis is also increasing in Eastern Europe, Russia, and Africa. If you have lived in any of these countries and are now receiving prenatal care in the US, it is especially important that you are screened to prevent damage to yourself and your baby.

Infection with chlamydia and gonorrhea (see p.414), two other sexually transmitted diseases, is more likely to cause problems with infertility. However, chlamydia infections are also the cause of serious eye infections in newborns, so if you think you may be at risk, tell your doctor so that they can help you prevent problems from developing.

Autosomal recessive disorders

Some genetic disorders are passed on only if both parents have copies of the faulty gene, and you may not know if you are only a carrier (in other words, if

SICKLE CELL SCREENING
If you are at a high risk of being a sickle cell carrier, you will be offered a screening test.

RHESUS NEGATIVE PREGNANCIES

▶ **If your blood group is Rhesus negative**, problems can arise in pregnancy if your baby inherits Rh-positive status from your partner. Rhesus status is rarely a problem in a first pregnancy, but if you are Rh-negative and exposed to some of your baby's Rh-positive blood during childbirth, you may develop anti-D antibodies that could cause problems in a subsequent pregnancy. The anti-D antibodies attack the next baby's blood, causing anemia and distress for the baby in the uterus and anemia and jaundice (see p.435) after the birth.

▶ **All mothers (both Rh-positive and negative)** are checked for anti-D antibodies during their first prenatal-visit blood tests. The blood tests are repeated at 28 weeks.

▶ **If you have developed antibodies**, you will be given more blood tests every four weeks, and your baby will be carefully monitored for signs of anemia or heart failure.

▶ **Even if you have not developed antibodies**, (but are Rh-negative), doctors now offer a routine preventative anti-D injection at 28 weeks, which mops up Rh-positive fetal blood cells and prevents the development of destructive maternal antibodies. These are offered in the first pregnancy, unless the father is also Rh-negative, in which case there is no risk that the baby's blood will be incompatible. This anti-D program has made rhesus hemolytic disease of the newborn relatively rare.

▶ **All Rh-negative mothers** who give birth to a Rh-positive baby are given an Rhlg injection within 72 hours of delivery. A blood test establishes the level of fetal cells present in her circulation, and if the concentration is high, another dose of Rhlg may be needed.

▶ **Rh-negative women** who have amniocentesis (see pp.140–43), chorionic villus sampling (see p.140), external cephalic version (see p.271), or those who have vaginal bleeding or abdominal trauma during pregnancy are given an injection of Rhlg within 72 hours. Those who miscarry and require surgical evacuation of the uterus or who undergo termination of pregnancy or who have an ectopic pregnancy are also treated with Rhlg.

KEY

− mother's blood

+ baby's blood

▲ antibodies

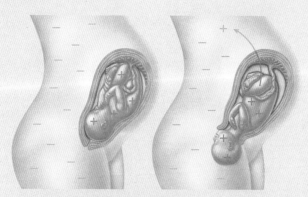

FIRST PREGNANCY Maternal and fetal circulations do not usually mix during pregnancy but during birth the mother may be exposed to her baby's blood.

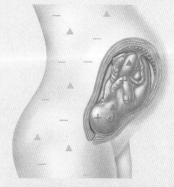

SUBSEQUENT PREGNANCY If the mother has developed antibodies to her baby's red blood cells they may cause problems in a future pregnancy.

you have just one copy of the mutation). Among white people of North American and European origin, the most common of these disorders is cystic fibrosis, which affects all the fluid- and mucus-producing glands in the body, leading to thick, abnormal secretions. Among the Ashkenazi Jewish population, Tay-Sachs disease, in which harmful chemicals accumulate in the brain, is most common. If you are of African or Mediterranean origin, sickle-cell anemia and thalassemia, both abnormalities of the oxygen-carrying hemoglobin, are most common, and you will be offered a special electrophoresis test of your hemoglobin to determine whether you have sickle-cell or thalassemia trait.

If you carry the sickle-cell trait, it is important that your partner's sickle-cell status is established early on, since there is a chance that your baby could inherit a double dose of the trait and develop the full-blown sickle-cell disease (see p.418 and p.425). Similarly, if you carry the A or B thalassemia trait, you will need to arrange for your partner to be tested too. A baby with full-blown thalassemia (see p.425) suffers from very severe anemia and iron overload, which eventually leads to multiple organ failure.

Hepatitis B and C

These viral infections cause liver disease but it is unusual for women to be infected for the first time during pregnancy. Both are more common if you have used intravenous drugs, had multiple sexual partners, or have been exposed to infected blood. If you have received a blood transfusion in a country where the screening of blood products is not rigorous, you may be at risk. The hepatitis B virus does not cross the placenta during pregnancy, but if you carry the virus, your baby will be at risk of infection at the time of delivery. The virus is not transmitted in breast milk, but babies are occasionally infected by blood if the mother's nipples are cracked and bleeding. As many as half of all babies infected with hepatitis B will develop cirrhosis or cancer of the liver in later life. This is why it is so important to know whether you are hepatitis B positive during pregnancy; if you are, your baby can be protected by IgG immunoglobulin treatment at delivery and immunized with hepatitis B vaccine soon afterward.

Hepatitis C infection is an important cause of liver disease worldwide, but it is only rarely transmitted to the baby during pregnancy or delivery. However, the risk is greatly increased if you are HIV positive (see p.130 and p.414). It is recommended that all pregnant women be tested for HIV. Screening for hepatitis C is not routine but may be offered to you.

> 66
> *...it is essential that your exact blood group is known, included in your maternity records, and available at any time of the day or night.*
> 99

HUMAN IMMUNODEFICIENCY VIRUS (HIV)

Screening for HIV infection is now offered to all pregnant women. If you are found to be positive, you will be offered the best advice and treatment for you and your baby; If found to be negative, you will be greatly reassured.

HIV is a retrovirus capable of incorporating itself into the genetic code, especially of white blood cells, which are responsible for fighting infection. HIV infection (see p.415) has now become an epidemic worldwide and, until very recently, being HIV positive often led to the development of full-blown autoimmune deficiency syndrome (AIDS). Today the situation has changed dramatically: people who are HIV positive can now receive antiretroviral drugs, which can protect them from the onset of AIDS.

HIV IN PREGNANCY

For pregnant women, knowing that they are HIV positive can improve their personal survival, thanks to the new drugs available. It will also significantly reduce the risk of their babies becoming infected. Cesarean delivery and avoiding breastfeeding further reduce the risk if your viral load remains high. These measures coupled with maternal antiretroviral drug treatment at and around the time of delivery cut the chance of a baby being infected with HIV from 20 percent to less than two percent.

However, these medical advances can only be offered if your doctors know that you are HIV positive.

OPPOSITION TO SCREENING

Thirty years ago I was involved in trying to introduce routine screening for HIV in our prenatal clinic at St. Mary's Hospital, London. We met considerable opposition from patients, midwives, and doctors, who were all concerned that the implications of being found to be HIV positive were so devastating that they should not be revealed as part of prenatal screening. The Royal College of Midwives actually advised members that routine screening was an invasion of a pregnant woman's privacy.

A CHANGE IN ATTITUDE

Several events had to occur before this attitude changed. First was the publication of European studies showing that the measures described above could reduce the transmission of HIV from mother to baby.

Next came the introduction of the antiretroviral drugs used singly and in combination, which appeared to offer a greatly improved chance

of halting the progression of HIV to full-blown AIDS.

The tragedy in our hospital was that it was only after two six-month-old babies were admitted with life-threatening infection caused by AIDS that the real importance of prenatal screening for this disease began to be fully understood.

The mothers of these two babies had not been offered testing for HIV during pregnancy by their local hospital. Having witnessed the horror of their children's suffering they wanted to support the introduction of routine screening.

POSITIVE APPROACH

During the months that followed, the uptake of HIV screening in our prenatal clinic rose from 30 percent to over 95 percent.

Much of the credit for this change in practice must go to the two midwives who introduced the service. Their supportive approach toward HIV screening was crucial in changing the attitudes of patients and their medical colleagues. Today HIV screening is offered as a routine part of prenatal care. Consent, though, is required.

Your blood test results

The results of your blood tests will be available in about two weeks and are usually filed in your maternity records. The exact date of your next visit will depend on the type of prenatal care you have chosen;. In my own practice we usually see women at about 17 weeks—after their serum screening blood test result is available—in order to discuss the results of these tests and ensure that any necessary action is taken. Any important abnormal test results may be duplicated and filed in a separate set of hospital-based records, with your permission, for the few women who have been identified as having high-risk pregnancies.

Common concerns

You may not have many concerns at this point. The risk of miscarriage is diminishing; you have survived the early pregnancy complaints; and you are not yet pregnant enough for the later ones.

Although your risk of miscarriage is now considerably reduced, any bleeding in early pregnancy invariably raises the fear of miscarriage occurring, especially if you have suffered one in the past, or experienced problems earlier in this pregnancy. So if you do experience some vaginal bleeding, see your doctor and have an ultrasound scan at the earliest possibility—in the majority of cases it will reassure you that nothing is wrong and that the bleeding is insignificant. If this is the case, it is a good idea to make sure that your doctor examines you internally and looks carefully at your cervix. In early pregnancy the flood of hormones can make its surface very fragile and prone to bleeding, particularly if you develop a mild infection, such as yeast. If there is any concern about the appearance of your cervix, your doctor will probably take a pap smear.

Varicose veins

Although varicose veins tend to be much more common in later pregnancy (see p.235) some women, particularly those who had varicose veins in a previous pregnancy, will have symptoms of aching and discomfort much earlier. If this applies to you, do make sure that you start wearing good support hose every day. Prompt attention to varicose veins in pregnancy can greatly reduce the problems that tend to develop at a later stage.

In early pregnancy the flood of hormones can make the surface of your cervix very fragile and prone to bleeding…

Your sex life

Some women find pregnancy brings added sexual fulfillment. The marked increase in vaginal secretions, together with the greater blood flow to all of the genital organs may mean that sex is more pleasurable than before. For many couples, there is the added thrill of being able to have unprotected sex, which can be a highly erotic sensation after years of using contraception. Added to this is the wonderful closeness that arises from the knowledge that you and your partner have created a life together. For many couples, this is a very powerful emotion, which enhances their lovemaking.

However, for many women, sex suffers. In spite of the many "sexy" images of famous women in late pregnancy that we are subjected to in newspapers and magazines, the fact remains that many mothers-to-be do not consider themselves to be very sexually appealing during their pregnancy and may start to feel this way as early as the third month when their shape starts to change. Being pregnant may change your perception of yourself and your partner's perception of you as a sexual partner and there are also physical and emotional reasons why your sex life may be different or diminished during pregnancy. During the first three months a woman may experience a lower libido because she feels nauseous or exhausted; her breasts might be too tender to touch; she might have had some bleeding or have reason to worry about the risk of miscarriage; or she might simply not feel like having sex. Later on, heartburn, indigestion, fatigue, sheer size and inability to get comfortable mean that women may have sex less often and find it less enjoyable. Even though loss of libido is extremely common, it can still come as a shock to both partners.

If you are feeling insecure about these issues, it is vital that you talk to your partner. You need to reassure each other that your feelings for each other have not changed just because one or both of you is not very interested in making love.

Clothes and hair

Most likely your clothes are starting to feel tighter around the waist now, but it is sensible to resist the temptation to rush out and buy a completely new

In spite of the many 'sexy' images of famous women in late pregnancy, the fact remains that many mothers-to-be do not consider themselves to be sexually appealing...

ENROLLING FOR CHILDBIRTH CLASSES

▶ **You will not start your childbirth classes** until your sixth or seventh month, but many classes get booked months in advance, so start researching now to find out what is available in your area. Some are called parent education classes but most cover similar ground.

A good childbirth class should explain the physiology of pregnancy; describe the best breathing techniques for labor advise you how to remain as active as possible.; help you prepare yourself for an active birth; and provide exercises that will make you stronger for delivery. Your instructor should discuss labor fully and honestly, giving a full description of pain-relief options. She should advise on basic baby care, including feeding, for the days and weeks following the birth. Some classes have followup groups after the birth.

▶ **All prenatal classes encourage fathers-to-be to attend**, although your partner should not feel pressured to attend every session. Some courses arrange classes for men so that they can discuss their feelings in a more open way and learn how they can prepare to be supportive during labor.

▶ **All hospital maternity units run prenatal courses**, as do midwife teams. These can be very useful if you are planning a hospital birth. You will have ample opportunity to put questions to the midwives and other obstetric personnel and this will make you all the more familiar with the way the labor ward works. By the time you arrive for the birth, you will be even more relaxed about your surroundings.

▶ **Questions to ask** abefore you enroll include who sponsors the classes; is the instructor certified by any organizations; can the instructor provide references; how many classes are in the series; and what topics are covered (see pp.437 and p.438).

▶ **Even if this is not your first baby**, enrolling in an childbirth class is still a good idea. You may be a little rusty when it comes to remembering the details of labor or the correct breathing technique.

▶ **One of the best spin-offs from attending an childbirth class** is the chance to meet women who are having babies at around the same time as you and who live in your area. Many women find that they make wonderful, enduring friendships in their childbirth classes.

wardrobe of clothes at this early stage. If this is your first pregnancy you can usually get by with some looser tops from your existing wardrobe perhaps with one or two pairs of pants or skirts with an elasticated waist. Wearing your partner's jeans for a few weeks may be an option. If you have been pregnant before, you may well find that you need to change into roomier garments earlier than you did in your first pregnancy. Once the muscles of the abdominal wall have been stretched by pregnancy, they are never quite as tight as they used to be.

Trips to the hairdresser are a good way to boost morale but many women tell me they are concerned that the chemicals in hair dyes might be potentially dangerous. There is no evidence that this is so; most permanent and nonpermanent dyes contain chemicals that have been in use for many years and are unlikely to be toxic in the kind of doses used to color hair every few months. However if you are worried, you could stick to highlights, which only color the hair shaft rather than expose the whole scalp to dye, and if you color your own hair, wear gloves and do it in a well-ventilated space.

PRENATAL TESTS

Screening and diagnostic tests for fetal abnormalities make up one of the most complex areas of prenatal care. I can appreciate that if you have been sailing through pregnancy so far, these may be topics you feel reluctant to consider—but I urge you to read on. I hope the information here will help steer you through the options available and enable you to base your choices on the most up-to-date evidence and information. This will give you confidence in any decisions that you may need to make.

All screening tests are optional and you and your partner have the choice of whether or not you want to have them. There is no right or wrong answer; each couple will have to decide what is right for them after considering all the available information. Although the vast majority of babies are born healthy, when problems do occur they tend to do so during fetal development because of an inherited genetic condition, an acquired problem due to, for example, an infection or a medication, or for no known reason. If an abnormality is detected through screening, it gives you and your partner the chance to prepare yourselves emotionally and practically for the prospect of caring for a child with a disability or for the possibility of terminating the pregnancy, depending on the extent of the problem. (For a fuller discussion on congenital abnormalities see pp. 144–47.)

Do I need to have tests?

All women have a chance of delivering a baby with an abnormality but a number of factors increase the risk and these should be taken into account when you decide which, if any, tests to have. You are at increased risk if:

• you have had a previous pregnancy that was affected by an abnormality

• you or your partner has a family history of genetic disorders or abnormalities

• you are over 35 years old

• you take or have taken medications that are known to harm a developing baby.

What is available?

First trimester screening has changed considerably in the last few years, due to the introduction of noninvasive prenatal testing (NIPT). Although provisions vary from one practice to the next, many more practices are now exploring the possibility of introducing NIPT to high-risk groups of women even if they must travel to a different location to have testing. NIPT is still considered a screening test and the difference between this and a diagnostic test is important. The combined screening test is offered to everyone between 11 and 14 weeks and this will identify most, but not all, of those at risk of carrying a baby with an abnormality. The combined test cannot diagnose a problem; it simply establishes a risk figure on which women then have to base their decision whether to go on to have more tests. The combined tests include a serum (blood) test and an ultrasound scan (see p.124 and p.174), which measures the Nuchal Fold Translucency Scan (behind the neck of the fetus). For those women who have missed the window of opportunity to have the combined test, a quadruple test, which is just a blood test, can be performed from 14 to 20 weeks' gestation. With every test you need to consider the detection rate and the false positive rate (see chart below). A false positive is a positive test result that is later shown to be negative. A positive screening test prompts doctors to offer an invasive test to confirm the diagnosis, so if the false positive rate is high, many women with pregnancies that are problem-free are exposed to more tests they do not need.

Diagnostic tests give a clear answer as to whether or not a fetus has an abnormality but these invasive tests are not routine, since they all require samples of the amniotic fluid, placenta, or fetal blood to be taken from inside the uterus and these procedures carry a small additional risk of miscarriage. The only definitive tests for Down syndrome (see p.147) are chorionic villus sampling (CVS) (see p.141) and amniocentesis (see p.143). Whether or not you choose to have one of these tests will depend on your views on abnormalities and how you would want to act if you knew for sure that your baby was affected by one.

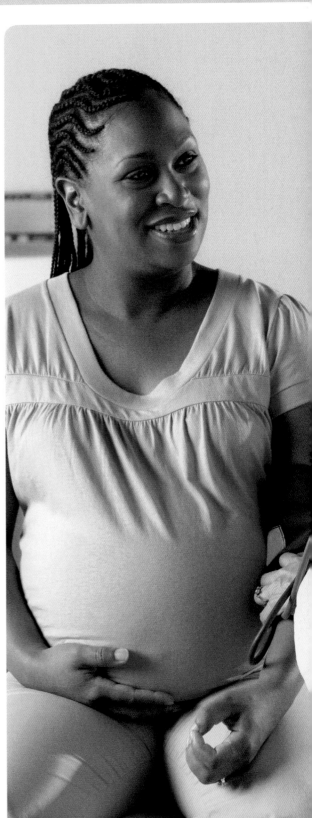

OLDER WOMEN Being over 35 years old puts you at a higher risk of having a baby with an abnormality, but the vast majority of babies are born healthy.

How accurate are the tests?

The detection rate of Down syndrome has effectively tripled over the last 15–20 years thanks to the work of committed obstetricians and scientists specializing in the field of first trimester screening.

The most important practical problem in prenatal screening is that it is difficult to achieve a high detection rate of affected babies and at the same time maintain a low false positive rate. In the past, we only had maternal age of 35 years or more to identify women who are at greater risk. Doctors will offer these women amniocentesis. However, if the mother's age alone is used to

BEING INFORMED Speak to your midwife or obstetrician about the prenatal tests available to you. It is your choice whether or not to have them.

decide whether an invasive test such as amniocentesis or CVS is performed, only 30 percent of babies affected by Down syndrome will be identified. If maternal age is combined with serum screening, the detection rate rises to 65 percent, but that is still only two-thirds of affected babies. However, when Nuchal Fold Translucency Scanning (NTS) is measured at 11–14 weeks the detection rate rises to 80 percent, and when the

NTS measurement is combined with a blood test that measures free beta-hCG and PAPP-A hormone levels, the detection rate is as high as 90 percent.

Nuchal Fold Translucency Scanning (NT)

This screening test was developed at King's College Hospital, London, in the 1990s to improve the detection of Down syndrome as early as possible in pregnancy. The scan is performed between 11 and 14 weeks by hospital-based specialists and is based on an ultrasound measurement of the depth of fluid present under the skin behind the neck of the fetus (see box, p.139). An NT is usually offered to every pregnant woman. If it is not offered, you can request the test. Check with your insurance carrier to see if the cost is covered.

An increasing number of physical markers for Down syndrome are being identified. For example, Professor Kypros Nicolaides at the Fetal Medicine Centre in London has shown that these babies often lack a nasal bone, have protruding tongues, a single hand palmar skin crease, or have abnormal flow patterns in their heart valves or blood vessels. By examining the baby's profile during the NT scan, he believes that the detection of Down syndrome can increase to over 95 percent, and the false positive rate be reduced to less than one percent.

Serum screening test (Quadruple test)

The Quadruple serum screening tests that is available measures four substances in your blood to predict whether your baby is at risk of Down syndrome, certain other chromosomal (genetic) abnormalities, or an open neural tube defect such as spina bifida (see p.146 and p.419). Serum screening tests are probabilities, or estimates of risk; they do not provide

---

SCREENING TEST COMPARISONS

The development of new tests over time has raised detection rates for abnormalities such as Down syndrome from 30 percent to 99 percent. The most recent advance in screening—NIPT—has the lowest false positive rate at just one percent, resulting in fewer invasive tests.

Method of screening	Timing (weeks)	False positive rate	Detection rate	Number of babies found to be affected after test result is positive
Maternal age		5%	30%	1:130
Nuchal Scan	11–14	5%	80%	1:47
Combined test	11–14	5%	85%	1:56
Quadruple test	16–20	5%	80%	1:50
NIPT	>10	0.1%	99%	9:10

TIMING OF TESTS

Date	Test
10 weeks	Noninvasive prenatal testing (NIPT)
11–14 weeks	Combined test (including measuring nuchal fold)
16–20 weeks	Quadruple test
18–22 weeks	Fetal anomaly scan (see p. 174) (diagnostic)
11–14 weeks	Chorionic villus sampling (CVS) (diagnostic)
From 16 weeks	Amniocentesis (diagnostic)

a definite answer. When you have a test, it is important to remember the following:

• An abnormal serum screening test (screen positive) does not mean that your baby has one of these abnormalities, but it does mean you are at greater risk, and may want to have more testing to get more information.

• If the baby is found to have a chromosomal disorder, you will be faced with the choice of whether or not to continue with your pregnancy.

Performing the test

Serum tests are usually performed at about 15 to 16 weeks at your obstetrician's office. The Quadruple test measures four substances in the blood: alpha-fetoprotein (AFP), uncoagulated estriol, inhibin A, and free beta-HCG. In babies with Down syndrome, the AFP and estriol levels tend to be lower and the HCG higher. The blood sample is sent to a special lab and analyzed for the different substances. The results are fed into an algorithm along with your age and the exact gestational age of your baby. The risk assessment is usually available within five working days and, depending on the result you will either be contacted by your obstetrician or it will be mailed to you. In the vast majority of cases, the test results will be low risk, but for the few results that show a high risk, it is very important that you receive the news as early as possible in order to provide you with the maximum amount of time to arrange more tests if you want to have them.

Your risk assessment

Your risk may be described as 1 in 45 or 1 in 450. Translated into simple terms these numbers mean that for every 45 or 450 pregnancies, one baby is likely to be affected by the abnormality and either 44 or 449 babies are unlikely to be at risk of the abnormality. Therefore, large numbers are good news, while lower numbers will raise the alarm that a baby may have a problem.

It sounds simple, but the reality is quite complicated. Some couples will consider a risk of 1 in 45 worrying, while others will interpret it as a risk that they are prepared to accept. At the other end of the scale, most couples and their obstetricians will consider a serum screening result of 1 in 450 as very reassuring while for others, the possibility that

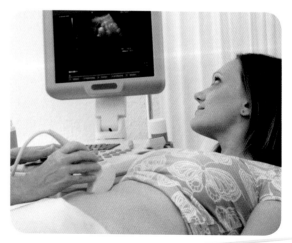

MEASURING THE NUCHAL FOLD This simple, noninvasive test performed in the first trimester is highly accurate in detecting Down syndrome.

MEASUREMENT OF NT

During an ultrasound scan at 11–14 weeks, the technician measures the fetus and the depth of the fluid under the skin behind the neck of the fetus (the nuchal translucency).

This screening test can only give you an indication of risk and is not a conclusive answer as to whether or not your baby has Down syndrome.

▶ **If the fluid measurement is below 3mm**, your baby is unlikely to have a problem. This will be the result for 95 percent of women.

▶ **If the measurement is 3mm**, there is a probability that your baby has Down syndrome or other chromosomal abnormality. The higher the measurement, the higher the chance. A measurement between 3 and 4mm is borderline.

▶ **If your NT measurements are high or borderline**, you will be counseled about the implications and offered the option of undergoing an invasive prenatal

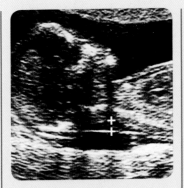

LOW RISK This fetus has only a small depth of fluid behind its neck and is at low risk of Down syndrome.

test, such as a chorionic villus sampling or amniocentesis. Only five percent of pregnant women will find themselves in this situation.

▶ **If you decide not to have an invasive test** following an NT measurement that was high or borderline, I strongly advise you to make sure that you have a detailed anatomy scan at 20 weeks. This

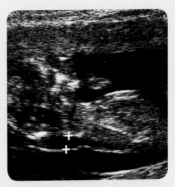

HIGH RISK The larger depth of fluid marked here puts this fetus at a higher risk of Down syndrome.

is because babies with a thicker nuchal translucency have an increased chance of heart, gut, and other structural abnormalities that will be visible on an ultrasound scan. If some or any of these are identified, your doctor can arrange appropriate pediatric help and advice before, during, and after the delivery if you decide to continue with the pregnancy.

their baby may be affected by a small but finite chance of an abnormality prompts them to have more tests. Of course, what the numbers cannot possibly take into account is your past experience and your personal views about having a baby affected by an abnormality, or alternatively, electing to terminate a pregnancy with a proven abnormality. There are no right or wrong answers to these dilemmas, which is why you need to discuss openly and honestly with your partner, obstetrician, and genetic counselor how you would want to act upon having an abnormal test result.

Noninvasive Prenatal Testing (NIPT)

This screening test analyzes the cell-free DNA (DNA fragments) in the mother's blood that originates from the baby. It offers a strong indication as to whether the fetus is at high or low risk of an abnormality such as Down syndrome, but avoids the risk of miscarriage from invasive tests.

In 1997 the presence of cell-free fetal DNA (cffDNA) in the maternal circulation was reported. Fetal DNA comes from the placenta. It can be

NIPT The mother's blood is screened for cell-free DNA, some of which comes from the baby. This test is more than 98 percent accurate at detecting Down syndrome.

detected from the first 4 to 5 weeks of pregnancy onward and is rapidly cleared from the maternal circulation within the first hour after birth. Maternal blood is therefore a reliable source of material for prenatal diagnosis (about 80 to 90 percent of the DNA fragments originate from the mother and the remaining 10 to 20 percent come from the baby).

Recently DNA sequencing technologies have become so precise that the detection of the extra material resulting from a trisomy of the fetal chromosomes within the mother's blood can now be performed accurately. Male fetal sex can be determined using cffDNA in maternal plasma by the identification of Y chromosome sequences. In the UK, this is being used to determine the baby's gender in women whose baby is at risk of an X-linked disorder, where early identification of a male baby indicates the need for an invasive diagnositc test to determine whether the affected X chromosome has been inherited. On the other hand, an invasive test is not required if the baby is female. In pregnancies at risk of sex-linked diseases, noninvasive prenatal testing (NIPT) (with ultrasound) has been shown to reduce the use of invasive diagnostic testing by nearly 50 percent.

In Rhesus-negative mothers, cffDNA testing has replaced amniocentesis for fetal blood grouping. This is important because if the baby is RhD-positive, amniocentesis for blood grouping risks boosting the level of antibodies, in turn converting a mild disease to severe. Determination of other fetal blood groups such as Kell, C, c, and E using cffDNA has also been reported.

To date, the published data indicate extremely good results for predicting trisomy 21 and trisomy 18. Initially these tests were less reliable for the

HAVING A CVS TEST

If you have a chorionic villus sampling test, you will first be given an ultrasound scan to identify exactly where the placenta is lying. The test requires a small sample of tissue from the placenta, and the amniotic sac has to be avoided during the procedure.

▶ A local anesthetic is injected into your abdominal wall to numb the area before a fine double-barreled needle is introduced into your uterus at the correct site to access some chorionic villi—the frondlike projections in the placenta.

▶ A syringe containing some special culture fluid is attached to the end of the needle and the placental cells are sucked up into the syringe.

▶ The tissue obtained is fresh living placenta, which means that when

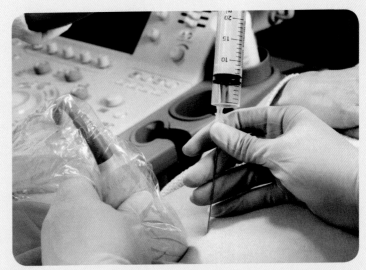

CVS TEST A sample of cells is taken from the placenta for analysis.

it reaches the cytogenetics laboratory culture and analysis is much quicker than from fetal skin cells that have been cultured from an amniocentesis sample (see overleaf). You will probably receive a provisional test result within 72 hours and a conclusive result in about 10 days.

detection of trisomy 13, but this has been evolving and subsequently better results have been reported. For trisomy 21 and 18, the detection is now close to 100 percent. Possible sources of error in the test are when the levels of DNA fragments are not high enough to be detected (too early, i.e., less than 10 weeks' gestation or when the mother is severely overweight), placental mosaicism (abnormal cell lines only present in the placenta and not the baby), and in multiple pregnancies.

Chorionic villus sampling (CVS)

This prenatal diagnostic test is usually performed at 11–13 weeks and involves obtaining a small tissue sample (a biopsy) from the placenta (see box above). Since the baby and the placenta develop from the same cells, the chromosomes in placental cells are the same as in the cells from the baby. The majority of women who undergo a CVS test do so to exclude Down syndrome. However, it is also used if a specific gene disorder is suspected, such as sickle-cell anemia or thalassemia major (see p.418). Fresh tissue can be collected and tested before 12 weeks, so a pregnancy with a severe chromosomal abnormality can be diagnosed in time for a suction termination of pregnancy to be performed, if this is what the parents choose.

Disadvantages of CVS

No invasive prenatal test is completely problem-free, so consider the following:

• The risk of miscarriage after CVS appears to be slightly higher than for amniocentesis, affecting approximately one percent of women who undergo CVS. This higher risk from the CVS procedure may be because the test is performed earlier in the pregnancy (at 11–13 weeks) than amniocentesis, and the pregnancy may have been going to miscarry anyway, but it is still a cause for concern.

• There is also evidence that when the CVS is done very early, the baby may develop an abnormality in the growth of its limbs.

• Occasionally, the placental tissues contain mosaic cells (abnormal cell lines only present in the placenta but suggest that the baby has a chromosomal abnormality). In this situation you will probably be advised to have an amniocentesis at a later date to check the results.

HAVING AMNIOCENTESIS

This procedure takes about 20 minutes and is performed using ultrasound guidance to find the best point to insert the amniocentesis needle. The optimal spot is where the needle can enter through the uterine wall and reach a pool of amniotic fluid, without touching either the placenta or the baby.

▶ The operator may inject some local anesthetic into your skin at this spot to reduce any discomfort, but the amnio needle is so thin that you will find this procedure is even less painful than a blood sample being taken from your arm.

▶ Once positioned in the right place, a syringe is attached to the outer sheath of the needle and the sample of amniotic fluid (about 10-20ml, the equivalent of 4 teaspoonfuls) is sucked into the syringe.

▶ The needle is then removed and the baby is scanned carefully to ensure that all is well.

▶ You will be advised to rest and avoid strenuous activity for 24 hours. Some women feel slightly sore for an hour or two afterward. A few have spotting of blood or leaking of amniotic fluid from the vagina. This usually stops within a short time.

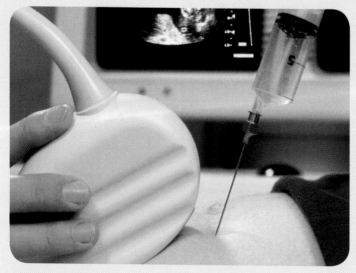

AMNIOCENTESIS A sample of cells is taken from the amniotic fluid for analysis.

LABORATORY TEST The chromosomes in the sample are analyzed.

A positive aspect of amniocentesis is that it is extremely rare for the results to be wrong and the risk of miscarriage is low.

Amniocentesis

Amniocentesis is the most commonly performed invasive test. It involves taking a sample of the amniotic fluid around the baby, which is then sent to the laboratory for analysis. The test is usually performed at 16 weeks of pregnancy, but can be performed at any gestational age. You will be offered an amniocentesis test if:

• you are more than 35 years old
• you have a personal or family history of a baby with Down or another chromosomal abnormality
• you have had an abnormal nuchal translucency scan result or a high-risk serum screening result in your current pregnancy.

When the amniocentesis sample reaches the cytogenetics laboratory, the fluid is spun hard to collect cells from the baby's skin together in a pellet. These cells have floated off the surface of the skin into the amniotic fluid in much the same way that our skin cells float off into bathwater. These fetal skin cells then have to be encouraged to grow in tissue culture, which takes time (usually one to two weeks) and much expertise. They then have to be brought to a stage where they are actively dividing (called the metaphase stage) to perform the chromosomal analysis. Sometimes the cells cannot be encouraged to grow or are very slow in growing, which can delay your result. Rarely, the cells that are cultured are found to be your own and not your baby's; in these circumstances you would need to have another amniocentesis.

Amniocentesis is not usually performed until 16 weeks of pregnancy because before this time, the number of fetal skin cells may be insufficient to set up a culture and obtain a result. Removing fluid from the amniotic pool at too early a stage may also cause problems with the development of the baby's lungs and limbs.

Pros and cons of amniocentesis

A positive aspect of amniocentesis is that it is extremely rare for the results to be wrong and the risk of miscarriage is low. Although generally quoted as one percent, in certain practices that are performing many of these procedures the risk is much lower— on the order of 1 in 300 or 0.3 percent. The risk of miscarriage is highest within two weeks of having the amniocentesis performed. It is relevant that only pregnancies with potential problems are going to have this test, so later complications may not be directly related to the amniocentesis and may have been going to happen anyway.

The downside to having an amniocentesis procedure is that you will not have the results of it before about 17–18 weeks of pregnancy. This means that if the result shows that an abnormality is present and you choose to terminate the pregnancy, you will have to undergo an induced labor to deliver the baby vaginally (see pp. 294–97).

CONGENITAL DISORDERS

Congenital means "born with." The term includes all genetic disorders and any physical or structural abnormality that may be present in a baby. This section will help you understand how and why many congenital disorders arise and give a context to many of the prenatal tests that have been described already. An increasing number of abnormalities can now be diagnosed prenatally, enabling parents and doctors to plan treatment, although some abnormalities become apparent only after a baby is born.

Genetic disorders are caused by abnormalities in our genetic material that are either inherited or arise when a previously normal gene mutates (undergoes a change that makes it function abnormally). Some genetic disorders in the fetus are known to be due to the presence of a single or several abnormal genes. Others result because the number, shape, or arrangement of one of the chromosomes is abnormal. Still more are due to a complex interaction between environmental factors and genes, which is not fully understood. Spina bifida and cleft lip/palate are two examples. Only a few, mainly Down syndrome and spina bifida, are screened for regularly in pregnancy, although diagnostic tests and genetic counseling are available to couples who have particular risk factors or a family history of genetic diseases or other congenital abnormalities.

Chromosome abnormalities

Before, during, and after fertilization the two sets of chromosomes (one from your egg and one from your partner's sperm) that make up your baby's complement of 23 pairs undergo a complex series of divisions and rearrangements. If one of the chromosomes is abnormal, or too many or too few chromosomes are left in the fertilized egg, an abnormal embryo or fetus may develop. Most are miscarried early, but sometimes the pregnancy continues and a baby with abnormalities is born.

About 6 babies in every 1,000 are born with a chromosomal abnormality. In stillborn babies the figure rises to 6 in every 100. The most common abnormalities are disorders in the number of chromosomes (either too many or too few), and the names reflect the number of the pair of chromosomes they affect.

Trisomies occur when three copies of one chromosome are present. Most trisomies are due to abnormal cell division (meiosis) in the egg, which occurs before fertilization. They are much more common in older women, because older eggs are more likely to be abnormal. The most common trisomies are Down syndrome/Trisomy 21 (see p.147); Patau syndrome/Trisomy 13 (see p.415); and Edward syndrome/Trisomy 18 (see p.415).

Monosomies arise when one chromosome is completely missing. The most common type of monosomy, Turner syndrome (see p.416), is the loss of an X chromosome in girls.

Triploidy is when the embryo has an extra set of 23 chromosomes (see p.415).

Extra sex chromosomes occur in disorders such as Klinefelter syndrome (see p.416), in which boys have an extra X chromosome.

Translocations (see p.415) are abnormal arrangements of the correct number of chromosomes. When the transfer of genetic material occurs between two chromosomes, genetic material can be lost, augmented, or simply exchanged.

Dominant genetic diseases

In these inherited diseases, only one abnormal gene is necessary for the disease to develop. Males and females are equally affected and have a 50 percent chance of passing the gene and the disease on to their children. Unaffected individuals cannot pass on the gene or the disease. Dominant diseases are rarely fatal in early life, as affected individuals would die before passing on the genes.

There is invariably a family history of the disease, but because dominant diseases are expressed to a greater or lesser degree in different individuals this may be difficult to establish without the help of a geneticist. In familial hypercholesterolemia (see p.416), for example, babies of affected parents can be tested at or after birth for high blood levels of cholesterol. Some of the dominant neurological disorders, such as Huntington's Disease (see p.416) and myotonic dystrophy, can now be diagnosed prenatally, thanks to advances in gene mapping technology, which can pinpoint the abnormality in DNA samples obtained by amniocentesis or CVS.

INCIDENCE AND CAUSES

▶ Major congenital abnormalities (such as heart or neural tube defects) are present in 4% of newborn babies and cause 1 in 4 perinatal deaths.
▶ Minor congenital abnormalities, such as an extra finger or toe, are present in at least 6% of newborns.
▶ About 40% of congenital problems are inherited as a result of genetic factors.
▶ Around 7% are acquired due to damage during development by infection (5%), exposure to drugs (2%), chemicals, X-rays, or metabolic disorders, such as uncontrolled diabetes.
▶ About 50% of congenital disorders are unexplained, although it is likely that the majority of these will be found to have a genetic origin or result from a mixture of environmental and genetic factors.

Recessive genetic diseases

In these diseases two copies of the abnormal gene (one from each parent) are needed for the disease to develop. The recessive gene is usually masked by a normal dominant gene, and therefore there may be no family history of affected individuals. However, when both parents are carriers, all of their male and female children have a 1 in 4 chance of inheriting two recessive genes and developing the disease, and a 2 in 4 chance of becoming a symptomless carrier of the disease.

Many recessive disorders can be diagnosed prenatally. Cystic fibrosis, sickle-cell anemia, and

❝ The most common congenital abnormalities are caused by disorders in the number of chromosomes— there can be either too many present or too few. ❞

thalassemia (see p.417) are detected by DNA analysis from amniocentesis or CVS samples, whereas biochemical disorders such as Tay-Sachs disease (see p.416) and phenylketonuria (see p.417) are diagnosed from blood samples.

Sex-linked genetic diseases

Diseases such as hemophilia, Duchenne's muscular dystrophy (see p.417), and Fragile X syndrome (see p.418) are caused by a recessive gene located on the X (female sex) chromosome. The disease only affects men, because women have a second X chromosome to mask the effect of the recessive gene. Women are carriers of the disease, which means that their children have a 50 percent chance of inheriting the abnormal gene. A daughter may not inherit the gene at all, or she may become a symptomless carrier because the second X chromosome will prevent her from developing the disease. A son has a 50 percent chance of developing the disease, because the Y chromosome inherited from his father will be unable to mask the disease. There is no male-to-male transmission of X-linked disorders, but occasionally an X-linked disorder may arise from a new random gene mutation.

Neural tube defects

Neural tube defects (see also p.418) are one of the most common serious congenital abnormalities. In the absence of prenatal screening, about 1 in every 400 babies is affected. Although the exact gene or genes have not been identified, these disorders tend to run in families. Incidence varies widely from one region to the next and has a strong connection with diet. Babies born with neural tube defects such as spina bifida are often severely disabled and require frequent surgical procedures and hospitalization. Disability typically consists of weakness or paralysis of the legs and urinary and fecal incontinence.

PRENATAL DIAGNOSIS

You may need genetic counseling and/or prenatal diagnosis if you have or have had:
▶ A child with a birth defect, chromosome abnormality, or genetic disorder
▶ A family history of either of the above
▶ A child with undiagnosed developmental disability
▶ An abnormal prenatal serum screen result
▶ A fetus with suspected abnormal ultrasound findings
▶ A maternal medical disorder that predisposes your baby to congenital abnormalities
▶ Exposure to an environmental hazard (teratogen) in your current pregnancy
▶ A parent known to be a carrier of a genetic disorder
▶ A history of recurrent miscarriage or fetal loss
▶ A previous neonatal death

Genetic counseling

Couples who know they have a family history of genetic disease or have already had a child with an inherited disorder are strongly encouraged to have genetic counseling when they are planning to get pregnant. They may want to have prenatal diagnostic procedures such as chorionic villus sampling, amniocentesis, or specialized ultrasound scans during pregnancy. Furthermore, for certain conditions it is now possible to offer them preimplantation genetic diagnosis (PGD). PGD is a technique in which eggs are fertilized in vitro and one of the cells of the tiny embryos are analyzed to ensure that it is free of the genetic disorder before being implanted into the mother's uterus.

DOWN SYNDROME

Although Down syndrome (Trisomy 21) is the most common chromosomal abnormality seen in live-born babies, numbers have fallen from 1 in every 600 births to 1 in 1,000 in recent years, due to an increased uptake in prenatal testing.

In 95 percent of cases of Down syndrome there is no family history. In three percent of cases, the extra chromosome 21 is attached to another chromosome (a translocation) and is inherited from one parent, who usually shows no signs of the problem. In the remaining two percent, mosaicism is present, which means that some cells in the body contain a third chromosome 21, whereas others have the normal two.

The risk of Down syndrome increases sharply with maternal age (see below), but since older pregnant mothers are routinely offered screening, most babies with Down syndrome are now born to women under 35. As a result, screening for all pregnant women is becoming routine.

Although about 50 percent of babies with Down syndrome miscarry, 9 out of 10 full-term babies with Down syndrome survive the first year of life. These babies are, however, at high risk of abnormalities of the heart or intestine and problems with hearing and eyesight, and usually have reduced muscular tone and are floppy. Physical features include slanting eyes, a single skin crease on the hands and feet, and a protruding tongue. The bridge of the nose is shallow or absent, which often means that the child is snuffly and susceptible to colds and chest infections.

All children with Down syndrome are mentally disabled, but the severity of the disability is variable and difficult to predict before birth. Recent advances in the way children with Down syndrome are educated has led to many leading relatively independent lives as adults. The average life expectancy is about 60 years old, although leukemia is common in childhood and thyroid disease and a form of Alzheimer's disease are common in adults.

The Nuchal Translucency scan is an early screening test for Down syndrome (see p.137). Routine ultrasound scans may detect other markers, such as the absence of a nasal bone, or heart, kidney, and gut abnormalities. Skin creases on the hands and eyelids are also indications that a baby may be affected by Down syndrome. Some babies with Down syndrome, however, have no obvious structural signs, and are not detected until after birth.

RISK OF DOWN SYNDROME

Maternal age at EDD Risk of Down syndrome

Under 25	25	26	27	28	29	30
1:1500	1:1350	1:1300	1:1300	1:1100	1:1000	1:900
31	**32**	**33**	**34**	**35**	**36**	**37**
1:800	1:680	1:570	1:470	1:380	1:310	1:240
38	**39**	**40**	**41**	**42**	**43**	**44**
1:190	1:150	1:110	1:85	1:65	1:50	1:35
45	**46**	**47**	**48**	**49**	**50**	
1:30	1:20	1:15	1:11	1:8	1:6	

*EDD= expected delivery date

▶ **WEEKS 13–26**

The second trimester

In the second trimester, your baby will grow steadily and the basic structures and organ systems that were formed in the early weeks will be more developed and consolidated. The size overall of the fetus will increase three- to four-fold, and its weight a dramatic 30-fold. Although over the coming weeks you will start to look noticeably pregnant, this is often a time of renewed energy, good health, and a sense of well-being.

CONTENTS

Your baby

IN THE SECOND TRIMESTER

WEEK 14
The eyes are now positioned at the front of the face with the eyelids tightly closed.

WEEK 16
The difference between the male and female genitalia is increasingly obvious. This male fetus now has a solid scrotum and a rudimentary penis.

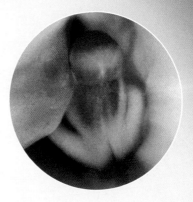

Your fully formed baby grows rapidly during the second trimester and can soon be felt kicking in the uterus.

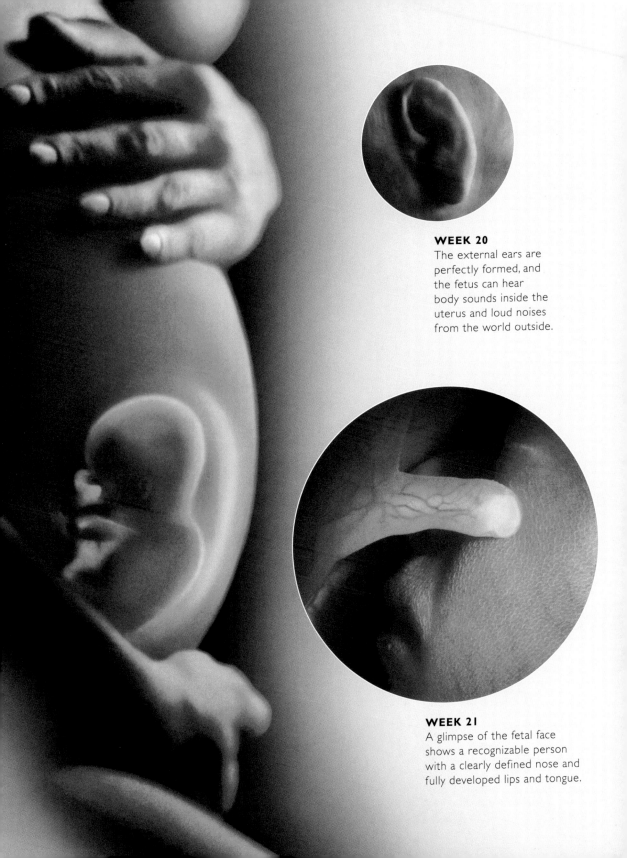

WEEK 20
The external ears are perfectly formed, and the fetus can hear body sounds inside the uterus and loud noises from the world outside.

WEEK 21
A glimpse of the fetal face shows a recognizable person with a clearly defined nose and fully developed lips and tongue.

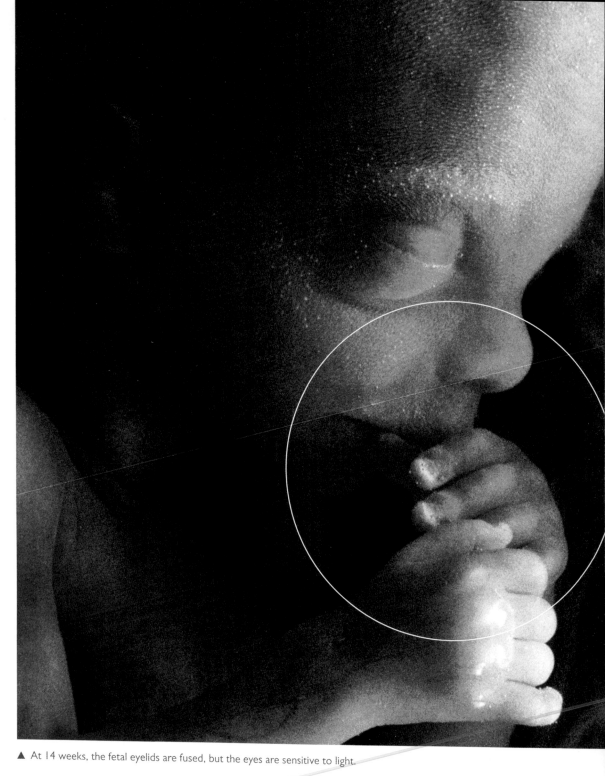

▲ At 14 weeks, the fetal eyelids are fused, but the eyes are sensitive to light.

| 1 | 2 | 3 | 4 | 5 | 6 | 7 | 8 | 9 | 10 | 11 | 12 | 13 | 14 | 15 | 16 | 17 | 18 | 19 | 20 |

▶ WEEKS 0–6   ▶ WEEKS 6–10   ▶ WEEKS 10–13 ▶ WEEKS 13–17   ▶ WEEKS 17–21

▶ FIRST TRIMESTER   ▶ SECOND TRIMESTER

▶ WEEKS 13–17
The developing baby

YOUR BABY IS LOOKING MORE and more like a human being. **Although the head is still relatively large, the length of the body is increasing rapidly. The development of the legs is catching up with the arms and very quickly overtakes them in length. The limbs now appear to be in better proportion with the rest of the body.**

Fingernails can be seen and toenails will start to develop in a few weeks. The trunk has straightened out, but the body still looks thin and is only covered by a layer of fine translucent skin through which the underlying blood vessels and bones can be seen clearly. Very soon, a protective layer of brown fat will start to form, which will help keep the baby warm.

The facial bones are complete, and the facial features are more delicate and much easier to recognize. The nose appears more pronounced and the external ears now stand clear of the sides of the head. The tiny bones in the inner ear have hardened, which allows the fetus to hear sounds for the first time. The eyes are looking forward, although they are still quite widely spaced, and the retina at the back of the eye has become sensitive to light. Although the eyelids are fully formed, they will continue to remain closed for most of the second trimester. However, your baby has already started to be aware of bright light beyond your abdominal wall. The recent development of facial muscles means that your baby can now make—but not yet control—facial expressions. If you happen to have an ultrasound scan at this stage, you may see your baby frowning, grimacing, or even squinting at you. Eyebrows and eyelashes start to develop, and the downy hair on the head becomes coarser and now contains some pigment. Inside the mouth, taste buds are appearing on the tongue.

life size

At 13 weeks, the fetus is about 3in (8cm) long and weighs about 1oz (25g). By the start of the 17th week, its size has increased dramatically to about 5in (13cm). Its weight is now about 5oz (150g).

Intricate movements

Perhaps the most important step forward in your baby's development is that all the connections between the brain, nerves, and muscles have been made. The nerves linking the muscles to the brain begin to develop a fatty coating of

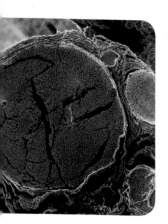

NERVE FIBRES
Signals pass quickly from the fetal brain to the muscles and limbs now that the nerve fibers are coated with fatty myelin sheaths.

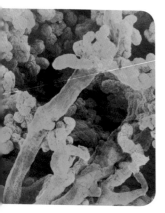

CHORIONIC VILLI
The frondlike villi (colored green) in the placenta allow the exchange of gases and nutrients with the mother's blood.

a substance called myelin, which helps transfer messages to and from the brain. As a result, the fetus is now capable of a wide range of quite intricate movements. The limbs can now move around their joints because the muscles that control this movement are now able to contract and relax. The arms are now long enough for the hands to meet together over the fetal body. When the hands touch they grasp each other and anything else they encounter—such as the umbilical cord. The fingers now curl and the arms and legs flex and extend. The fetus can make a fist or suck its thumb.

Despite all this activity, most first-time mothers are not aware of the fetal movements because the amniotic fluid acts as a cushion and the baby is still not large enough to directly stimulate the nerve endings in the wall of the uterus. Some second-time mothers who know what to expect report that they can feel "quickening"—fluttering sensations in their abdomen—but definite fetal movements are not usually recognized until about 18 or 20 weeks.

The placenta

The placenta continues to grow in size and produce essential hormones (see pp.158–59), which are needed throughout pregnancy to ensure that the baby's growth is on target and the mother's uterus and breasts continue their growth and development. In addition to providing the oxygen and nutrients that your baby needs until the time of delivery, the placenta has now formed a very sophisticated barrier that will help combat the risk of many infections for the rest of your pregnancy. In addition, the placenta will dilute the effect of any medical drugs on the baby, as well as nicotine and alcohol taken by the mother. By the end of the 16th week of pregnancy, the placenta has thickened to a depth less than ½ inch (1cm) and now spans around 3in (7–8 cm).

The amniotic fluid

The amniotic fluid, which fills the sac surrounding the fetus, plays a very important role in fetal development at this stage. It allows freedom of movement and the development of essential muscular tone, while protecting your baby from knocks and bumps. During the first trimester, the amniotic fluid was absorbed through the fetal skin, but in the early weeks of the second trimester the fetal kidneys start to function. From now on your baby will swallow amniotic fluid and excrete the fluid back into the amniotic cavity. Although the amount of amniotic fluid remains relatively constant, it is being absorbed and replaced continuously. The presence of the right amount of amniotic fluid is particularly important for the development of the fetal lungs. Although your baby will continue to obtain all its oxygen and nutrient supplies from the placenta until birth, the lungs must be able to float in a full bath of amniotic fluid in order to expand and develop optimally in preparation for breathing in the outside world. At this stage the bath contains about 180–200ml of amniotic fluid, which is equivalent to the contents of an average paper cup. During this period, the fetus starts to shed some of its own skin cells into the amniotic fluid. This is an important milestone because these cells can be used to determine the chromosomal status of your baby if you decide or are advised to have an amniocentesis test (see pp.140–42). Until now the skin cells available were too few to be a reliable source of information about your baby, which is why amniocentesis is usually not recommended until 15–16 weeks of gestation.

17 WEEKS OLD On a 3-D ultrasound the fetus is seen wiggling and floating, but its movements are cushioned by the amniotic fluid and may not yet be felt.

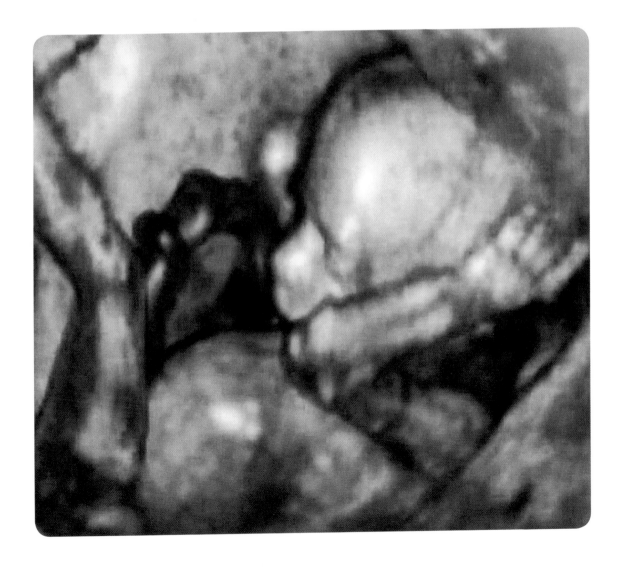

Your changing body

You will probably have noticed that your waist has become a little thicker by now and your belly has become rounder but the exact time in weeks at which you become noticeably pregnant to other people depends very much on your prepregnancy weight and shape.

Nevertheless, during the next few weeks you will be conscious of the fact that you are starting to "show" and that work colleagues and friends who are not in the know are starting to look quizzically at your abdomen. At the beginning of the second trimester, your uterus has grown to the size of a small melon and as a result can now be felt rising out of your pelvic cavity. From now on, its size can be easily assessed by gentle abdominal palpation.

...work colleagues and friends who are not in the know may now be looking quizzically at your abdomen.

Skin pigmentation

Increased skin pigmentation is very common in pregnancy and usually starts to be noticeable at the end of the first or the beginning of the second trimester. The extra estrogen secreted by your body stimulates cells in your skin called melanocytes to produce pigment that darkens the skin. The areola around your nipples may be the first noticeable change; in addition to becoming darker, the areola usually increases in size as well. Moles, birthmarks, and freckles are also likely to enlarge and darken, as will any areas of scar tissue. Most women develop a linea nigra or "black line" of pigmentation, which stretches down the center of their enlarging abdomen. The linea nigra can be prominent in some women from early in the second trimester, whereas in others it does not appear until a little later in the pregnancy. All these changes in color are entirely normal and usually fade away after the baby is born.

Increased blood flow

None of these changes to your body would be possible without an increase in your circulating blood volume and important adaptations in the way that your heart and blood vessels work. The water content of your blood increased early in your pregnancy, but now the volume of red blood cells is becoming measurably larger. Your cardiac output (the amount of blood that is pumped through your heart per minute) continues to increase. Your stroke volume (the volume of blood pumped by your heart at every beat) and heart rate will also increase but thanks to the action of progesterone, your blood vessels will cope with these dynamic changes by becoming more dilated and relaxed.

At the beginning of the second trimester, 25 percent of your blood is being directed to the uterus in order to support your growing baby and placenta, which is an enormous increase compared to the two percent that your uterus used to receive before you became pregnant.

The blood flow to your kidneys will continue to rise until the 16th week, after which it levels off. The filtering capacity of your kidneys, which started to increase in the first trimester, is now 60 percent higher than it was before you were pregnant and will stay at this level until the last four weeks of your pregnancy, when it will fall again. However, the tiny tubules in your kidneys, responsible for reabsorbing the substances that pass through them, are now working overtime. Hence, it is not uncommon for your urine to contain small amounts of sugar and protein.

How you may feel physically

Now that your pregnancy has entered its second trimester, you will be much more confident that this baby is going to become a reality. On a physical level, you will most probably be feeling less nauseous and be regaining some of your usual vitality.

You have probably started to tell friends, family, and colleagues that you are pregnant and find you have become the focus of much attention, congratulations, and advice about what you should and should not be doing as a pregnant woman. Advice is usually offered freely, not just by close family members and friends, but also by people you hardly know, and although it is almost always well meaning, there are times when it may be confusing and even distressing, not to mention entirely irrelevant to your pregnancy. There is absolutely no need for you to be elated one moment and desperately worried the next by such potentially erroneus information, so be careful who you discuss your pregnancy with at this stage. Remember to treat unsolicited advice and cautionary tales with a large grain of salt.

Minor irritations

There are some minor problems that may become noticeable in the early weeks of the second trimester. For example, some of you will notice that your nose is permanently stuffy, although you are not suffering from a cold. You might even be experiencing nosebleeds, blocked ears, and bleeding gums. These symptoms are nothing to worry about, but because they are due to the increased blood

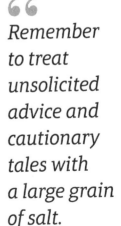

Remember to treat unsolicited advice and cautionary tales with a large grain of salt.

THE MAJOR PREGNANCY HORMONES

From the first day of your pregnancy most of the subtle and more dramatic changes in your body and the way it is functioning are under the control of key hormones. These are produced from existing sources and glands within your body but also, increasingly as the pregnancy progresses, by the placenta and the developing baby.

HORMONE	WHAT IT DOES	WHERE PRODUCED
HUMAN CHORIONIC GONADOTROPIN (HCG)	Maintains secretion of pregnancy hormones estrogen and progesterone by the corpus luteum in the ovary until the placenta takes over.	Produced in large quantities by the young placenta peaking at 10–12 weeks, then declining rapidly.
ESTROGEN	Increasingly high levels in pregnancy boost the blood flow to body organs and promote the dramatic growth and development of the uterus and breasts. Softens collagen fibers in connective tissue to allow ligaments to become more flexible.	Over 90 percent is a type called estriol, which is produced by the placenta; fetus is also involved in the estrogen production process.
PROGESTERONE	Relaxes blood vessels to cope with increased blood flow. Has similar relaxing effects on digestive and urinary tracts. Hypnotic effect may induce placid feelings in pregnancy. Relaxes muscles and helps ligaments and tendons loosen to accommodate the growing uterus and prepare the birth canal for delivery. Prevents contractions until birth is due. Prepares the breasts for lactation.	Until week 6–8, corpus luteum produces progesterone to maintain the pregnancy. By the end of the first trimester, progesterone is produced entirely by the placenta.
HUMAN PLACENTAL LACTOGEN (HPL)	Similar to growth hormone, HPL accounts for 10 percent of the placenta's protein production. Diverts mother's glucose reserves to the fetus; also has effects on maternal insulin production and uptake to help the transfer of nutrients to the fetus. Has a role in breast development and in milk secretion after delivery.	Produced by the placenta from five weeks onward; levels rise throughout pregnancy.

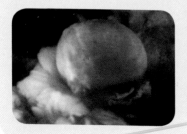
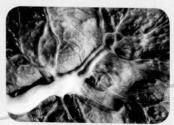

Hormone factories The key site of hormone production in early pregnancy is the maternal ovary (far left), but by 12 weeks the placenta and fetus take control. The placenta both produces hormones and martials the fetal and maternal resources to produce estrogen.

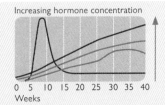

Increasing hormone concentration

0 5 10 15 20 25 30 35 40
Weeks

KEY

—— Human chorionic gonadotropin
—— Estrogen
—— Progesterone
—— Human Placental Lactogen

Key hormones A graph illustrates the surge in HCG hormone early in pregnancy and the steady rise in levels of estrogen, progesterone, and HPL throughout pregnancy.

HORMONE	WHAT IT DOES	WHERE PRODUCED
PROLACTIN	Stimulates breasts to produce milk. Prolactin levels increase during pregnancy, but their effect is blocked until after the birth.	Produced in the anterior lobe of the pituitary gland in the brain.
RELAXIN	An insulin-like substance in the blood that helps soften the pelvic ligaments for delivery and also aids cervical ripening (softening and thinning) ready for dilatation and childbirth.	The ovaries produce relaxin.
OXYTOCIN	Causes the muscles of the uterus to contract. Levels rise in the first stage of labor and are further stimulated by the widening of the birth canal. Oxytocin helps the uterus contract after childbirth and is stimulated by the baby sucking on the nipple during breast-feeding.	Produced by the posterior lobe of the pituitary gland. Receptors in the uterus increase in late pregnancy, enabling oxytocic drugs to be used to induce and augment labor.
CORTISOL AND ADRENOCORTICO-TROPIC HORMONE (ACTH)	Production increases from the end of the first trimester onward. Effects contribute to stretch marks and high blood glucose levels. Cortisol has an important role in helping mature the fetal lungs.	Cortisol secreted by the maternal adrenal glands and the placenta. ACTH from the pituitary gland prompts release of cortisol.
ANDROGENS (TESTOSTERONE AND SIMILAR HORMONES)	Vital building blocks for the production of estrogen during pregnancy. Some testosterone is needed for the development of the male external genitalia.	Produced largely by the fetal adrenal gland. Fetal testes also produce testosterone.

Hormone effects The expansion of your uterus is made possible by a boost in blood flow due to estrogen and the softening effect of progesterone on your ligaments and muscles. After birth, prolactin and oxytocin stimulate your breasts to produce milk.

> *You may find that you are increasingly forgetful—a symptom that is jokingly referred to as maternal amnesia.*

flow to the mucous membranes of your nose, mouth, ears, and sinuses, they are likely to remain with you for the rest of the pregnancy. So it makes sense to think about ways to alleviate these symptoms, even though it is unlikely that you will be able to get rid of them entirely until the baby is born.

Try to avoid spending too much time in heated, dry environments at work and home, including those with air conditioning, both of which dry out the air. You can improve the atmosphere by using portable humidifiers in the rooms you use most frequently. If you experience repeated nosebleeds, you should discuss this with your doctor, who may suggest that you visit a specialist.

You may find that you are increasingly forgetful—a symptom that is often jokingly referred to as maternal amnesia. I think it is simply a reflection of the fact that women are so preoccupied with the excitement of pregnancy during this period that other things seem less important and fail to register. Even if you find yourself forgetting things and unable to multitask in your usual way, rest assured that this vagueness often abates later in pregnancy and you will almost certainly be back to your normal self after the birth.

Your prenatal care

Even though most women remain physically fit and healthy throughout pregnancy, it is usual to see your prenatal doctor every four weeks during the second trimester.

The exact timing of your visits depends upon when you began prenatal care, the results of your blood and urine tests, the type of care you have chosen, and whether you have any particular problems that need attention, but the procedure at each routine checkup is usually much the same.

• Your urine will be tested for protein and glucose. It is important to exclude any infection if protein is persistently present by sending a sample for culture It is not unusual to find small amounts of glucose in your urine on isolated occasions. If it is found at several visits, an early glucose tolerance test to check for gestational diabetes may be advisable (see p.212).

• Your blood pressure will need to be measured with the correct size of cuff to make sure that you are not at risk from some of the more common complications of pregnancy.

• Your doctor will most likely examine your hands and feet to make sure that you have no swelling or edema.

• Your abdomen will be examined and the distance between your symphysis pubis (pubic bone) and the fundus (the dome-shaped top of your uterus) will be measured to ensure that the uterus is growing steadily by about 1cm per week. This measurement, which is always expressed in centimeters, is known as the symphyseal-fundal height (SFH). From 13 weeks onward, it is usually possible to feel the fundus just above the symphysis pubis, but regular measurements of the SFH are not usually taken until 26 weeks. Of course, measurements vary depending on a woman's height and build, together with the number of babies she is carrying and the quantity of amniotic fluid. If you are carrying twins or more, your fundal height will be much higher.

• Your doctor will also listen to your baby's heartbeat by placing a special monitor called a Doppler ultrasound over your uterus. This uses Doppler sound waves (which are completely safe for use in pregnancy) to record your baby's heart rate, which at this stage in pregnancy will be about 140 beats per minute—approximately double your own.

• Most doctors weigh pregnant women at every office visit (see p.408).

Prenatal records

Sometimes it is helpful to request a copy of your prenatal records sometime between the initial visit and the 20-week scan. This provides you with clear documentation of what has happened in the past and during your current pregnancy, which you can immediately share with any doctor that you may need to consult. Having your records available also makes it possible for you to review your progress in the journey through pregnancy anytime you want to. Most pregnant women are very good at remembering to bring their records to prenatal appointments; consequently, lost records are rarely a problem.

YOUR PRENATAL VISITS

Whether you receive your prenatal care from a single doctor or midwife or your caregivers are part of a larger practice, your prenatal care is likely to follow a similar pattern. Your appointments may not correspond exactly to this chart, but they will follow the same general pattern.

5–7 weeks	First prenatal visit
16–18 weeks	First trimester serum screening tests, Nuchal Fold Translucency Scan (NT)
16–20 weeks	Results of blood tests
20 weeks	Detailed fetal ultrasound scan
24 weeks	See doctor/midwife
27–28 weeks	Blood tests for anemia, glucose screening, receive Rhogam injection if blood group is Rhesus negative
30 weeks	See doctor/midwife
31 weeks	See doctor/midwife
34 weeks	Blood test for anemia; discuss birth plan, vitamin K injections for newborn baby, and infant feeding
34–41 weeks	Weekly visits to doctor/midwife
35–37 weeks	Group B Streptococci (GBS) screening. Go to the labor and delivery unit at the hospital for non stress test (NST), which is a recording of the baby's heart rate, and an ultrasound scan to assess fluid volume and fetal well-being
41 weeks and 3 days	Visit with doctor. Discuss possible induction of labor

Things to consider

At this stage of pregnancy your immediate concern is probably finding something appropriate to wear to work. You may also be wondering if some of the side effects of pregnancy can be prevented—for example, by taking extra care of your skin.

Stretch marks

The vast majority of pregnant women will develop some stretch marks during pregnancy. These are caused by the collagen beneath the skin tearing as it stretches to accommodate your enlarging body. Stretch marks can make an appearance at an early stage in pregnancy, and they tend to occur initially on the breasts, because these are the first parts of the body to start expanding, and then on the abdomen, hips, and thighs. The number and extent of stretch marks varies greatly from one woman to the next and is determined mainly by your genes and your age. As you get older your skin loses its elasticity, making stretch marks more likely. There is some evidence that if you are physically fit and well toned before pregnancy and ensure that your weight gain during pregnancy is gradual, you can limit their appearance.

If you do develop stretch marks—and remember, few women do not—rest assured that, with time, their pink appearance (and any itching) will fade. They will not disappear entirely but will become a lighter silvery shade that makes them less visible. There are numerous anti stretch mark creams on the market, but despite what the manufacturers would like us to believe, I'm afraid that no cream applied to your skin can have much effect on what is happening to the deeper layers of collagen that lie well below the surface. That said, massaging creams into your skin to keep it smooth and supple is very pleasurable—and any good moisturizing cream will do.

STRETCH MARKS
If you are prone to stretch marks there is nothing you can do to prevent them. However, their dark appearance will fade with time and they will be much less noticeable.

What to wear

You will find that some of your favorite clothes no longer fit, but try to resist the temptation to go out and buy a whole new wardrobe. You are well advised to wait and buy clothes for the last trimester, by which time there will be very little in your current wardrobe that you will be able to wear. You also need to consider the change in seasons between now and the final months of pregnancy. So look carefully at your current wardrobe, put away anything that has become out of the question or feels even slightly uncomfortable,and concentrate on what still fits and can be worn for the next few weeks.

It is very tedious to have only a limited selection of clothes, so this is a good time to review some of your partner's T-shirts, sweatshirts, and jeans to see if any are suitable. Borrowing larger sizes from your partner or friends is a great way to get through this transition period. You will then be able to save your clothes budget for the later stages of pregnancy, when you are likely to be desperate for a few new things to wear.

Cosmetic effects

If you feel that having tanned legs and arms and a brown belly are essential to your self-esteem, be aware that, in general, you will tan more quickly during pregnancy because the amount of pigment in your skin has increased. It is a good idea to avoid long periods in the sun to minimize the risk of burning, and prematurely aging your skin. Tanning beds should be avoided for much the same reason. When you are in the sun, make sure that you use a high-SPF sunscreen on your face, neck, and shoulders. Alternatively, try self-tanning products, all of which are safe to use in pregnancy.

If you want to get rid of unwanted hair, depilatories are safe to use but shaving or waxing may be your preferred option if you want to minimize contact with chemicals. Avoid waxing over any heavily pigmented moles or varicose veins.

Those of you with tattoos and piercings may have a few questions now that you are pregnant. A belly ring or bar will become uncomfortable as your abdomen grows and is unlikely to stay in place when your navel begins to protrude in late pregnancy, so have it removed sooner rather than later. Nipple rings can be left in place for now but will need to be taken out if you want to breastfeed. Tattoos on any area of your body that expands may change beyond recognition. You should not have any new piercings or tattoos during pregnancy because of the risk of infections such as hepatitis B and C (see p.129), and HIV (see p.415).

TIGHT FIT Put away anything that no longer fits and see if you can borrow a bigger size.

DIET AND EXERCISE

During the second trimester you will probably find that you have regained your appetite and are able to resume normal eating habits. You are also likely to feel more energetic—ready to pick up your old exercise routine or start a new one.

EATING WELL

Make the most of this trimester by eating well because by the next, you may well find that your appetite and digestion go awry again, this time because your growing baby is compressing your digestive system. For most women, lack of time rather than inclination leads to a less than ideal diet during pregnancy. One solution is to stop worrying about producing elaborate nutrient-balanced meals and concentrate instead on regular snacks chosen from a range of healthy foods. You can supply yourself and your baby with everything you need by keeping some of the following foods available:

▶ whole-grain bread and pita and fillings such as grated cheese, lean meats, hard boiled eggs, mashed sardines, salmon, tuna fish, humus, baked beans, salad, tomatoes, and vegetables
▶ chopped fresh vegetable sticks—carrots, cucumber, and celery
▶ fresh fruit—washed and prepared
▶ fruit juices, skim milk, mineral water, herbal and fruit tea, or decaffeinated tea and coffee
▶ unsweetened whole-grain breakfast cereals, oatmeal
▶ low-fat yogurt and cream cheese
▶ dried fruits—apricots, prunes, raisins, and figs
▶ nuts and seeds (sunflower and sesame seeds)
▶ whole-grain crackers
▶ cereal snack bars
Keep some of the more portable items in a refrigerator

GOOD POSTURE

Posture becomes increasingly important now that your shape is changing fast. Try to stand tall, as if your head is being stretched upward by a thread, aligning yourself in a straight line from the top of your head, down through your pelvis and perineum to your feet.

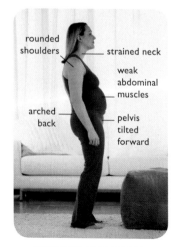

rounded shoulders
strained neck
weak abdominal muscles
arched back
pelvis tilted forward

INCORRECT POSTURE

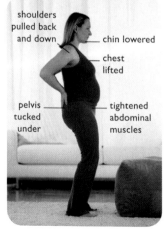

shoulders pulled back and down
chin lowered
chest lifted
pelvis tucked under
tightened abdominal muscles

CORRECT POSTURE

at work and carry crackers, fresh and dried fruit, nuts, and a small bottle of mineral water or fruit juice in your handbag.

PREGNANCY PROBLEMS

One of the fears that women have during pregnancy is that a bad diet will give rise to a range of complications. Conditions affected by diet include preeclampsia (see p.426), gestational diabetes (see p.427), and intrauterine growth restriction, IUGR (see p.429). However, I must stress that many women who develop complications have perfectly good diets—no worse than others who remain untouched by these problems. Women are generally quick to blame themselves—and, typically their diets—as being responsible for other problems such as premature labor and high blood pressure. Let me reassure you that these conditions are not related to diet.

EXERCISING SAFELY

High-impact sports, such as jogging, skiing, and horseback riding, are progressively inadvisable over the second trimester. Jogging is not dangerous for your baby but it puts pressure on your joints, tendons, and ligaments, and can lead to long-term damage. Be aware, if you are skiing or horseback riding that your growing belly alters your center of balance, making you more likely to fall. If any of the above is your preferred method of keeping fit and healthy, consider switching to other forms of exercise—bicycling, walking, and swimming are among the best.

KEGEL EXERCISES

This is one set of exercises that you should do on a regular basis— I cannot emphasize enough how important they are. Your pelvic floor resembles a hammock of muscles, which support the bladder, uterus, and bowel and surround the urethra, vagina, and rectum. Loss of tone in these muscles and/or damage from a prolonged vaginal delivery can lead to stress incontinence—small leakages of urine when you cough, laugh, or sneeze. This often continues after the birth and later in life, especially after menopause, when lack of estrogen compounds the problem.

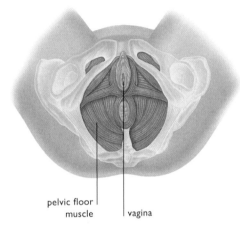

Your pelvic floor During pregnancy, your pelvic floor muscles soften, relax, and become stretched under the pressure of the enlarging uterus.

pelvic floor muscle | vagina

Practice Kegel exercises every day throughout pregnancy, and continue them after the birth. One way to remember is to practice them at particular times in your schedule, for example, when you brush your teeth or wait for the bus or train—after a while they will become automatic.

▶ **Empty your bladder** before starting the exercises. (It is no longer recommended to try to stop urine flow midstream.)

▶ **Tighten and release the muscles** around your urethra, vagina, and anus. You should feel a lifting sensation in your pelvic floor muscles. Hold for a few seconds, then relax slowly. Build up to 10 squeezes, 10 seconds for each squeeze and resting briefly between squeezes.

▶ **Now squeeze and relax the muscles more quickly,** holding for about a second at a time. Repeat 10 times.

▶ **Now squeeze and relax** each set of muscles around your urethra, vagina, and anus in turn from the front to the back and from the back to the front. You can perform these regularly during the day.

▲ At 19 weeks, the facial features are already remarkably refined.

| 1 | 2 | 3 | 4 | 5 | 6 | 7 | 8 | 9 | 10 | 11 | 12 | 13 | 14 | 15 | 16 | **17** | **18** | **19** | **20** |

▶ **WEEKS 0–6**   ▶ **WEEKS 6–10**   ▶ **WEEKS 10–13**   ▶ **WEEKS 13–17**   ▶ **WEEKS 17–21**

▶ **FIRST TRIMESTER**   ▶ **SECOND TRIMESTER**

▶ **WEEKS 17–21**

The developing baby

THE GROWTH OF YOUR BABY'S BODY AND LIMBS continues very rapidly during this time and, as a result, the head now appears in better proportion to the rest of the body. By the end of week 20 the head makes up less than one-third of the total length of the fetus.

Your baby's legs in particular have undergone an amazing growth spurt and are now longer than the arms. From this point on, the rate of growth of the body and limbs will start to slow down, although the fetus will continue to gain weight at a steady pace until the time of delivery. This relative slowing in physical size is an important milestone because it marks the fact that the baby is now developing in different ways. The lungs, digestive tract, and nervous and immune systems are all starting to mature in preparation for life in the outside world. The skeleton can now be seen clearly on an X-ray because more calcium has been deposited in the bones to harden them.

Your baby's sexual organs are now well developed and the difference between the male and female external genitalia is increasingly obvious. Inside the body of a baby girl, the ovaries contain the three million eggs that she will be born with; the uterus is fully formed and the vagina is starting to become hollow. A baby boy's testes have not descended from the abdominal cavity into the scrotum, but a solid scrotal swelling alongside a rudimentary penis can often be seen between the boy's legs on an ultrasound scan. An ultrasound scan at 20 weeks should be able to determine the sex of your baby, provided she or he is facing the right way. On the chest wall of both boys and girls, early breast tissue (mammary glands) has developed and nipples can be seen on the skin surface.

Improving senses

Although your baby's eyelids are still generally closed, the eyeballs can roll from side to side, and at the back of the eye, the retina is light-sensitive because nerve connections to the brain have been established. Your baby's taste buds

life size

At 19 weeks the fetus now measures about 6in (15cm) from crown to rump. Its weight is now about 8oz (225g). By the end of 21 weeks, its length is around 7in (17cm) and it weighs about 12oz (350g).

The fetus may move around vigorously if exposed to loud noises such as music at a rock concert.

are so well developed that they can now distinguish between sweet and bitter flavors (although these are unavailable in utero) and many of the first "milk" teeth have developed inside the gums. The mouth opens and closes regularly, and an ultrasound scan may capture your baby sticking its tongue out. Although it still has no conscious thoughts, it now hears sounds very clearly such as those of your heart beating, blood pulsing through the vessels in your lower body, and your digestive system churning around. It is often suggested that one reason why newborn babies usually stop crying when they are placed over their mother's left shoulder is that they recognize the comforting memory of her heartbeat.

The fetus can also hear sounds from outside your body and may jump or move around vigorously if exposed to loud noises such as music at a rock concert. The pattern of the fetal heartbeat, now clearly audible if an electronic monitor or Doppler ultrasound is placed on your abdomen wall in the correct place, also changes in response to loud external sound. The skin has also become responsive to touch and, when firm pressure is placed on your tummy, your baby will move away from the intruding stimulus.

New nerve networks

All of these sophisticated developments in your baby's senses are due to the fact that the nervous system is developing rapidly and maturing steadily. New nerve networks are being formed continuously and acquiring fatty, insulating myelin sheaths, which enables them to transmit messages to and from the brain at great speed. A fibrous sheath starts to grow around the nerve bundles in the spinal cord to help protect them from mechanical damage. These

FETUS AT 19 WEEKS ·····

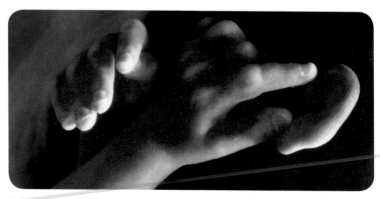

A fine covering of downy hair forms on the eyebrows and upper lip.

Your baby now hears sounds very clearly—your heart beating and the rumblings of your stomach.

adjustments to the nervous system help your baby become much more active. Even though you may not be able to feel the movements yet, your baby is constantly moving, twisting, turning, stretching, grasping, and doing somersaults. This increase in muscular activity allows movements to become more refined and purposeful, improving motor skills and coordination and helping develop stronger bones.

Skin and hair

The fetus starts to look a little plumper and less wrinkly during this stage of pregnancy as thin layers of body fat begin to form. Some of this fat is insulating brown fat, which starts to be deposited in pockets at the nape of the neck, behind the breastbone, around the kidneys, and in the groin areas. Babies that are born prematurely or are underweight have very little brown fat, which is why they have such difficulties trying to maintain their body temperature and become cold very quickly.

There is still very little fat beneath the skin, so the blood vessels, particularly those around the head, are clearly visible and the skin continues to look red and translucent. However, your baby's entire body is now covered with a fine downy layer of lanugo hair, which was first seen around the fetal eyebrows and upper lip at about 14 weeks. Lanugo hair is thought to be one of the mechanisms that help the fetus keep warm until it has sufficient fat reserves. This is probably why babies born before 36 weeks are usually still covered in lanugo, while those born at full term have shed most of the hair during the last few weeks in the uterus. This fine covering of skin hair also helps ensure that the baby remains covered in the thick white waxy coating, called vernix, that starts to be secreted from the sebaceous glands in the skin during this second trimester. The vernix protects the fetal skin from fingernail scratches and prevents it from becoming waterlogged during the many weeks that it is immersed in amniotic fluid.

Support systems

The placenta continues to be the fetal life support system and is now fully developed functionally. However, it will continue to grow, tripling in size by the end of a normal pregnancy. Up until now, it weighed more than the fetus, but from this point on, the fetal weight overtakes that of the placenta.

It is quite common to find the placenta lying low in the uterus during the 20-week ultrasound scan, but this is not a cause for concern at this stage. Although the placenta is firmly attached to the uterine wall, the uterus surrounding it will grow considerably both upward and downward during pregnancy. The lower segment of the uterus starts to form in preparation for delivery at around 32 weeks, with the effect that the majority of placentas are no longer low lying when visualized on an ultrasound scan. Of course, the placenta does not change its position; instead, the uterus grows around the placenta at a different rate and at different times in the pregnancy. In fact, the uterus continues to grow until about 37 weeks and, at term, less than one percent of women have a low-lying placenta (see p.240 and p.428).

The pool of amniotic fluid around the fetus continues to increase: by the end of the 20th week it contains about 11fl oz (320ml). This is a dramatic increase when compared to the 1fl oz (30ml) of amniotic fluid that was present in the uterus at 12 weeks. The temperature of the fluid remains at 99.5°F (37.5°C), which is slightly higher than the mother's body temperature. This is another factor that keeps the fetus warm.

Your changing body

For some women, weight increases gradually, while others notice a big spurt one week and no apparent increase during the next. Overall, you will gain, on average, 1–2lb (0.5–1kg) per week during this period, and by week 21 you will definitely look pregnant.

…your blood pressure has not shot through the roof because most of your blood vessels have become more dilated and flexible…

At 18 weeks the fundus of your uterus can be felt midway between your pubic bone and navel (umbilicus) when your doctor gently palpates your belly; by week 24 it will probably be at or just below your navel. The symphyseal fundal height (the distance between the top of your uterus and your pubic bone) will be 9½in (24cm). Although this measurement is not as accurate as a scan, it is a quick way to establish that your baby is growing satisfactorily.

Feeling the heat

The volume of blood in your circulation continues to increase steadily and by 21 weeks it will measure nearly 11 pints (5 liters). This increase is needed to supply the many organs in your body that are now working much harder than usual. The uterus is receiving the biggest share of this extra blood flow, which is vital to perfuse the placenta and provide sufficient oxygen and nutrients for your baby; an extra 17 fl oz (0.5 liters) of blood will continue to be pumped to your kidneys every minute for the rest of your pregnancy. A higher than normal proportion of your blood flow is also going to your skin and mucous membranes and their blood vessels have become dilated to accommodate it. This is one of the reasons why pregnant women have blocked and stuffy noses, feel the heat, sweat more profusely, and feel faint during pregnancy.

Extra blood volume

To deliver this extra blood to your organs, your cardiac output must continue to increase gradually and by 20 weeks, your heart is pumping about 15 pints (7 liters) of blood per minute. However, your heart rate (the number of times your heart beats every minute) cannot be allowed to rise too much or you would start to experience palpitations. The extra blood volume and the stronger pumping activity of your heart should logically result in your blood pressure rising dramatically, but important changes that are taking place in the blood vessels, which usually prevent these problems from developing.

One of the major reasons why your blood pressure has not shot through the roof is because most of the blood vessels in your body have become more dilated and flexible (known medically as a fall in their peripheral resistance). They contain much more blood than they used to, thanks to the action of progesterone and other hormones, and this extra blood volume and your dilated blood vessels can lead to unwelcome physical symptoms such as varicose veins (see p.235) and hemorrhoids (see p.217). This fall in peripheral resistance ensures that, in most cases, blood pressure alters very minimally during the first 30 weeks of pregnancy, unless complications such as pregnancy induced hypertension develop (see p.426). After 30 weeks there is a normal tendency for the blood pressure to rise, but this increase should never be fast or excessive.

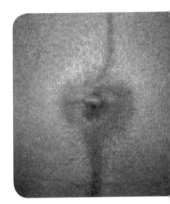

LINEA NIGRA This line of pigmentation down the abdomen is usually more noticeable in women with dark or olive skin.

Changes to your skin

As a result of the dilated blood vessels in your skin and the high levels of estrogen in your body, you are likely to notice tiny red marks called spider nevi appearing on your face, neck, shoulders, and chest.

The pigmentation in the skin around your nipples, genitalia, and the linea nigra down your belly will continue to be more noticeable. This line, which is more prominent in some women than in others, marks the point where the right and left abdominal muscles meet in the midline. These strap muscles will start to separate from now on to accommodate your growing uterus, but why this should be accompanied by increasing pigmentation on the surface of your abdominal skin is a bit of a mystery.

Some women develop chloasma—also called the mask of pregnancy—on their face. In fair-skinned women the chloasma appears as darker, brownish patches mainly on the bridge of the nose and cheekbones and sometimes

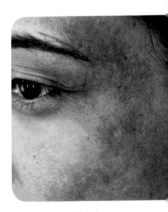

CHLOASMA This brown pigmentation usually develops symmetrically on the cheeks and other areas of the face.

around the mouth. On darker skin, the patches appear lighter than the normal skin tone. Since all of these skin pigmentation changes are due to pregnancy hormones, they usually disappear or fade quite quickly after the baby is born.

How you may feel physically

By now, you are probably rediscovering your prepregnancy energy levels. Indeed, many women find that, although they now look pregnant, they don't feel very different physically from the way they did before their pregnancy began.

FIRST FLUTTERS It may take a while before you are certain that you are feeling your baby move.

Your appetite will return—indeed it may increase—and you will find that you can once again enjoy normal meals. Make the best of this time in pregnancy and make sure that you follow a well-balanced nutritious diet (see pp.43–49). It will not be very long before your appetite is upset again, this time because of heartburn, indigestion, reduced stomach capacity, constipation, and all of the other gastrointestinal problems that develop in later pregnancy as the baby grows inside your abdominal cavity and your digestive system becomes sluggish. Your sex drive may also have returned and for many couples this is a reassuring sign that pregnancy has not permanently changed their physical relationship. Indeed, some find that their sex life becomes particularly enjoyable during this trimester. A variety of factors contribute to this. Physically you are generally feeling better, which is important since nausea and fatigue are real passion killers, and your hormone levels are less likely to be swinging all over the place. In addition, both you and your partner may be feeling more relaxed in this trimester, having had time to get used to the idea of you being pregnant.

The first time you become aware of your baby's movements is a physical experience that is almost inseparable from the emotional response that follows it. If you have not had a baby before, you may dismiss these first fluttery sensations as gas, but after a while you realize they are unrelated to your digestion; they feel quite different. No one can predict the exact date when you will feel your baby moving around, but it is likely to happen now or very soon.

Your emotions

By now you have probably told most of the people with whom you have regular contact that you are expecting a baby but, even if you have been keeping the news to yourself, your changing shape will have given other hints that you are pregnant. Whether you are at home or at work, you are likely to find that everyone around you wants to talk about your pregnancy. Like many women, you may enjoy this new-found closeness with relative strangers and welcome the chance to share your excitement about the pregnancy. There are very few occasions in life when we embark on a deeply personal conversation with a complete stranger and then allow them to pat our belly or place a paternalistic arm around our shoulders. This happened all the time when I was pregnant and I confess that, after some initial feelings of surprise, this generous display of warmth and goodwill confirmed for me that I was going through a very special time, and that everyone around me recognized this fact. Having said that, I do appreciate that some women view this outside interest in their pregnant state as an unwanted invasion of their privacy. If you find yourself in the latter group you may need to find ways to protect yourself from becoming resentful and angry.

Overall, this is one of the most enjoyable stages in pregnancy and you are probably feeling more serene and calm than usual. Try to enjoy the relative peace of the second trimester because the third trimester will probably bring with it some emotional highs and lows, not to mention some physical discomfort.

> *…you may enjoy this new-found closeness with relative strangers and welcome the chance to share your excitement about the pregnancy.*

Your prenatal care

You will see your doctor every four weeks during the second trimester for checks on your urine, blood pressure, and the height of your fundus. Your baby's heartbeat will be listened to using a Doppler ultrasound monitor or, more rarely, a fetoscope (see page 161).

If you have not already been given the results of your early blood tests and serum screening tests, this will be done at one of your prenatal appointments during this stage of pregnancy.

The fetal anatomy or anomaly scan

Most maternity units in the US offer women an ultrasound scan at about 20 weeks (see pp.174–75), although the exact timing varies from one practice to another to complement additional prenatal tests such as serum screening blood tests. However, it is usually performed between 18 and 22 weeks and is frequently referred to as the fetal anatomy scan because at this stage of pregnancy, your baby's organs and major body systems are sufficiently developed for it to be possible to detect the vast majority of potential structural

YOUR 20-WEEK SCAN

By around 20 weeks your baby is sufficiently developed
for most of the major organs and body systems to
be viewed on ultrasound and checked for signs of
a problem. For most women, this scan will reassure
them that all is progressing well.

Below is a summary of the most common checks during this scan but they will not necessarily occur in this order, because your baby will be moving around constantly and may not always be in the right position. Measurements and observations of your baby will be taken as the opportunity allows. The sonographer will only check off the boxes on the checklist when everything has been seen clearly. So if the position of your baby makes this impossible at first, you may be asked to walk around and then return for further measurements or even come back in a week or so for another attempt.

▶ **The fetal heartbeat** is usually the first item to be checked. The sonographer will also look for the four chambers of the heart. If a specialized heart scan is needed, this usually takes place at 22–24 weeks.

▶ **In the abdominal cavity**, the shape and size of the stomach, intestines, liver, kidneys, and bladder will be examined and the sonographer will check that your baby's intestines are now completely enclosed behind the abdominal wall.

The diaphragm, the muscular shelf that separates the chest from the abdominal cavity, should be complete and the baby's lungs should be developing.

Although abnormalities in any of these organs are uncommon, you will find information about some of the problems that may be identified on pages 416–22.

▶ **Your baby's head and the spine** will be examined, starting with the covering bones of the skull, to check that they are complete. The fetal spine has straightened, which means that the sonographer can move the scanning probe up and down, checking each bone, or vertebra to make sure that there is no evidence of spina bifida (see p.419).

▶ **The brain** contains two ventricles or fluid-filled cavities, positioned on the left and right side of the midline, which are lined with a special system of blood vessels called the choroid plexus. Rarely, the ventricles are enlarged. In the

HOW YOUR BABY MEASURES UP

The measurements taken during the scan help establish whether your baby is the right size for your dates. They are recorded as abbreviations in your notes, with measurements in millimeters, which are checked against charts of normal values. This provides an estimate of the size of your baby in terms of weeks out of 40.

BPD	Biparietal diameter (the distance between the side bones of your baby's head)	45mm = 19+/40
HC	Head circumference	171mm = 19+/40
AC	Abdominal circumference	140mm = 19+/40
FL	Femoral length (length of the thigh bone.)	29mm = 19+/40

These measurements suggest that this fetus is appropriately grown for 19 plus weeks of gestation. Assuming that the scan was performed at 19–20 weeks, the baby is the right size for dates.

DETAILS FROM A 20-WEEK SCAN

chin   lung   liver   bowel

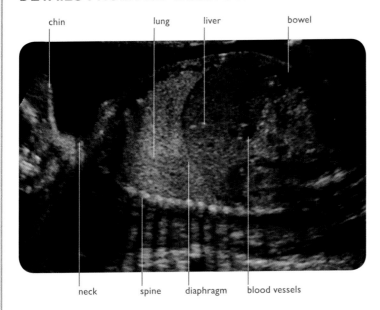

neck   spine   diaphragm   blood vessels

UNDERSTANDING YOUR BABY'S SCAN This right-side image of the fetus shows the chin and neck (far left in image) and the lung is visible as a pale area above the diaphragm, the dome-shaped sheet of muscle that separates the chest from the abdomen. The liver appears as a large shadow below the diaphragm, punctuated by two blood vessels, which appear as black circles. The dark M shape (top right in image) is the fetal bowel.

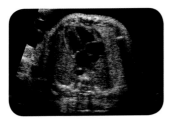

FETAL HEART The four chambers of the heart and blood vessels are clearly shown.

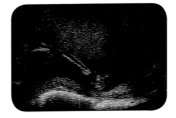

LEG AND FOOT The length of the femur (thigh bone) is a good growth indicator.

SPINE The skin over the spine is checked to make sure that it is complete.

unlikely event that this problem is identified, you will be offered further scans.

▶ **The position of the placenta** will be recorded as lying on the anterior or posterior wall, or at the fundus (top) of the uterus. It is very common for the placenta to be low-lying at this stage and you may be advised to have another scan at about 32 weeks to check that it is positioned higher.

The umbilical cord will be checked and the place where it inserts into the placenta may be noted. Most of the time it is in the center, but if it is at the very edge of the placenta (known as velamentous cord insertion, see p.430), there may be problem.

▶ **The volume of amniotic fluid** will be assessed, to ensure that there is neither too much, a condition known as polyhydramnios (see p.427), nor too

little (oligohydramnios, see p.427). In either case you will need to have more test that include a scan in a few weeks' time.

▶ **The sex of your baby** can sometimes be determined if the scan reveals a penis between the fetal legs. But it is not always possible to conclude that you are carrying a girl because the penis can be hidden. For this reason, some practices are unwilling to disclose these findings.

UNDERSTANDING YOUR SCAN Your doctor will be able to help interpret exactly what you are seeing.

abnormalities. The majority of anatomy scans will demonstrate that your baby appears to be developing normally. Far from being a cause for concern, it will offer you and your partner great reassurance, not to mention the thrill of seeing your baby in great detail. For many couples, this is the magical moment when they realize that their pregnancy is going to result in a real human being. Most doctors will offer you a printed picture for your family album; my twin girls are still entranced by the photo of themselves at the 20-week scan.

Confirming results

For some women the scan result may show one or more causes for concern. Many abnormalities are accompanied by physical signs which are described as markers for a disorder or syndrome. For example, about 70 percent of babies with Down syndrome have structural abnormalities in their heart or gut, a single skin crease in the palm of their hands, and slanting eyes with epicanthic folds that can often be seen on an ultrasound scan. When a scan picks up a physical sign that suggests Down syndrome, I almost always advise the parents to have an amniocentesis test (see pp.140–43). They then have a diagnosis they can be sure of and can be given full counseling and advice on their next step.

Although the 20-week scan provides vital information, I should make it clear that a normal scan cannot guarantee that a baby does not suffer from an abnormality. About 30 percent of babies with Down do not have an obvious structural marker. It is also impossible to prenatally diagnose problems such as autism or cerebral palsy. Scans do have limitations, even when the sonographer is an expert and the machinery is state-of-the-art.

Common concerns

Most of the complaints that crop up in the second trimester are minor, and many of them are due to your blood vessels dilating to cope with the extra volume of blood circulating in your body.

Feeling dizzy or faint is a very common symptom during the second and third trimesters of pregnancy, and is rarely a serious problem. As I explained earlier, the blood vessels around your body, and particularly in the veins of your pelvis and legs, now contain a much larger volume of blood. When you stand up quickly, you may feel momentarily dizzy because all that extra blood in your leg veins needs a bit of extra time to be redistributed to your head and other organs. Headaches are another common complaint. Most of the time they are due to tension and anxiety or to your nasal congestion. However, always seek prompt advice if you find that you are suffering repeated headaches because they can be the first sign that you are developing high blood pressure (see blood pressure problems, p.426).

Skin rashes

Pregnant women are very susceptible to a wide range of skin rashes, which are usually due to the massive hormonal changes occurring in your body. However, always check with your doctor if you experience persistent problems.

It is not unusual to develop dry patches of flaky and highly itchy skin, most commonly on your legs, arms, and abdomen. Applying simple moisturizing creams usually helps, but occasionally an antihistamine or even a low-dose steroid cream may be needed to relieve the irritation. The dilated blood vessels in your skin may make it look pinker, or even blotchy, which can sometimes be mistaken for a rash.

Since your skin needs to get rid of the extra heat from these dilated blood vessels, it is normal for you to sweat more. As a result, you may develop a sweat rash under your arms or breasts and in the groin areas, where sweat accumulates and does not have much chance to evaporate quickly. Wearing looser clothes, preferably cotton, and avoiding synthetic fabrics close to your skin will reduce the problem, as will careful hygiene and using unscented soaps and deodorants. Occasionally, a sweat rash can become infected with a fungal organism (for example, yeast see p.216). These are not serious, but can make you itchy and uncomfortable and you may need a prescription topical antifungal cream to help clear the infection.

APPLYING CREAM
Using moisturizing cream will help improve itchy, flaky skin on your legs, arms, and abdomen.

Things to consider

The chances are that the clothes you need to wear now are quite different from those that you imagined you would be wearing at this stage. Feel pleased if you resisted the temptation to buy new outfits at the first sign of your growing belly.

66

...many women feel the urge to go on vacation—a sort of 'make hay while the sun shines' type of break.

99

Now is the time to give some thought to what clothes you really need to be comfortable at work and at home. Think of this exercise as being similar to packing for a vacation when your baggage allowance is restricted. There is no shortage of sources for maternity clothes—they can be found in the big department stores, in some malls, in maternity stores, catalogs, and online stores—but careful selection is the key.

Mix and match separates are usually the best solution, since they offer the greatest flexibility and you can accommodate repeated changes in your size relatively easily. The main difference between special pregnancy tops and normal jackets or shirts is that the front is longer than the back so that, once your belly has been taken into account, both are the same length. However, the back of some jackets have a way of expanding to fit your enlarging shape, so if you need to wear suits or jackets to work, you could well find that buying yourself one good jacket and changing the skirt or pants as and when your growing abdomen demands it, is the best way to deal with your changing size.

When it comes to skirts and pants, the priority is to find clothes with a comfortable, expanding waistline. In the short term, buying pants a couple sizes larger than usual works well, but you will soon find that this is not the long-term answer, since the elastic waist starts to dig into you as the weeks go on. True maternity skirts and pants have clever button elastics or paneling at the top, both of which allow for your expanding waist and provide comfort for the later stages of your pregnancy. You might also want to invest in some drawstring pants or a dress with no waist to see you through a good part, if not all, of the remaining months.

Maternity pantyhose provide more support than ordinary pantyhose and even though you will probably have to buy different sizes as your pregnancy continues they are a good investment. I can remember how much more comfortable I felt when I was wearing them and how conscious I was of my tired and aching legs when I did not. Avoid knee socks, which can block circulation at the top of your calf and encourage the formation of varicose veins. Cotton-rich ankle socks allow your skin to breathe and do not impede circulation.

If swimming is part of your exercise routine, buy a swimsuit that will see you through the whole nine months. You will feel a lot more comfortable in a special maternity bathing suit, many of which can be bought for around the same price as an average swimsuit.

A 40-week pregnancy spans several seasons and most of your major clothing purchases need to be for the season when your body shape has altered most significantly. If your first and second trimesters fall in the summer months, you may have several items of loose clothing in your wardrobe to tide you over until your shape demands a formal shopping expedition. If your third trimester is in the winter months, resist the temptation to buy an expensive winter coat in a larger size, since you are unlikely to want to wear it next winter after your baby has been delivered. In the meantime you could borrow a coat from a friend, or buy an A-line design with no front fastenings, which can be used indefinitely.

A change of scenery

During the second trimester many women feel the urge to go on vacation or a weekend away—a sort of "make hay while the sun shines" type of break. Inexpensive flights can be found to all corners of the globe and the question of traveling in pregnancy and how safe it is crops up continually. Most of the answers are to be found in the Staying Safe in Pregnancy section, which includes a full discussion about travel in pregnancy, safety of immunizations, and the various precautions you should take (see pp.36–37).

As far as advice specific to this stage of pregnancy goes, if you have had previous pregnancy complications, think carefully about traveling abroad at this time; the last thing you need is to find yourself far from home, if something unpredictable happens. Managing an unforeseen problem is distressing enough without having to deal with a foreign language, a different health-care system, and the uncertainty of knowing whether your health insurance will cover the situation. However, if you have no history of problems and your doctor has said that you are fit to travel, then it is highly unlikely that you will encounter any pregnancy-related problems during your vacation.

TAKING A BREAK This stage of pregnancy can be a good time for travel as long as you have no pregnancy problems.

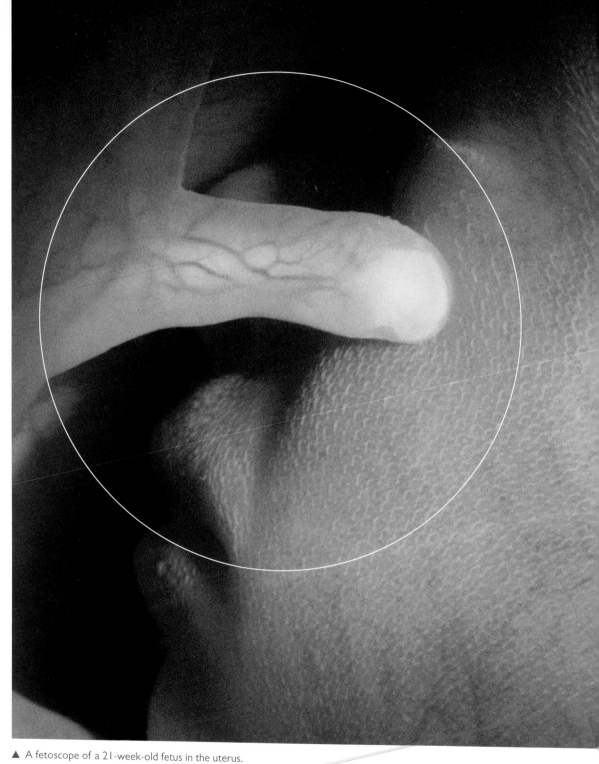

▲ A fetoscope of a 21-week-old fetus in the uterus.

| 1 | 2 | 3 | 4 | 5 | 6 | 7 | 8 | 9 | 10 | 11 | 12 | 13 | 14 | 15 | 16 | 17 | 18 | 19 | 20 |

▶ **WEEKS 0–6** ▶ **WEEKS 6–10** ▶ **WEEKS 10–13** ▶ **WEEKS 13–17** ▶ **WEEKS 17–21**

▶ **FIRST TRIMESTER** ▶ **SECOND TRIMESTER**

▶ WEEKS 21–26
The developing baby

YOUR BABY IS CONTINUING TO GROW in length and to put on weight steadily, although it will be some time before it develops a plump, chubby appearance. The facial features are now very well developed and eyebrows, eyelashes, and head hair are clearly visible.

The skin still looks pink and wrinkly, but no longer appears completely translucent, because some subcutaneous fat is being laid down. Two distinct layers of skin have developed: the surface layer or epidermis and a deeper layer called the dermis. The epidermal skin layer now carries a surface pattern on the fingertips, palms, toes, and soles of the feet, which is genetically determined and gives rise to a unique set of finger- and toe prints for this future human being. The underlying dermis grows little projections containing blood vessels and nerves. The surface of the skin remains covered with a fine layer of lanugo hair and a thick coating of white vernix. This waxy protective coat will remain until just before birth: babies that are born prematurely are often still covered with a thick layer of vernix, whereas babies born late or post mature, have lost all their vernix and, as a result, their skin is dry and flaky.

Inside the body

Several important developments are occurring inside your baby's body. The nervous and skeletal systems continue to mature, so that, instead of just wiggling or floating, movements become more deliberate and sophisticated, including kicks and somersaults. The fetus practices thumb-sucking and starts to hiccup. Any object that the hands encounter is firmly grasped and, amazingly, this grip is strong enough to support the baby's entire body weight.

The brain is developing and its activity can be monitored electronically on an electroencephalogram (EEG). By 24 weeks, the fetal brainwave patterns are similar to those of a newborn infant. The brain cells that have been programmed to control conscious thought are beginning to mature and research suggests that from this time onward, the fetus starts to develop a

life size

The 21 week-old fetus measures around 7in (17cm) and weighs about 12oz (350g). By the end of the second trimester, it will have increased in size to about 10in (25cm) from crown to rump and weighs just about 2lb (1kg).

21	22	23	24	25	26	27	28	29	30	31	32	33	34	35	36	37	38	39	40

▶ WEEKS 21–26 ▶ WEEKS 26–30 ▶ WEEKS 30–35 ▶ WEEKS 35–40

▶ **THIRD TRIMESTER**

primitive memory. Certainly, your baby can now respond to noises in your body and also to loud outside sounds and your physical movements. It is thought that a baby can distinguish between his mother's and father's voices at this stage, and is able to recognize them after birth. Studies have shown that babies are able to recognize a particular piece of music that they have "heard" repeatedly in utero. This theory might explain why my young daughters always seem to feel comfortable and relaxed when they hear Italian opera or Nina Simone—both are acquired tastes, but the twins were exposed to them repeatedly while I was pregnant.

The eyelids will be open toward the end of this period. Although many babies have blue eyes at birth, the final color of their eyes is not known until many weeks after delivery.

A cycle of sleeping and waking has started to develop. Unfortunately, this is not always synchronized with your daily pattern and you may find yourself worrying about the lack of movements during the day, only to find that you are kept awake for most of the night because the baby is buzzing around.

The fetal heart rate has slowed considerably from 180 to 140–150 beats per minute by the end of the second trimester. From this point on, monitoring the heart rate pattern on an electronic fetal monitor becomes one of the most useful methods of assessing your baby's well-being.

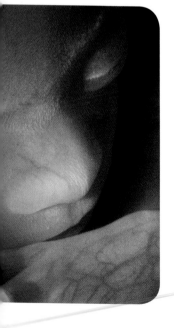

TOUCHING The hands move toward the face and touch and grasp anything they encounter.

The fetus is opening and closing its mouth regularly as large quantities of amniotic fluid are ingested. The fluid is digested and waste products from the metabolic processes in the fetal body are passed across the placenta, via the umbilical cord, to be disposed of in the mother's blood. The remainder, which is effectively excess water, is then excreted back as urine into the pool. At 26 weeks the volume of amniotic fluid has risen to about 16fl oz or almost 1 pint (500ml) and the entire pool is changed or recirculated every three hours.

The fetal lungs are still relatively immature and it will be several weeks before they can breathe unaided. Nevertheless, your baby has started to make breathing movements and will practice them until the time of delivery. The lungs are full of amniotic fluid, which helps the fetus develop more air sacs (alveoli); these need to be buoyant in the amniotic fluid in order to multiply properly and later to expand. A network of tiny blood vessels around the sacs is forming; this will be essential for the transfer of oxygen to the rest of the baby's body after birth. If the water breaks before the end of the second trimester, the baby's lung development is almost always compromised and breathing difficulties after birth are invariably a problem.

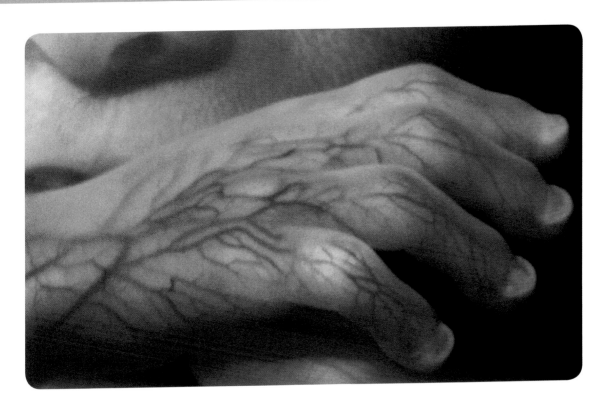

A viable baby

Even though the lungs are still relatively immature, the fetus has now reached a stage of potential viability, which means that it may be able to survive outside the uterus, albeit with the help of neonatal intensive care medical expertise, such as perinatologists, and a ventilator to assist with breathing. The limit of potential fetal viability is currently approximately 24 weeks. Babies that are born before the age of 24 weeks are very unlikely to survive. Generally, any baby that is less than 20 weeks old when born is very unlikely to survive and will be considered either a miscarriage (see p.341) or a stillbirth (see p.342). After 24 weeks, the chances of survival begin to increase although, until the age of approximately 30 weeks, there is still a high chance of the baby suffering from a physical or mental disability (see p.339). Nevertheless, each day that passes after 26 weeks is accompanied by an increase in lung maturity and a reduced risk of other problems as well. As I will explain in the next section, the first few weeks of the third trimester of pregnancy are crucially important in terms of potential fetal survival.

SKIN AND NAILS
A layer of underlying fat has been laid down, but blood vessels are still visible. Fingernails have developed and also a unique surface pattern on the fingertips.

Your changing body

During the next few weeks, most women will gain about 1lb (half a kg) in weight every week, although, the exact amount varies between women and between weeks. Ideally, you should gain 13–14lb (6–6.5kg) during this second trimester.

…you may notice that your skin is looking particularly rosy and your hair is thick and glossy.

If you find you are putting on quite a bit more weight than is recommended during this last part of the second trimester, remind yourself that only about one pound of this weight will have gone to your developing baby. The rest can be attributed to your growing uterus and breasts and the increased blood and fluid volume that your body now contains, but also to your maternal fat reserves (see p.42). Lots of extra pounds gained now will be difficult to shed after the birth. If your weight gain continues to be excessive as you move into the third trimester, you will be at greater risk of developing gestational diabetes (see p.427) and preeclampsia (see p.426), not to mention feeling unnecessarily tired and being troubled more than usual by back pain. So try to eat a sensible balanced diet and restrict your sugar and carbohydrate intake. If you do need to control your weight, reducing your calorie intake gradually will not harm your baby's growth. Some women put on very little weight in pregnancy, but as long as their diet contains all the necessary nutrients, this is no reason for concern.

The expanding uterus

Your uterus continues to expand, and between 21 and 26 weeks will rise above your belly button (umbilicus). This growth has been achieved by an increase in the size of the muscles of the uterus, which are still attached to the same supporting ligaments. It is not surprising that many women experience stitchlike pains down the sides of their abdomen as the uterus expands and stretches these ligaments to their limits. The height of the fundus measures approximately 7½in (22cm) at 22 weeks; 9½in (24cm) at 24 weeks; and 10in (26cm) at 26 weeks.

To accommodate the increasing size of your uterus and baby, several other changes occur. As the uterus moves upward in your abdominal cavity, your rib cage will move upward too, by as much as 2in (5cm), and the lowest ribs begin to spread sideways. This frequently causes discomfort or pain around the rib cage and may mean that you start to feel breathless. The stomach and other digestive organs become compressed and progesterone continues to relax the muscles in the gut. As a result, it is common to experience heartburn, indigestion, and constipation at this stage in your pregnancy (see p.187).

A pregnancy bloom

Between 21 and 26 weeks the increase in your cardiac output continues to rise slowly but steadily, along with the volume of blood in your circulation, but your stroke volume and heart rate are leveling off and will not increase more. To deal with these cardiovascular changes, your peripheral resistance must reduce even more to ensure that your blood pressure does not alter significantly.

The up side to all of this extra blood flow and the enormous quantities of pregnancy hormones is that you will probably notice that your skin is looking particularly rosy and your hair thick and glossy. This is because women shed less hair during pregnancy and the increase in their metabolic rate means that hair also grows faster than usual. After the birth, you will start to lose hair in much greater quantities than usual—but to a large extent you are simply losing post-birth what you would normally have lost during the nine months.

Changing center of gravity

Your posture will have changed by this stage in your pregnancy. The enlarging uterus and baby are slung forward in the middle of the body, and consequently, a pregnant woman is forced to find ways of restoring her altered center of gravity. Added to this mechanical load, the ligaments of the pelvis have been softened by the pregnancy hormones contained in the vastly increased blood flow to the pelvis. This is an essential change since the pelvis needs to relax sufficiently in order to be able to accommodate the passage of a 6½lb-plus (3kg) baby through its previously rigid walls. However, this means that your pelvis can no longer function as the stable girdle that it once was and your pregnant body must find a way of compensating for these changes to their stability. The simplest way to deal with this mechanical challenge is to lean backward, arch your back, and use a wider gait than normal, but this change in posture is frequently accompanied by back pain, because the ligaments of your abdomen, back, and pelvis are strained.

Your fast growing belly is beginning to affect the way you move, sit, and lie. You will start to notice that you feel unstable wearing high-heeled shoes and that certain chairs are less comfortable than others. You may need to support the small of your back when you are sitting, and when you are lying down, there will be certain positions that are much more comfortable than others. These changes and effects are especially noticeable if you are expecting twins. You will find some practical advice on adopting a good posture and preventing back pain later in this section (see p.193) and more specific advice on backache later in this trimester (see p.218 and pp.243–44).

GOOD HEALTH Through the second trimester of pregnancy, many women appear to be glowing with good health.

How you may feel physically

Even first-time mothers will now be very aware of their baby's movements every day and left in no doubt that these sensations are the work of a lively, growing baby. I think this is one of the most exciting milestones reached by women in their journey through pregnancy.

You will also find your baby's movement very reassuring, because you will now be able to monitor the welfare of your baby yourself, instead of having to rely solely on the information you receive from your doctor, midwife, or the latest scan report. I remember vividly my own astonishment and overwhelming joy when, during a meal one evening, a commotion in my belly occurred, which resulted in my plate being bounced across the table in front of me.

Unwanted advice

Every woman carries her pregnancy differently, and however you are carrying yours, it is more than likely that your baby is exactly the right size. Yet it can be difficult to remember this advice when everyone seems to have an opinion as to whether your pregnancy is too big or too small for your dates. Throwaway lines such as "Goodness, you have put on a lot of weight" or "You're getting as big as a house" can be truly upsetting, particularly if you are already feeling self-conscious about the extra pounds you have put on. Conversely, "Are you sure that you are eating properly?" and "You do look small. Is everything all right?" may be well intended, but are guaranteed to alarm you at a vulnerable moment. If this constant analysis of your size in pregnancy begins to bother you, I suggest that you explain as calmly and as tactfully as you can that you find it distressing and you would prefer family, friends, and colleagues to stop. After all, people do not usually pass judgment on your body shape when you are not pregnant.

Another problem may arise if you are so engrossed in the subject of pregnancy that you find yourself inadvertently encouraging people to open up and recall all kinds of things about their own pregnancies that might be better left unsaid. Some of their anecdotes will be encouraging and useful (particularly if you have a specific problem that you are worried about) but others can be downright frightening. I think there is a tendency for people who have had children to forget how worrying scary stories can be for those who are currently going through a pregnancy. Again, my advice is to be honest. Gently but firmly explain that you would rather not hear another tale of a premature birth or an excruciating labor. Far from being offended, most people will understand and

...there is a tendency for people who have had children to forget how worrying scary stories can be...

PROBLEMS WITH DIGESTION

At this stage of pregnancy, problems with your digestion become much more common. You may well have started to experience mild or even severe symptoms of heartburn and indigestion and suffer episodes of constipation, too.

INDIGESTION

As your enlarging uterus begins to compress your abdominal organs, the capacity of your stomach is reduced and your whole digestive system begins to slow down. Food remains in your stomach and intestines longer, making you prone to indigestion—a sensation of a heavy lump at the bottom of your stomach. At times you may have a constant dull or stabbing pain in your abdomen, sometimes with back pain.

HEARTBURN

You may also have heartburn due to the fact that the valve between your esophagus and stomach has become relaxed and is no longer very efficient at preventing food mixed with acid gastric juices from being regurgitated back. This irritates the lining of your esophagus, causing a searing or burning sensation behind the front of your rib cage.

As long as the symptoms pass within a couple of hours, they are not a cause for alarm. In the meantime, there are several ways to minimize your digestive problems:

▶ **Eat little and often,** avoiding heavy, fatty, highly spiced, or pickled foods, which make symptoms worse.

▶ **Drink a glass of milk** or eat a plain yogurt before meals and before bedtime. This can help relieve heartburn by neutralizing stomach acid.

▶ **Sit upright** when you are eating to reduce the compression of your stomach.

▶ **Avoid lying down** for at least an hour after you have eaten and keep your head well propped up on several pillows when you go to sleep at night to reduce the problem of heartburn.

▶ **If your symptoms** are severe, ask your doctor to prescribe you an antacid such as Mylanta or Gaviscon.

CONSTIPATION

Your sluggish digestion can also make you constipated, leaving you feeling heavy and irritable. Try some of the following remedies:

▶ **Boost the fiber** in your diet by eating more fresh fruit and vegetables and whole-grain bread and cereals.

▶ **Increase your fluid intake** by making sure you drink at least 4 pints (2 liters) of water per day.

▶ **Exercise regularly.** As little as a 20-minute walk each day can help relieve constipation.

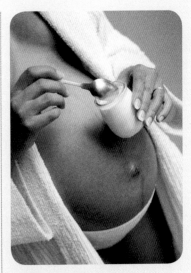

EASING HEARTBURN Plain yogurt may help relieve symptoms.

▶ **Bulking laxatives** such as Metamucil or Colace are effective because they contain complex sugar compounds, which the human gut cannot digest. They absorb water, helping produce a bulkier, softer stool that can be passed without straining and discomfort.

▶ **Laxatives** containing senna are not recommended in pregnancy because they irritate the gut, which has the potential to trigger uterine contractions.

some will also realize (and most probably deeply regret) that they have caused you unnecessary anxiety.

Your prenatal care

You will continue to have prenatal appointments every four weeks at your doctor's office and by now will know what to expect in terms of the various checkups. For the majority of women, this stage in your prenatal care is a quiet and enjoyable time.

Although I feel it is important to cover the other tests and procedures that may form part of prenatal care here, I want to add that it is very unusual to encounter any of these serious complications during this stage of pregnancy, either with your health or with that of your baby.

It is unlikely that you will have another routine ultrasound scan during this period unless problems that need more investigation were identified at the 20-week scan. If the scan suggested anything abnormal in the development of your baby's organs—for example, an intestinal blockage or problems with the kidneys or urinary tract, you will have additional scans between 21 and 26 weeks, possibly at a special center. If the abnormality is confirmed, you may be advised to have a late amniocentesis (see pp.140–3) or fetal blood sample test (see cordocentesis, p.143) to see whether or not the baby is affected by a

PREDICTING PREMATURE BIRTH

Today, the majority of babies that are born prematurely survive and develop normally, but those that are born very premature (before 30 weeks) still have a significant risk of disability if they survive (see p.339). Therefore, any test that helps predict which babies are at risk should always be carefully considered.

Some two percent of women have a very short cervix and it is thought that half of these may give birth very prematurely as a result. Some doctors are now using vaginal ultrasound scans throughout the second trimester to help identify these pregnancies and offer preventive treatments.

If your doctor is participating in this research, you will have the length of your cervix checked using a vaginal ultrasound probe. If your cervix is found to be shorter than average, you will be closely monitored.

Some researchers believe that inserting a stitch in the cervix to lengthen and close it may be of help, although this procedure is not without risk. Some doctors give progesterone treatment to prevent contractions and/or steroid treatment to reduce the risk of breathing complications if the baby is born prematurely.

Screening for premature birth is not universal, but most doctors will provide extra surveillance.

chromosomal or genetic problem. The initial results from the amniocentesis usually take three working days. For more unusual conditions, results can take up to three weeks because the skin cells in the amniotic fluid have to be cultured before they can be examined for chromosomal abnormalities. This will seem like an eternity for worried parents-to-be, but new molecular biology techniques at some specialist centers can greatly reduce the time taken to establish the baby's genetic makeup (see p.143).

If you have had a previous baby with a heart abnormality or if there is a family history of cardiac problems, early referral to a fetal cardiologist is advised, since at times early cardiac scans at 12 or 16 weeks can offer great reassurance to the expectant mother. This is followed by a detailed cardiac scan at 22 weeks. At this stage the four chambers of the fetal heart and its connecting pipes can be seen more clearly; this improves the accuracy of the scan and the advice that can be offered to you.

The other type of ultrasound scan you may be offered is a Doppler blood flow scan (see p.257), which examines the way blood is flowing in the vessels of the uterus, placenta, and umbilical cord. Research has suggested that reduced blood flow through the uterine arteries at this stage in pregnancy may be a way of identifying women at high risk of developing high blood pressure (see blood pressure problems, p.426) or problems with the growth of their baby, (see p.214 and intrauterine growth restriction, p.429). As a result, some doctors now screen women with Doppler scans at 24 weeks. The small minority of women (5 percent) in whom the blood flow is reduced can then be carefully monitored for changes in their blood pressure. The Doppler blood flow scan can also be used to look at the flow of blood in many of the baby's arteries and veins, which is a good indicator of the baby's general well-being.

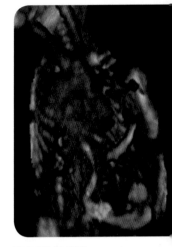

BLOOD FLOW
A Doppler scan shows blood flowing through the major fetal vessels. The heart is seen as a large red stucture center left, and the yellow-colored blood vessels, bottom right, lead to the umbilical cord.

Common concerns

Concerns in pregnancy are usually a mixture of nagging complaints and anxieties about things you may have heard or the way your pregnancy is progressing. I hope the advice here sets your mind at rest on a few issues but remember that your doctor will also be happy to talk through any concerns that you may have.

It is quite common to feel dizzy or light-headed when you change your position suddenly at this stage of pregnancy because of the massive changes in the way your increased blood volume is distributed in your body. A significant proportion is being directed to the uterus to support the placenta and baby and large quantities of blood are being stored in your pelvic and leg veins because of the reduction in peripheral resistance. When you stand up quickly it takes a few minutes for the blood in your pelvic and leg veins to redistribute; meanwhile, there is a shortage of blood to the brain leaving you light-headed and even inclined to faint. Similarly, if you have been standing for long periods, the extra pooling of blood in your legs may leave your brain short of blood, especially when it is hot and your blood vessels dilate even more to cool you down.

There are steps you can take to reduce these dizzy spells or, worse, fainting episodes, both of which can be quite frightening and unpleasant.

• Make sure that you do not get up too quickly from a sitting or lying position. Try to let the blood flow adjust gradually.

• Try to avoid becoming overheated, particularly in hot weather. One of the most common times for feeling dizzy or faint is when you try to get out of a hot bath too quickly because this leaves your circulation completely unable to deal with the shifts in blood volume needed to prevent you from feeling light headed.

• Be sure to eat regularly and choose foods such as complex carbohydrates (see p.44) that release energy gradually, so that your blood sugar levels are prevented from rising and then falling too quickly.

• If you feel faint or dizzy, sit down and try to put your head between your knees or lie down and raise your feet above your head or at least above your pelvis, so that the blood in your leg veins returns to your brain as quickly as possible.

Even if you suffer from dizziness regularly, your baby is not in any danger because the blood supply to the uterus and placenta is being maintained at your expense. However, when you lie flat on your back the weight of the uterus can end up pressing against blood vessels in the pelvic area, depriving your placenta (and hence your baby) of oxygen, so avoid this position.

The main danger of dizziness is if you suddenly start seeing stars while driving or getting on a train. For this reason, interrupt your travel regularly. If you have to stand for long periods, make sure that you keep shifting weight from one leg to the other. Better still, walk around if you can.

Checking fetal movements

Many women I meet in my practice are worried about how many fetal movements they should be feeling each day or night—anxieties that can be made worse by the doctor asking at each visit "Is the baby moving well?" If this

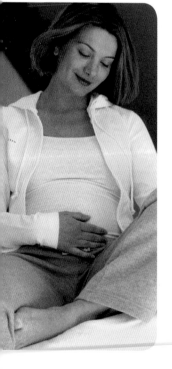

PATTERNS OF MOVEMENT You will begin to recognize your baby's patterns of movement and become alert to any changes.

is your first baby it is obviously difficult for you to know, so I owe a big debt to the distressed pregnant woman who pointed this out to me many years ago when I was at the start of my training. As a result of this lesson, I changed my methods of finding out about fetal movements.

The issue of fetal movements is important in pregnancy for the simple reason that they are one of the best ways for you and your doctor to assess the well-being of your baby. Having said that, I don't intend to offer rigid rules about how many movements you should be feeling per day or night, at any stage during your pregnancy journey. This is because every pregnancy is different and every baby develops its own pattern of movements, which can change as the pregnancy progresses. Some babies are more active than others generally and all babies go through periods in the day when they are quieter or are engaged in more vigorous pursuits. Over the weeks, you will get a feel for your own baby's pattern of activity, perhaps noticing that your baby responds with a kick when you are in a certain position or becomes still at certain times of the day. So, instead of advising that you should feel 5, 10, 20, or 50 movements every 12 or 24 hours, you should look out for any dramatic changes to your baby's pattern of movements and should one occur, consult your doctor promptly. In particular, if you do not feel your baby move during a 24-hour period, you should seek advice immediately.

> *Over the weeks, you will get a feel for your own baby's pattern of activity…*

Abdominal aches and pains

Abdominal pain is always worrying and, for pregnant women, it invariably raises concerns that the baby is at risk. During the second trimester, the most likely explanation for aches in the lower abdominal area that virtually every woman experiences at this stage in pregnancy is that the ligaments supporting your enlarging uterus are being put under enormous strain. However, if you develop regular abdominal pain, or notice that your entire abdomen has become too tender to touch, tell your doctor immediately so that it can be investigated promptly. There are several possible causes and, although uncommon, they can have potentially serious consequences. The most serious is pain in the uterus itself, which may be the first signs that you have had some bleeding behind the placenta from a placental abruption (see p.428) or are at risk of premature labor (see p.340). The pain may be sharp and stabbing or a dull, constant ache and may or may not be accompanied by vaginal bleeding. Whichever, your doctor will want to examine your uterus.

A uterine fibroid (see p.423) is a benign mass of muscle in the uterine wall, which may cause problems in the second trimester because high levels of estrogen and progesterone encourage the fibroid to grow along with

You should always try to avoid lifting heavy weights in pregnancy but if you have young children, this is likely to be impossible at times.

the rest of the uterus. Occasionally, this rapid growth causes the center to degenerate, resulting in severe pain in the uterus and abdomen, localized to a particular spot. Pain from fibroid degeneration is very unpleasant, but usually disappears with bed rest and pain medication without causing any problems to the baby. Occasionally, large fibroids positioned in the lower part of the uterus or beside the cervix will lead to problems near the time of labor if the baby's head does not have sufficient room to descend into the pelvis.

Nausea, vomiting, and/or diarrhea accompanied by abdominal pain is uncommon in pregnancy, but is almost always due to food poisoning or viral gastroenteritis. It is usually self-limiting and, although very unpleasant, disappears quickly with no harm to you or your baby. There is no need for medical treatment apart from making sure to drink plenty of fluids to replace those you have lost. If the nausea and vomiting persists or is accompanied by abdominal pain, consult your physician. Very occasionally, symptoms may be due to listeria infection (see p.50 and p.413), which is a potential cause of late miscarriage and intrauterine death and is best treated with penicillin antibiotics.

Appendicitis is another rare, but important cause of persistent abdominal pain in the second trimester of pregnancy. However, it can be very difficult to make the diagnosis in pregnant women, since the appendix is no longer in its usual position in the lower right corner of the abdomen but has been displaced by the growing uterus.

Urinary tract infections are another important and common cause of abdominal pain in the second trimester. It is usually felt in the lower part of your abdomen, above the pubic bone and is likely to be accompanied by discomfort when urinating. Remember that you may not notice the early symptoms of a urinary tract infection, such as cystitis, before organisms have traveled up the dilatedurinary tract to the kidneys and developed into pyelonephritis. In mid- and late pregnancy, urinary tract infections can lead to irritability of the uterus and premature contractions, not to mention long-term damage to your kidneys if left untreated. For this reason, if you have abdominal pain in pregnancy, your urine will be tested and you will probably begin a course of antibiotics while you wait for the results. These medications will not affect or damage your baby. If you do have a urinary tract infection, it is essential to finish the course of antibiotics, even if symptoms disappear quickly, and have another urine sample examined to make sure the infection is gone. Urinary tract infections that are inadequately treated will recur and, worse still, may then become resistant to the usual antibiotic treatments.

PREVENTING BACK PAIN

Back pain is so common in pregnancy it is unusual for a woman not to suffer from it. You are likely to begin to notice it around the end of this trimester, so here are some tips that may help reduce its severity and prevent it from becoming worse.

▶ **When you are upright**, adopt good posture: stand tall, and hold your shoulders back (your back should always be in a straight line). Remember that if you slouch over your expanding belly, your back will arch and this will aggravate your lower back pain. Try not to stand for long periods of time.

▶ **Invest in some good flat or low-heeled shoes**, preferably with support for the arches of your feet and sturdy soles. Wearing high heels from this point on will make you feel more unstable and increase the strain on your back.

▶ **Sit well: this is particularly important** if you spend long periods working at a desk. Make sure that both shoulder blades and the small of your back are against the chair back and that the seat supports your thighs. The chair should be at the right height for you to keep your feet flat on the floor and the computer screen should be at eye level.

▶ **When you are driving** check that the car seat supports the small of your back and that you can reach the hand and foot controls easily. Your seatbelt may feel uncomfortable but you must wear it for every trip.

▶ **When you are resting**, raise your feet and legs to take the pressure off your back and pelvis. Later on in pregnancy you may need a firmer mattress to help support your back. You will also find that sleeping on your side will help reduce the strain in the ligaments of your back.

▶ **When you get out of bed**, first turn on your side and, keeping your back straight, swing your legs over the side of the bed. In this position you will be able to push yourself upright using the strength of your arms without placing any strain on your back.

▶ **Regular gentle back exercises** will help the muscles and ligaments of your back stretch and become more supple. Pelvic tilt exercises are especially helpful, as are exercises that help strengthen the muscles of your back (see p.219).

▶ **Try not to gain too much weight**—every extra pound places more strain on your back.

LIFTING HEAVY WEIGHTS

You should always try to avoid lifting heavy weights in pregnancy but if you have young children, this is likely to be impossible at times. When your toddler needs to be carried use this method. Squat down and hold him close to you. Keep your back straight and use the muscles in your legs to push yourself upward as you stand. Use the same technique to lift any heavy weight.

Dental health

In addition to visiting the dentist regularly, it is often a good idea to see a dental hygienist for some general advice on gum care. Sore gums that bleed when you brush or floss your teeth are very common; they become soft and spongy thanks to the increased blood supply and hormones of pregnancy. To help firm your gums and prevent bacteria from infecting the broken skin, increase the number of times you brush your teeth. Recent research has suggested that gum disease in pregnant women may contribute to the development of problems such as late miscarriage and premature labor. Although the mechanisms are not completely understood, it is possible that a constant focus of inflammation or infection in the mouth can result in complications in other areas of a pregnant body.

Things to consider

I hope that you have already enrolled in a childbirth or parent education class because the best ones tend to fill up well in advance. If you haven't found a place yet, make it a priority to do so now.

The more knowledgable and prepared you are for labor, the more relaxed and confident you will feel; in addition to this, you will find that the opportunity of sharing concerns, feelings, and experiences with other parents-to-be is a source of comfort and relief. Even if this is not your first baby, sign up for a refresher course—you will be surprised how quickly you have forgotten the details of labor and breathing techniques. Also, there is a constant evolution in how a hospital deals with labor and what sort of pain relief it can offer, so even if you had your first baby in the same hospital, there may have been changes of which you are unaware.

Parents and in-laws

If you are one of those lucky individuals who usually enjoys an uncomplicated relationship with their parents and in-laws, or step-parents for that matter, you may be surprised to find that things are not quite as straightforward as they were before you became pregnant. There are so many different permutations of

" Your mother may be offering all sorts of advice about pregnancy and child-rearing already and may be put off if you plan to deal with your pregnancy and newborn differently. "

a family unit today that I could not begin to cover them all but I think it is useful to reflect on the fact that, although pregnancy invariably brings joy to a family, or extended family, it can also give rise to conflicting emotions that can reverberate upward through the generations as well as downward.

Your mother may be offering all sorts of advice about pregnancy and child-rearing already and may be a bit put off if you are taking the view that life has moved on in the intervening years and that you plan to deal with your pregnancy and newborn baby differently. Similarly, your mother-in-law will have her own views, too, that don't always jibe exactly with those of your own mother. Both are capable of loaded questions such as, "Why are you working so hard?" or, "Are you planning to quit work after the baby is born?" If these are beginning to bother you, ask your partner for help. Presenting a united front and demonstrating clearly that you are making decisions together about issues such as childcare and work may help silence the critics in the family.

Telling children about the baby

If you already have a child or children, the other delicate issue you will shortly have to address, if you have not already done so, is when to tell them that they will soon have a baby brother or sister. Of course the timing depends, to a large extent, on the age of the child. A two-year-old may not have noticed your changing shape and will certainly not have linked it to the fact that there is a baby growing inside you. However, older children will be aware of your altered appearance, so it is much better that they hear the news from you instead of accidentally from someone else.

If you have not told them already, reflect for a moment on why you have not. I suspect that this will be because you are worried that any existing children may fear they are going to lose some of your love and attention by having this new person in the family. This is a logical concern for a small child who cannot yet know that parents have limitless love when it comes to their children. So you need to constantly reassure them.

The other thing to remember about young children is that they have little perspective of time and have no way of understanding that your pregnancy is going to continue for another six months or so before they see the end product. Children absorb most information they are given but often choose to process it in their own time, usually at a later date. So, when you are trying to have a sensitive conversation about the future baby, do not worry if your

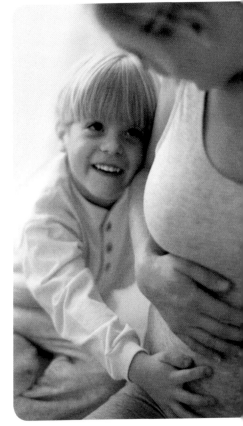

NEW ADDITION An older child may want to ask exactly how his new brother or sister will arrive in the outside world.

toddler appears to have no interest in the subject and suddenly interrupts to ask if he can have a snack. Equally, do not be surprised when the next day, or the next week, he returns to the subject of the future baby without any warning. Just pick up the threads of the conversation and continue to reassure him.

Your girlfriends

During the second trimester of pregnancy, your relationship with girlfriends who don't have children or are not planning to have them in the near future can become strained. Now that your attention is focused on visits to the obstetrician, how many times a day you can feel your baby kick, and which parenting class you want to join, your friends may find your new topics of conversation rather limited. They may be wondering what happened to the prepregnant you and whether that person will ever reemerge.

The reality is that you are moving on to a different phase of life, and this change in your relationship will be even more noticeable after your baby is born. But there is no reason why a close friend cannot be equally important to you, just because your lifestyle has taken on a new direction. If you value your friendships, there is every good reason to continue them.

As far as nights out are concerned, whiling away an evening in a crowded bar or club may no longer be an attractive option in pregnancy but you can make the most of meals out and get-togethers at friends' homes while you still have the freedom to do so. Reassure them that there will be nights out after the birth even if they end a little earlier, either because you need to get as much sleep as possible or because your babysitter will want to go home. Having a baby does not mean that you do not want to see your friends, or that you have to become such a bore that you are no longer able to involve yourself in their lives or listen to their current news.

What's in a name?

Before pregnancy, you might have assumed that the only possible difficulty in choosing a name would be to whittle down a long list of names to a few contenders. Now that you are pregnant, you may be surprised to find that the whole exercise is more complex than you had imagined.

Part of the problem may be that you find it difficult to relate to your baby at this stage. Some mothers-to-be start

GIRLFRIENDS Your relationship may change but you don't need to lose touch with your nonpregnant friends.

chatting away happily to their baby when it is only a few inches long, but believe me, there is no need to feel disturbed if it is taking you a little longer to develop the relationship. Knowing the sex of the baby at the earliest possible opportunity helps some parents think about their unborn baby as a real person and this may also help when it comes to choosing a name; others feel this takes away some of the excitement of the birth. For the latter group, choosing two sets of possible names for their baby boy or girl can be one of the magical parts of the pregnancy.

The next problem is that you may soon be feeling pressured by suggestions from family and friends. Some women deal with the issue by inventing a completely absurd name, which shocks their relatives into silence; others opt for a vague, noncommittal approach. Having said that, it is also common for parents to make their final choice during the birth, or days, or even weeks afterward when the legal requirement to fill in the birth certificate forces them to make a decision. There are only two outcomes I can guarantee: the first is that, as you watch your baby grow, you will not be able to imagine your child being called anything other than the name you chose. Second, when your child grows up, she will tell you how she wished you had given her a different name.

CHOOSING A NAME
Sometimes the perfect name does not become apparent until you meet your baby face to face.

The last name dilemma

If you are married or in a long-term relationship, the issue of which last name your baby uses may be open for debate. Although the assumption is that the child will take the father's last name, there is no legal reason why this should be so and you may have strong feelings that your own last name should be used. One increasingly favored option is either to combine the last names or to use the mother's name as an additional first name.

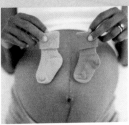

▶ **WEEKS 26–40**

The third trimester

Your baby is capable of surviving if delivered now, albeit with medical assistance, but the remaining weeks in the uterus are vitally important. Your baby's development is now focused on maturing the lungs, digestive system, and brain so that they can function in the outside world. As your abdomen expands to accommodate your growing uterus and baby, your thoughts increasingly turn toward the birth.

CONTENTS

Your baby

IN THE THIRD TRIMESTER

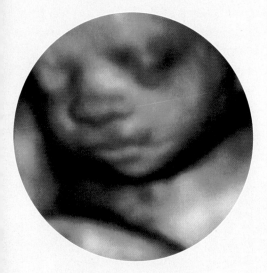

WEEK 27
The baby begins to develop
a pattern of rest and sleep
alternating with active periods.

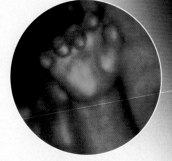

WEEK 28
Skin creases are visible on
a chubby hand with perfectly
formed fingernails.

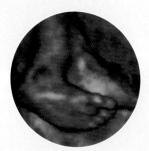

WEEK 29
Movements are strong and purposeful
and include hefty kicks, punches, and
rapid changes in position.

66 *During the weeks between now and
delivery, the baby's body systems are
becoming ready for life outside the uterus.* 99

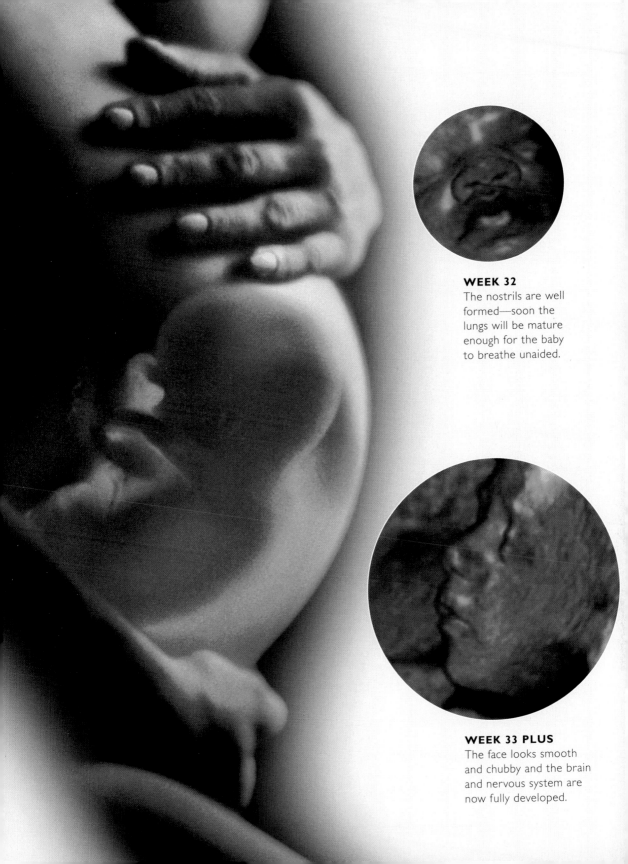

WEEK 32
The nostrils are well formed—soon the lungs will be mature enough for the baby to breathe unaided.

WEEK 33 PLUS
The face looks smooth and chubby and the brain and nervous system are now fully developed.

▲ By 27 weeks, the eyelashes and eyebrows are fuller and the fetus can blink.

| 1 | 2 | 3 | 4 | 5 | 6 | 7 | 8 | 9 | 10 | 11 | 12 | 13 | 14 | 15 | 16 | 17 | 18 | 19 | 20 |

▶ **WEEKS 0–6**   ▶ **WEEKS 6–10**   ▶ **WEEKS 10–13**   ▶ **WEEKS 13–17**   ▶ **WEEKS 17–21**

▶ **FIRST TRIMESTER**   ▶ **SECOND TRIMESTER**

▶ **WEEKS 26–30**

The developing baby

DURING THE NEXT FEW WEEKS, your baby continues to grow in length, while the body weight increases quite significantly because white fat is being deposited under the skin. The baby now looks plumper as the abdomen and limbs fill out and the skin starts to lose its wrinkly appearance.

Subcutaneous fat helps your baby regulate its own body temperature, an essential development for life after delivery, although this ability is only partially acquired in the uterus, and newborn babies still lose heat very quickly. As the fat is laid down, the lanugo hair becomes sparser and soon only a few patches over the back and shoulders will remain, although the coating of white vernix will stay until about 36 weeks. The hair on the scalp will lengthen and the eyebrows and eyelashes will become fuller. Skin creases can be seen on the hands and feet, and little fingernails and toenails are also clearly visible. The testes of male babies start to descend into the scrotum.

Now that the eyelids are open, your baby will start to blink and will become much more aware of differences in light. This new sense allows the baby to be more responsive to external stimuli; many mothers notice their baby has developed a distinct pattern of rest alternating with activity. The baby is also able to start focusing its eyes at this stage, although the distance is limited to about 6–8in (15–20cm) until after birth.

Getting ready to breathe

From now until the end of the pregnancy the further development of the lungs is vitally important. By about 29 weeks, most of the smaller airways (bronchioles) are in place and the number of alveoli (little air sacs), which lie at the end of the bronchioles, are increasing. The formation of alveoli continues throughout pregnancy and after birth. In fact, the lungs do not become fully matured until a child is eight years old, which is why many childhood respiratory problems disappear or improve as children get older.

not life size

At the start of the third trimester, an average fetus measures about 10in (25cm) from crown to rump and weighs just about 2lb (1kg). By 30 weeks, its length is about 11in (28cm) and it weighs about 2–3lb (1–1.5kg).

| 21 | 22 | 23 | 24 | 25 | **26** | **27** | **28** | **29** | **30** | 31 | 32 | 33 | 34 | 35 | 36 | 37 | 38 | 39 | 40 |

▶ **WEEKS 21–26**　　　　▶ **WEEKS 26–30**　　▶ **WEEKS 30–35**　　　　▶ **WEEKS 35–40**

▶ **THIRD TRIMESTER**

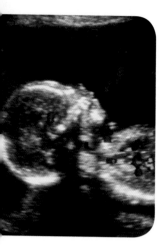

PRACTICE BREATHING
The flow of amniotic fluid in and out of the baby's mouth as it practices breathing movements is seen as red patches on this color Doppler scan.

The next step in lung maturity is the production of a lipid called surfactant, made by the lining cells of the lung, which coats the air sacs in a very thin film. The surfactant works by reducing the surface tension inside the alveolar air sacs, in much the same way that dish soap disperses grease on dishes. This process is important because when the baby takes a first breath of real air, the air sacs need to be as elastic as possible to be able to expand successfully. When the first breath is exhaled, the air sacs must be prevented from collapsing in readiness for the next lungful of air. A baby born before 35 weeks cannot yet produce sufficient quantities of surfactant and this, coupled with the inadequate development of the bronchioles and alveoli, the newborn baby's lungs are too rigid to deal with the constant flow of air entering and leaving them. If you are unlucky enough to have a premature birth, you will probably be given an injection of steroids before the baby is born; this helps to stimulate the lungs to produce surfactant (see p.342). The neonatal doctors may also decide to spray artificial surfactant into your baby's lungs after he or she is born in order to make them more elastic.

Of course, your baby is not breathing air at present because the placenta is still supplying all its oxygen requirements. Nonetheless, your baby will have started to make rhythmic breathing movements as it continues to develop its lungs in preparation for birth. These movements of the chest wall can be seen on an ultrasound scan and explain the "hiccups" you sometimes feel—short jerky movements that are quite different from other activity you can feel going on inside you.

Your active baby

You will be very aware of fetal activity between 26 and 30 weeks. Although conditions inside the uterine cavity are becoming cramped, there is still enough room for the occasional somersault and complete change of position. The amniotic fluid is not being produced at the same rate as a few weeks ago so movements are no longer well cushioned and hence more noticeable than previously. It is very common to see dramatic changes in the shape of your abdomen when the baby heaves itself into a different position. Some women worry that they will be damaged or that their baby will suffer an injury as a result of all this vigorous activity. Neither will occur. There is still enough amniotic fluid to protect the baby and the thick muscular wall of your uterus is more than enough to prevent damage to your internal organs. Before leaving the subject of movements, I want to remind you that there is no correct number of movements or kicks that you should experience every day (see pp.190–1) but any sudden changes in your baby's regular pattern of movement needs to be reported to your doctor immediately, just in case your baby is in trouble.

There is no correct position for the baby to be in at present but many babies at this stage are still lying with their head uppermost. This often results in the mother feeling the head butting against her rib cage, which can be quite unpleasant and can sometimes be the cause of quite harp pain. Like every other problem in pregnancy, it will not last forever. Most babies turn around to embark on their journey into the outside world head first.

Maturing systems

Your baby's nervous system continues to become more intricate and sophisticated. Constant movements of the muscles help make movements and reflexes more coordinated. Your baby will practice the sucking reflex on a thumb or fingers whenever the opportunity presents itself but the ability to suck from the breast will not be fully developed until 35–36 weeks.

The fetal bone marrow has now taken over as the main producer of the baby's red blood cells. This will help your baby become independent after delivery because these cells will transport oxygen around the blood stream. A simple immune response to infection is now in place.

As your baby reaches 30 weeks, its ability to survive in the outside world has improved dramatically and, as a result, the vast majority of babies that are born at this gestation will manage extremely well, with a bit of help from the special care baby doctors. From now onward, every day spent in utero reduces the time that your baby would need to spend in a neonatal unit. Even though the baby's physical size is not changing dramatically, its functional maturity is taking a considerable leap forward.

By 30 weeks, the placenta will weigh just under 1lb (450g), which is a massive increase compared to the 6oz (170g) it weighed at 20 weeks. Every minute, it receives about 16fl oz (500ml) of blood from your circulation.

YOUR BABY'S BRAIN

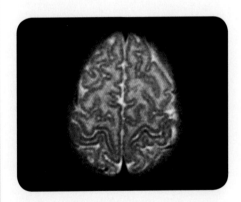

The brain grows in size and now starts to fold over to fit inside the bony skull. An anatomical slice through the upper part (cortex) of the fetal brain now looks like a walnut or a map of the Scandinavian fjords, with lots of little inlets and inward projections. The protective myelin sheath that began to form around the nerves of the spinal cord many weeks ago now extends to the nerve fibers entering and leaving the brain. As a result, nerve impulses can now travel much faster from the brain to the rest of the body. This ensures that, in addition to more intricate movements, the baby is now able to learn new skills.

Your changing body

Your uterus is continuing to expand at a steady pace. By 26 weeks it has reached above your umbilicus or belly button and during the next few weeks you will notice that your abdomen is enlarging both upward and sideways.

By the 30th week, the height of the uterine fundus measures approximately 12in (30 cm) from your pubic bone. I say approximately because there is much individual variation and it is important not to become worried if your fundal height is a few inches lower or higher than the standard textbook measurements. Your doctor will measure your uterus at each visit and if there is any serious discrepancy in size, they will arrange for you to have a series of ultrasound growth scans (see p.214) together with some other tests to ensure that all is well with your developing baby (see pp.256–9).

INSIDE YOUR ABDOMEN Your uterus grows both upward and outward, reducing the available space for your stomach and intestines.

Cramped inside

In order to accommodate this enlarging mass, your other body organs need to make a few adjustments so you may experience some new symptoms or an exaggeration of some previous ones. The intestines and stomach become more compressed upward because they can no longer fit themselves comfortably around the sides of the uterus. This upward displacement often results in heartburn and/or indigestion: look back to the feature on page 187 for the best ways to deal with this. Similarly, even if you had a healthy appetite during the second trimester of pregnancy, you will probably find that you can no longer manage to eat a large meal at one sitting.

Your bladder is also unused to this extra pressure in the abdominal cavity and can no longer hold the quantities of urine it used to be able to cope with; this causes its own set of irritations (see p.215).

You may also experience some rib pain or discomfort, since your rib cage is now pushed outward to make more room for the increasing contents of your abdominal cavity. Some women are lucky and go through the whole of their pregnancy without any rib pain. However, if your body frame is smaller than average or you are carrying twins or triplets then you are very likely to notice rib discomfort. It will be made worse if your baby is an especially strong

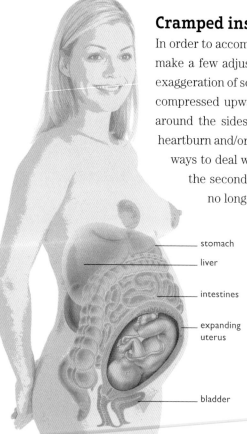

stomach

liver

intestines

expanding uterus

bladder

SHORT OF BREATH

You will notice a definite change in your breathing pattern during the third trimester. There are several reasons for this:
▶ The high levels of progesterone increase your body temperature and breathing rate.
▶ In addition, as your ribs flare outward, your diaphragm has to stretch further and in the process it becomes less flexible. This reduced movement of the diaphragm forces you to try to breathe more deeply.
▶ Last, your expanding uterus pushes the abdominal contents up against the diaphragm leaving your lungs with less room to expand when you try to take a deep breath.

With all these conflicting pressures it is hardly surprising that pregnant women frequently experience episodes of breathlessness, dizziness, and light-headedness toward the end of their pregnancy. If you look back to page 190 you will find practical advice on ways to reduce some of these symptoms.

kicker or spends a lot of time in the breech position (see p.269) because the fetal head will belly up against your diaphragm and rib cage. You may be particularly uncomfortable when you are sitting down, because this compresses everything even more. If you have a desk job, it is sensible to make a few adjustments to your routine. Try to make sure that you get up and walk around regularly. When you are feeling particularly uncomfortable, keep changing your sitting position until you find a better one. Make a determined effort to maintain good posture.

A surge in circulation

From 26 weeks onward, your circulatory system embarks on another surge. The total blood volume is now about 11 pints (5 liters)—a 25 percent increase over normal, although the maximum blood volume will not be reached until about 35 weeks. This increased blood volume means that your cardiac output (the quantity of blood pumped by the heart at each beat) continues to increase during the next few weeks. However, more relaxation of the blood vessels around your body is no longer an option because all your blood vessels are now at maximum capacity. In fact, from this point onward, your peripheral resistance will have to increase slightly and your blood pressure will start to rise, although this should be a small and very gradual change.

Your body tissues also become thicker because there is so much fluid on board that it has to be accommodated somewhere—so it is very common and perfectly normal for your fingers and legs to become slightly swollen. Having said that, if you notice that your face, fingers, or legs have suddenly become much puffier or swollen, these may be early signs of preeclampsia (see p.426) and you need to be seen urgently. Preeclampsia usually develops after

CIRCULATION CHANGES

The graph shows a steep rise in cardiac output and blood volume from mid-pregnancy onward and a corresponding dip in the peripheral resistance of the body's blood vessels.

KEY

— Cardiac output

— Stroke volume

— Heart rate

— Blood volume

— Peripheral resistance

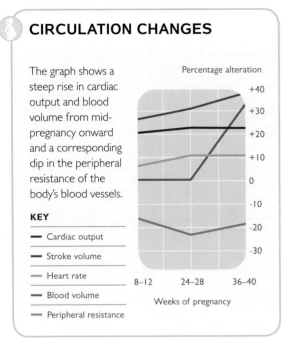

Percentage alteration

Weeks of pregnancy

30 weeks and then only in a minority of women. Rarely, it develops earlier and in such cases, it is likely to become severe.

The continuing changes in your circulatory system means that the blood flow to your skin and mucus membranes is increased. In response, the peripheral blood vessels dilate and this is why pregnant women in the third trimester notice that they "feel the heat" and sweat more easily, sometimes profusely. Many women also find that the palms of their hands and the soles of their feet are red in color and feel as if they are on fire. This is called palmar erythema. All of these skin changes are entirely normal and will disappear after you have delivered. They merely reflect the fact that you need to get rid of the extra heat that your increased metabolism and your baby's metabolism is generating. If the blood vessels in your skin did not dilate, you would not be able to maintain your body temperature or your baby's at a constant level and you would literally overheat—just like a car with a broken radiator.

Breasts and colostrum

Your breasts will feel a lot fuller by now, due to the combined and continued action of the hormones that are responsible for breast growth in pregnancy. The veins on the surface of your breasts become more prominent and more visible during this last trimester and the nipples and areolae continue to darken.

Under the influence of these hormones, the internal structure of your breasts has been changing and developing ready for lactation and breast-feeding (see p.396) While the placenta is still in the uterus, high levels of estrogen and progesterone block the key hormones that trigger milk secretion but you may notice from this stage onward that you are producing a clear liquid, which leaks from the nipple at all sorts of strange times, for example, when you are in the bath or making love. This liquid is called colostrum and it is the fluid that your baby will suckle for the first few days of life, before your real milk comes through. Colostrum contains sugar, protein, and antibodies—in fact, all your baby's nutritional needs—and the most likely reason it has started to be produced now is for those babies that decide to make an early appearance into this world. Having said that, please do not worry if you do not see any colostrum throughout your entire pregnancy. I can assure you it is there, it is just that you are one of the fortunate women from whom it has not actually leaked out.

How you may feel physically

You are likely to be actually enjoying your enlarged size by now, even if at times you feel rather compromised by it. Remember to walk tall and keep your back straight, since bad posture will put a lot of pressure on your lower back during these last three months.

I think it is useful to make a few suggestions here about how you can address the predictable fatigue and loss of energy that may have started by now. Many pregnant women tell me that, however much they rest, they are still tired and lacking in energy. The usual advice is to spend time with your feet up but this is more difficult to do when you have a job and/or other children, not to mention the rest of your life to run. So the message needs to be—think realistic, not idealistic. Try to find ways to reduce your commitments at work and at home.

Delegation is often the answer and I suspect that you will be pleasantly surprised by how receptive your colleagues and your family are if you just give them the opportunity to help you. So instead of trying to be superwoman, identify someone else to go to the extra committee meetings at work. Ask your partner to go to the school parents' evening, or send your apologies. Look at the household chores with a fresh eye. Do they really need to be done or could they be left to a later date when you are feeling more energetic? If the answer is yes, as may be the case with basic shopping and housework, then think about ways to employ someone to do them. If this is not possible, then you will need to ask your partner, another family member, or a supportive friend for help.

As any mother of more than one child will tell you, there can be real problems dealing with your first child or other children when you are heavily pregnant. When your two-year-old toddler is having a tantrum and refuses to take a bath, you need to step back a few paces and give yourself a break. Ask yourself—is this bath really necessary? If the answer is no, then skip the bath. If the answer is yes, go for a compromise of using a sponge on the face and grubbiest parts of your child. Remember that becoming fraught about daily domestic issues will help no one and you need to make sure that you conserve your physical and emotional energy for things that really do matter.

> *When your two-year-old toddler is having a tantrum and refuses to take a bath, you need to step back a few paces and give yourself a break.*

Your emotional response

You are definitely in the home stretch now—well past the halfway mark in this pregnancy journey. This is a real transition period in emotional terms, because the birth of your baby, which used to be a rather abstract notion, suddenly becomes very real.

Your baby has a very good chance of survival now, so you may be starting to become impatient, wishing away the next few months. At the same time you may be experiencing contradictory feelings of panic at the thought of having a baby to take care of in the near future. If this is your first baby, you are probably starting to worry about how qualified you are to be let loose on a newborn infant. After all, many women have never changed a diaper, let alone held a very young baby, before they are handed their own in the delivery room. If this is not your first child, you will understandably be concerned about how your other children will adapt to the challenge of another little person who will have demands on your time and attention. You may also worry about how this new baby will fit into your already hectic life.

Since the possibility of birth now exists for real, you are probably also starting to think about how you will cope with labor and delivery. If this is your first birth, you will be conscious that you are about to sail into uncharted waters. If you are reading this book stage by stage, I would suggest that now is the time to jump ahead and read the sections on pain relief, labor, birth, and life after birth to equip yourself with the practical and emotional aspects of childbirth and beyond. Like most other important events in our lives, the better informed you are, the more capable you will be at dealing with the challenge positively and confidently. Make sure that you start your childbirth or parenting classes now, if you have not already done this.

Some women who are normally slender and weight-conscious feel suddenly liberated by their large, rounded belly.

A positive body image

If you are normally a fit, healthy woman, the physical downside of pregnancy can come as something of an emotional shock. You may become increasingly frustrated by your growing bulk, which is preventing you from leading your life as you used to. On the other hand, you may be loving every minute of your new-found voluptuousness. Some women who are usually very slender and weight-conscious tell me that they feel suddenly liberated by, and enormously proud of, their large, rounded belly. They view their body as an affirmation of their sexuality, especially since it might be the first time that they have ever had a generous cleavage! Similarly, women who have previously been worried about their body size may for once be reconciled with their larger shape and positively enjoying it. The truth is that the way we feel about our heavily pregnant bodies has a lot to do with how well we feel on a day-to-day basis during the latter stages of pregnancy and the way in which our partners react to our distended bellies and swollen breasts (not to mention extra all-over weight). Some women become deeply attached to their growing belly and a little sad at the prospect of losing it, while others feel that the day it disappears cannot come quickly enough.

Involving your partner

Some men are closely involved in their partner's pregnancy from the beginning, but many more do not show much interest in the details of pregnancy, labor, and life after birth until rather late in the day. If this is the case with your partner, you may be starting to feel concerned that he is still not as involved in your pregnancy as you might have wanted. Indeed, some men are reluctant to become demonstrably involved at all. They are not necessarily being unsupportive, it is just that men and women tend to be on different wavelengths during this unique time.

Women tend to immerse themselves totally in their pregnancy, because it is both physically and psychologically part of them. It is not surprising that men are less able to do the same, since they are, by definition, physically detached. Many tend to carry on with their lives as if nothing has changed. Although they may recognize that life will change dramatically after the baby arrives, it may be difficult for them to translate this somewhat abstract notion into the day-to-day reality.

If you are first-time parents, pregnancy provides you with the opportunity to start shifting your relationship from the couple that you currently are, to the family that you will shortly be. However, I do think it is important to remember that trying to change your partner into your ideal of what he should be during your pregnancy is unlikely to be successful. He is going to need to adapt in his own way and in his own time—albeit with a few prompts from you. He is undoubtedly just as eager as you are that everything goes smoothly during labor and after, so perhaps the most important thing that you can do now is to make sure that he is sufficiently informed to offer you practical and emotional help throughout. Then he will be able to look back on the event and feel he was as involved in the birth of his child as he wanted to be.

POSITIVE BODY IMAGE
How you feel about your heavily pregnant body is usually a reflection of your overall health and well-being at this stage of pregnancy.

Your prenatal care

As long as there are no complications in your pregnancy, you will probably continue to have monthly checkups until about 32 weeks, which means that you will need to see your prenatal doctor only once during the next few weeks.

Prenatal schedules vary but it is usual to have a routine appointment at 28 weeks. A blood count and antibody test will be performed (see below) and your urine and blood pressure will be checked. Your doctor will examine your hands and legs and any sudden swelling will be investigated with blood tests to ensure that you are not developing preeclampsia.

You may already be experiencing mild practice contractions called Braxton Hicks' contractions (see pp.237–38) which travel down your uterus and cause it to harden momentarily, although these become more common after 30 weeks. However, if you are experiencing any prolonged or painful uterine activity, especially if it is accompanied by lower back pain, report it immediately.

THE GLUCOSE TOLERANCE TEST

Gestational diabetes (see p.427) is a common complication in pregnancy, largely because of the strain that pregnancy puts on a woman's kidneys and metabolic system. In severe cases, the symptoms are similar to those for diabetes, including extreme thirst, a need to urinate frequently, and fatigue. However, many pregnant women who develop gestational diabetes have no symptoms, which is why a routine glucose tolerance test is done between 24 and 28 weeks if you have a family history of diabetes, are of an ethnic background that puts you at a higher risk, a BMI of 30 or more, or if you are having twins. If you have a past history of

gestational diabetes, you will be asked to check your glucose levels at home on a regular basis. The test is very simple:
▶ You will be asked for a urine sample at each prenatal visit. It is tested for the presence of glucose.
▶ A glucose screening test is done at 24–28 weeks of pregnancy. The patient drinks a special sugar mixture and, one hour later, a blood sample is drawn and the level of glucose is measured.
▶ If the glucose screening test is abnomal, the patient is given a glucose tolerance test on an empty stomach. A baseline blood sample is taken, then the patient drinks a glucose solution. One more blood

sample is taken after two hours. These provide a valuable assessment of your sugar metabolism. The results of this test are usually available within five working days.

If you have gestational diabetes you will follow a low-sugar, low-carbohydrate diet for the rest of your pregnancy. If this does not control the problem, you may need to take pills to reduce your high blood sugar levels or possibly have regular insulin injections. Although only a small percentage of women have the problem after delivery, having gestational diabetes increases your risk of developing Type II or late onset diabetes by 30 percent.

Blood tests at 28 weeks

Your hemoglobin (blood count) will be tested between 26 and 30 weeks to make sure that you have not developed anemia (see p.424). If your hemoglobin is less than 10.5g/dL, you will probably be advised to take an iron supplement. It is important to build your red blood cell count now, since your hemoglobin is likely to drop more toward the end of pregnancy, thanks to the increased fluid content in your bloodstream. However, gastrointestinal upsets, constipation problems, and sometimes diarrhea are common side effects of iron supplements so if you have problems, ask for a different brand. Liquid preparations available in your local pharmacy may be kinder on your digestive system. Above all, aim to eat iron-rich foods, particularly those that contain lots of fiber, such as dried apricots and raisins.

The antibody screen uses a portion of the same blood sample to check your blood group again and make sure that you have not developed any red cell antibodies. This is particularly important if you are Rhesus negative (see p.128 and p.425). Although the time of greatest risk is at delivery, when a Rhesus-negative mother may become sensitized to blood from her Rhesus-positive baby, these women are now offered an injection of anti-D at 28 weeks and after delivery, even if this is their first pregnancy.

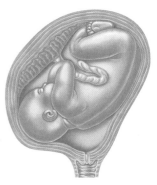

OBLIQUE LIE The baby is lying at an angle across the uterus.

Your baby's position

In addition to measuring the height of your fundus and listening to the fetal heartbeat, your doctor will palpate your abdomen and at this stage may be able to determine the position in which your baby is lying. From now on, your prenatal records will carry a record of the baby's position at each checkup. (For a full description of these different positions and the abbreviations in your records see pp.268–70).

The lie of your baby at this stage is most likely to be longitudinal (vertical), but it could also be transverse (lying horizontally from side to side in your uterus) or oblique (at an angle). The presentation refers to the part of your baby that is nearest to the pelvis. It can be a cephalic presentation (head down) or a breech presentation (head up). If the lie of your baby is transverse, there is no presenting part at the present time. This is nothing to worry about, since the lie and presentation of the baby can change many times between now and the onset of labor. Similarly, do not be concerned if your doctor cannot determine which way up your baby is lying between 26 and 30 weeks. Even the most skilled clinicians may find it impossible to decide whether your baby is head down or head up at this stage.

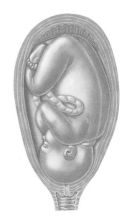

LONGITUDINAL LIE In this lie the baby is vertical with its head or bottom down.

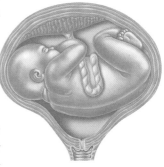

TRANSVERSE LIE The baby is lying horizontally across the uterus.

CHECKING YOUR BABY'S GROWTH

At each visit, your doctor monitors your baby's growth by palpating your abdomen and measuring the fundal height of your uterus. If a more detailed check is needed, it is usually done with a series of ultrasound scans.

GROWTH SCANS

Fetal growth scans are revealing because most problems at this stage in pregnancy are likely to affect the rate at which your baby is growing. The size of the baby's head, limbs, and abdominal girth will be recorded, and the relationship between the various measurements examined carefully because late pregnancy problems may not affect all aspects of growth equally.

IDENTIFYING THE PROBLEM

Intrauterine growth restriction (IUGR manifests itself in different ways depending on the cause (see pp.256–57 and p.428). For example, if the placenta is not working well (as can be the case if the mother has high blood pressure or preeclampsia) the growth of the baby's head will be maintained but usually at the expense of the growth of the baby's abdomen.

This is because the blood supply carrying oxygen and nutrients from the placenta will be diverted to the baby's brain, and the abdominal organs receive less. To compensate, the baby's liver will start using fat stores causing the liver (and abdominal girth) to become smaller. This growth pattern is known as "head sparing growth retardation"– a somewhat frightening term, but one that sums up a clever survival mechanism that ensures that the fetal brain is protected in a potentially difficult situation.

COMPARING MEASUREMENTS

Your baby's measurements will be compared with previous and future measurements because the rate of growth over time is what determines whether it is safe to leave the baby in utero or whether the baby needs to be delivered immediately.

If your baby is not growing very well but is not in distress, you will be asked to return for another scan in 2 weeks. This may seem a long time to wait, but it is difficult to interpret changes in measurements within a shorter interval of time.

INTERPRETING GROWTH CHARTS

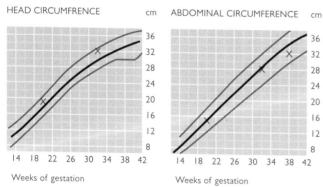

HEAD CIRCUMFRENCE   cm

ABDOMINAL CIRCUMFERENCE   cm

Weeks of gestation

Weeks of gestation

ON EACH GRAPH the 50th percentile (red line) is the average and the 90th percentile line (above) and the 10th percentile line (below) represent the upper and lower ranges of normal growth. On the head circumference graph, the baby's head is growing steadily. On the abdominal circumference graph, the velocity of growth of the abdomen is showing a decline, possibly because blood and nutrients are being directed to the heart and brain at the expense of organs in the abdomen.

Common concerns

This is a weightier section now, reflecting the fact that pregnancy-related problems and irritations are more numerous in the third trimester. Fortunately, there are remedies and strategies that will help reduce some of their effects.

Urinary frequency during the day is common—the usual reflex signals that start when the bladder is full occur much earlier when there is an increasingly heavy baby pushing down on it from above. Although there is nothing that can be done about this further mechanical design fault of pregnancy, remember that, if you need to urinate very frequently and can only expel a tiny amount on each occasion, you may have developed a urinary infection and need to make sure that your urine is properly tested (see p.192).

You may have started to leak small quantities of urine when you sneeze, cough, or laugh; this is called stress incontinence and is common toward the end of pregnancy. Renewed attention to Kegel exercises can reduce the problem as can cutting out tea, coffee, and colas, which have a diuretic effect.

Sleep patterns in the third trimester are often disturbed and your bladder will make a significant contribution to this by ensuring that you have to get up several times a night to use the bathroom. Several of my patients have suggested that these nightime interruptions are designed to help you adjust to the inevitable lack of sleep after the baby is born. They may be right, but when I was pregnant I would have preferred to sleep undisturbed at night between 26 and 40 weeks and find out about nightime vigils at a later stage.

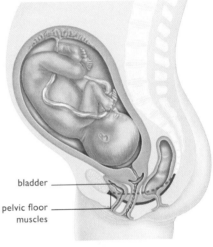

bladder

pelvic floor muscles

BLADDER PROBLEMS
During a cough or sneeze you may leak urine. The problem is due to the weight of the baby pressing on your bladder and also weak pelvic floor muscles, shown above as a solid line (with a dotted line indicating their pre-pregnancy position).

Vaginal infections

It is normal for vaginal discharge to increase from the second trimester of pregnancy onward. However, it should always remain clear and mucuslike, odorless or just mildly smelling—similar to the discharge you may have experienced before the start of your menstrual period. You may find that you need to use a thin panty liner. If the discharge becomes yellow-green, develops a stronger smell, or your vulva, vagina, and anal region become reddened and painful, particularly when you urinate, tell your doctor. She will take some swabs to check for vaginal infection, which if left untreated may increase your risk of going into premature labor.

DEALING WITH YEAST

Many pregnant women are troubled by yeast (candidiasis) in pregnancy. The following remedies may help relieve symptoms:

▶ **Cream and suppository remedies** can be bought over-the-counter or prescribed by your doctor. Suppositories, inserted into your vagina, are the most effective because they tackle the root cause of the infection by increasing the acidity of your vaginal secretions. They will not harm your pregnancy, and a single pessary may resolve the problem. Creams applied to the vulva may reduce the discomfort temporarily, but they won't cure the underlying problem.

▶ **Personal hygiene** is important. Make sure that you always wipe your anal region from front to back (rather than back to front) after a bowel movement. Bathe regularly and keep your vulva clean and dry. Avoid highly fragranced soaps and bubble baths, particularly if the vulval skin is reddened and sore.

▶ **Adding a few drops of vinegar** to your bath or bathing the vulval area with a weak solution of cider vinegar may relieve symptoms. Alternatively, you could try live yogurt to balance the body's natural bacteria. You could try smearing the yogurt into the vaginal entrance to relieve itching.

▶ **Wearing cotton underwear**, and avoiding tights or tight-fitting jeans will give the skin around your genital area air to breathe.

▶ **Reducing your intake of sugar and yeast** may be useful if you have recurrent yeast infections, since they can both aggravate the problem.

Most itchy vaginal infections are due to yeast (candida infection), which is an innocent, albeit uncomfortable, side effect of pregnancy (it is not sexually transmitted). You will notice an itchy, curdlike discharge (which looks a bit like cottage cheese) around your vagina. Yeast is not a cause of preterm delivery and most women experience at least one episode of it during pregnancy. It is largely due to the vaginal environment becoming less acidic during pregnancy thanks to the effects of hormones, and this encourages the growth of the yeastlike fungus *(Candida albicans)*, which is normally present in small numbers in the vagina and gut. Another common cause of yeast is antibiotic treatment because the antibiotics kill off some of the normal housekeeping bacteria in the gut and vagina, allowing the candida organisms to gain a hold. Some women suffer recurrent episodes of yeast during pregnancy.

Headaches

Headaches are common in pregnancy and are usually nothing to worry about. However, some women suffer from migraine episodes that can leave them quite debilitated. If you are suddenly experiencing severe headaches, report them to your doctor promptly. Please do not be tempted to wait until your next appointment, because severe headaches at this stage in pregnancy can be a sign that your blood pressure is too high (see p.426). Even if your headaches prove to be nothing serious, your doctors will be able to suggest some safe remedies.

Itching skin

By the end of pregnancy your skin will have stretched by an extra 150–290 square inches (77–155 square cm) and can become dry and itchy as it becomes increasingly taut over your enlarging belly. Stretchmarks often make an appearance around now and can make the problem worse. Expensive creams marketed specially for pregnant women, to prevent or reduce stretch marks, are unlikely to do more than relieve dry itchy skin temporarily. I can promise you that cheaper remedies—simple, unscented emollient creams or oil such as baby oil or olive oil—are just as effective at keeping your skin supple and well hydrated. You can also reduce itchiness by wearing cotton clothes to keep the skin cool. If the itching persists or becomes severe, particularly affecting your palms and the soles of your feet, you may need a blood test (see p.242 and p.424), so please tell your doctor.

Hemorrhoids

Many women become troubled by hemorrhoids by this stage in pregnancy. These are dilated veins around the inside and outside of the anus, or back passage, which are caused by the pressure of the baby's weight in your pelvis. Hemorrhoids frequently cause throbbing pain and itching around the anal area and they may also bleed. You may find that you can feel a swollen tender vein protruding out of your anus or notice some light red bleeding on the toilet paper after you have had a bowel movement. If you are constipated, you are more likely to strain in an attempt to empty your bowels and this can cause the hemorrhoids to swell more, so make sure you drink plenty of water each day, increase your intake of dietary fiber and exercise regularly. Lifting heavy weights can aggravate the problem. Over-the-counter creams that contain a lubricant and a light local anesthetic will help relieve the discomfort as can icepacks, particularly when you have had a long and tiring day.

Leg cramps

Many pregnant women suffer from leg cramps, particularly at night. You may find that you wake up suddenly, gripped by painful, violent spasms in one of your legs or feet. Some doctors think that the pressure of the uterus on certain nerves in the pelvis may be the trigger while others suggest they may be due to low calcium or salt levels, or an excess of phosphorus. However, none of these theories has been proven so don't even consider trying to adjust your levels of these minerals with supplements or dietary changes. When you get a cramp attack, simply flex the leg, calf, or foot in the opposite direction. So if your calf cramps, for example, stretch it out by straightening your leg and flexing your

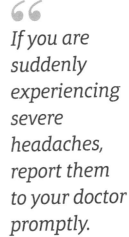

If you are suddenly experiencing severe headaches, report them to your doctor promptly.

RELIEVING BACK PAIN

Women used to be told that they had to put up with backaches in pregnancy because nothing could be done about it. In fact, there are several practical measures that help relieve symptoms—you don't have to put up with debilitating pain.

The kind of back injury that tends to occur at this stage of pregnancy is generalized back pain. Later in the trimester you may develop more specific back problems, such as sciatica, pubic symphysis dysfunction and sacroiliac joint pain—these are covered in detail on pages 243–44.

First, make sure that your doctor identifies exactly what the problem is: the back is such a complex area that pain could be due to any number of causes. It is not sensible to embark on treatments that are at best inappropriate, at worst dangerous.

Consider consulting a chiropractor, but make sure that you choose a trained practitioner (see pp.436–38). A skilled chiropractor will relieve symptoms of backaches (or joint pain) by gentle manipulation and massage but you should never agree to any realignment of vertebra—especially in the lower spine—by "clicking" them into place or using short, sharp, manipulations.

Once your doctor has made a diagnosis, back exercises designed for pregnancy may be helpful (some of are shown here). Referral to a physical therapist may also be helpful.

PROTECTING YOUR BACK

Because your belly now weighs more than usual, you may find that walking even short distances can pull on your abdominal ligaments or give you lower backache. Your pelvic ligaments are under more strain than ever and, and because they are more elastic than usual, it is inevitable that they will complain when made to work harder. Turn to the tips on page 193 to remind yourself how to protect your back when lifting and how to support your back while you sleep.

STRENGTHENING MUSCLES

Regular exercise will help build stronger back muscles and improve your posture and, as a result, support your spine and lumbar region and help reduce—if not prevent—back pain. Physical activity will also help you sleep better since you will feel calmer, thanks to endorphins released during exercise, which have a slight painkilling and mood-enhancing effect.

AN ORTHOPEDIC BELT

Another practical way to reduce back strain is to wear an orthopedic belt (these are usually advertised as

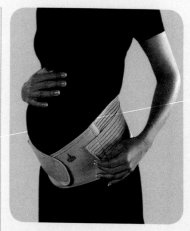

SUPPORT A belt may bring instant relief.

maternity belts in magazines and online). The belt is positioned just below your belly and straps around your pelvis with velcro fastenings. Wear it during the day and take it off at night. I remember how important my belt was to me at this stage in pregnancy. I was carrying twins, and am short (about 5ft 2in), which meant that by 26 weeks, I was feeling extremely unsteady on my feet and the backaches were awful. As soon as I put on my belt, the relief was immediate. Why they are not regularly recommended to pregnant women is a mystery to me.

BACK EXERCISES

If your back is giving you problems, try some of the exercises below. They will help strengthen the muscles that give support to your spine and pelvis and will keep you supple, which will be of great benefit during labor. As always, stop exercising if you feel any discomfort, and if you are unsure about any specific exercise, ask the advice of an obstetric physical therapist (contact him/her through your doctor).

▶ **Knee hug** Lie on your back with your arms hugging your knees (make space for your belly). Gently roll a little from side to side to release tension in your lower spine and pelvis. This is very soothing for your lower back.

▶ **Spinal twist** Lie on your back with your knees bent and feet together, arms out at shoulder height. Slowly drop your knees to one side, while turning your head to look in the opposite direction. Feel your spine twisting gently. Raise your knees back up and repeat on other side.

▶ **Spinal relaxation** Lie on your back, with your knees bent, shoulder-width apart, with your arms at your sides. Push up on your legs so that your thighs, pelvis, and back as far as shoulder blades are lifted off the floor. Lower your back down slowly, exhaling at same time. Repeat five times.

▶ **Pelvic tilts** Lying on your back with your knees bent, pull in your lower abdominal muscles, squeeze your buttocks, and press the curve of your back into the floor. Hold for 10 seconds (don't hold your breath) and release slowly. Repeat five times, building up to 10.

▶ **Knee squeezes** Lie on your back, with your knees bent and your feet together. Squeeze any object roughly the size of your fist between your knees (such as a can of beans). Hold the contraction for 10 seconds and repeat 10 times, twice a day. Progress to an object the length of your forearm (such as a paper towel roll) only when the previous exercise is absolutely painless throughout the contraction. This exercise is particularly good if you are suffering from pubic symphysis dysfunction.

▶ **Birth ball** Sitting upright on an inflated birth ball will help to promote good posture.

KEY BACK STRETCHES

SPINAL STRETCH Sitting down on bent knees, legs slightly apart to make way for your belly, stretch your arms out in front of you along the floor. Feel the stretch all along your spine.

CAT STRETCHES Kneeling on all fours, knees and arms shoulder-width apart, arch your back into a hump, clenching your buttock muscles and tucking your pelvis in. Hold, then release slowly until your back is flat again. Repeat five times.

You may be anxious to ensure that you are not regarded simply as a mother-to-be but also as a working woman and colleague.

foot toward you as you simultaneously massage the calf area until the pain fades away. Although leg cramps are uncomfortable, they are nothing to worry about, since they are temporary ailments that will disappear after your baby is born. However, constant leg pain should always be investigated because of an increased risk of deep vein thrombosis (DVT) in pregnancy (see p.424).

Carpal tunnel syndrome

Some of my patients become alarmed because their fingers sometimes feel tingly, as if they had pins and needles in them. Occasionally, they might even feel a little numb or weak, as if they have lost sensation. This common ailment is caused by fluid retention, which swells the band of tissue (carpal tunnel) at your wrist and puts pressure on the nerves and ligaments that lie in this tunnel, before they enter your hand. The symptoms will disappear after the birth of your baby, because you get rid of all the excess water that has accumulated. In the meantime, if you become seriously uncomfortable, your doctor will refer you to a physical therapist, who may prescribe a splint to support your wrist. You could also try sleeping with the affected arm propped up on a pillow, to help drain the excess fluid. Remember that diuretic drugs used to get rid of excess body fluid should not be used in pregnancy (see p.35).

Things to consider

If you are working you should make sure that you and your employer are agreed on all of the specifics of your maternity leave (see p.63). Your doctor can give you a form now that you have passed the 26th week of pregnancy, confirming. Your due date and enabling you to qualify for maternity pay and leave.

Like many women, you may be hoping to continue working as long as possible to maximize your maternity leave after your baby is born. However, if you need to stop work for medical reasons, you can obtain a letter from your doctor so that you will qualify for short-term disability.

If you work in an office where certain styles of clothing are not permitted or are frowned upon, clothes can be a major issue as you near the end of your pregnancy. Although anyone who sees you will probably instantly realize you are pregnant, you may be anxious to be sure that you are not regarded simply as a mother-to-be but also as a working woman and colleague. If slipping into comfortable leggings and baggy T-shirts is not an option, then you will have to struggle on with wearing a jacket or similar business casual attire for as long as you can bear it. Borrow or buy a few outfits that will fit the bill.

If you are heavily pregnant during the summer, it may be quite difficult to find clothes that can keep you both cool and decent. You feel like a furnace for much of the time and if it is summer or you are living in a hot climate, you may feel unbearably hot, puffy, and sweaty. Heat rashes form in all sorts of recesses of your body (under your arms, under your breasts, between your legs). If your feet and hands have swollen, shoes start to pinch, and finger rings may become so tight and uncomfortable that they are impossible to wear. There is not much you can do about this, other than avoiding situations where you will feel particularly hot, such as overcrowded restaurants or stuffy theaters, and by wearing loose, light clothes made of natural fibers such as cotton.

Starting childbirth classes

You will probably be starting your birthing classes now—most hospital-run classes have start dates every four weeks and recommend women start their course between 30 and 32 weeks. Monthly start dates are also the norm for other forms of childbirth classes. If you do have a choice of when to start, always opt for the earlier date since you never know what might happen. You need to make sure that you attend the all-important sessions on labor and pain relief, before you are forced to find out all about it first-hand in the delivery room. For exactly the same reason, you want to avoid missing the visit to the maternity ward, if you have not already been shown around. This is especially important if this is your first birth, if you are expecting twins (50 percent of twins are delivered before 35 weeks), or if you have had a previous premature delivery.

Your partner may not want to attend all of the classes, but make sure that he at least knows when the special session for fathers is and encourage him to attend this one. It is far better that he has prior warning about the role you would like him to play, rather than leaving him to find out for himself during labor.

PREPARING FOR BIRTH Breathing and relaxation classes will help focus your attention on and prepare you for the next stage—your baby's birth.

Your sex life

A couple's sex life is often revitalized in this period of pregnancy since most women are feeling both physically and emotionally well. You may be conscious that time is running out before the birth of your baby, with the inevitable disruption that this will bring to your night's sleep and, as a result, to your sex life. Tiny babies are not very considerate about their parents' need for intimate times together alone. The only thing that may hinder your sex life at this stage are nagging doubts so let me deal with a few of them.

• Although rarely discussed, the fact that you may feel the baby moving inside you while you are making love can make you feel inhibited, or it can make you laugh. Although the sensation may be disturbing, it is certainly not a sign that your baby is upset by your lovemaking.

• You may both be worried that you will harm your baby by having penetrative sex or concerned that sex could trigger labor inadvertently since your partner's semen contains prostaglandin (a hormone used to induce labor). In addition, orgasm causes the uterus to contract.

The reality is that no amount of sexual activity will harm your unborn child or trigger labor in a normal pregnancy, so you can continue an active sex life unless you have been told to stop because of a potential problem or complication. Examples include a previous premature delivery or a risk factor for premature labor such as a short or slightly dilated cervix (see p.188); threatened premature labor (see p.340); recent bleeding and/or a low-lying placenta (see placenta previa, p.240 and p.427); and ruptured membranes.

Disturbing dreams

On a different note, many women report that they experience strange dreams during late pregnancy. They may be sexually explicit dreams or disturbing dreams involving the death or illness of babies and children. Both are common and can cause great anxiety, since you will undoubtedly start to wonder about their meaning. So it is important to remember that they are not an omen of awful things to come. Like all good and bad dreams (most of which we do not remember because we do not wake through the dream phase of sleep) they are a method of coping with day-to-day concerns and fears. Think of them as a way of sifting through negative emotions without having to experience them in reality. One of the reasons these dreams seem more common during the last trimester is that you wake much more often (either because you need to go to the bathroom or because you are uncomfortable) and are more likely to recall them.

...disturbing dreams are a way of sifting through any negative emotions without having to experience them in reality.

DIET AND EXERCISE

In the third trimester, eating nourishing food and maintaining your fitness has a dual purpose. Both will help reduce fatigue and boost your well-being as you enter the last stretch of pregnancy and begin to build your strength for labor.

GOOD EATING

Your diet is less crucial to your baby's well-being than in the first trimester, and unless you are surviving on a diet of potato chips and sodas, your baby is likely getting everything it needs.

▶ **Your weight gain should be** around ½ lb per week during the final three months (see p.42) although it may be minimal in the final weeks of pregnancy.

▶ **Your daily calorie intake** in the third trimester of pregnancy should increase by only 200 calories, which is not very much at all. An extra healthy snack, such as an apple or an orange, each day will supply all you need.

▶ **You may need to eat more frequently** during the last few weeks because you feel the need to snack regularly. Your body is most likely laying down some final, extra reserves in preparation for labor, so choose foods that are nutritious and will give you vital energy. Since you never know when you will go into labor, the better your diet in the weeks ahead, the better your ability to physically manage the demands on the day.

▶ **Maintain your fluid intake** (at least eight glasses a day) to make sure that your body is fully hydrated; this will give you more energy. It may seem obvious, but if you haven't stopped smoking yet, you should now: smoking starves the placenta, and thus your baby, of oxygen.

FIT FOR LABOR

There is no reason why you should stop exercising until the day you deliver, unless your doctor has specifically told you to stop.

▶ **Certain activities will now be difficult or uncomfortable** The chances are that you will have given up white-water rafting, horseback riding, and running by this point!

▶ **If you are accustomed to a particular sport,** you may be able to continue for a little longer, albeit at a gentler pace, provided you feel well and your doctor gives you the green light (although I'm sure you will know very well when things are becoming more than you can manage).

▶ **If you haven't tried pregnancy swim classes, yoga, or exercise classes yet,** try to find time to do so soon. You will be surprised by how good they make you feel.

▶ **Whatever your chosen method of exercise,** make sure that you are doing your Kegel exercises and are paying attention to your posture.

SHOPPING FOR YOUR BABY

There are no fixed rules about what to buy for your baby, but from my own experience and that of the many pregnant women I have talked to over the years, there are certain items that are more essential than others. Broadly speaking, there are two main areas to think about when choosing what you need for your baby's first few months: clothes and equipment.

Choosing baby clothes

Young babies cannot regulate their temperature very well, so they need to be kept covered during the early weeks, but not so much that they start to overheat. The general rule of thumb is that, for the first two months, babies need one more layer than you would wear on any given day (although this varies depending on the time of year and on the baby, since some feel the cold more than others). Remember also that babies have little or no hair, so on cooler days they will need to wear a hat outdoors (never indoors) and, if it is sunny, they must have a hat that protects their head, neck, and face.

Comfort, practicality, and ease of washing are the main criteria when choosing baby clothes. Look for clothes that do not restrict your baby's movements; that can be put on and taken off easily without causing your baby discomfort; that do not have bows, ribbons, and lace that will catch fingers; and that allow the skin to breathe. Babies grow so fast that clothes for newborns may only last a couple of weeks and, if you and your partner are taller than average, you may even have a baby who at birth is already wearing clothes for older infants. So, other than two or three all-in-one bodysuits for newborns, I would advise you to go straight for the next size up when choosing your baby's clothes. Don't feel shy about borrowing newborn clothes if they are offered—they get very little use and can be handed back within a month of the birth.

Make sure that the clothes you buy are made of a breathable fabric that can be machine washed at a reasonable temperature (minimum 100°F)

THE RIGHT SIZE Do not buy too many newborn clothes—they will only be worn for a few weeks.

and tumble dried. Cotton is the best fabric for breathability, comfort and ease of washing; synthetic-only fabrics are less appropriate, particularly in the early weeks. Wool is good in winter, but it can be irritating worn next to a baby's delicate skin. Make sure all-in-one suits are quick and easy to undo—in the early months, you will be changing at least 10 diapers in a 24-hour period. Go for the styles that have snap closures around the bottom to make sure that your baby is not put through the unnecessary contortions of taking the whole garment off for every diaper change. All-in-one suits without feet have the advantage that they will not cramp the growth of your baby's toes.

If you are having an fall or winter baby, you will need some warm outerwear, too. All-in-one suits, with integrated hoods and booties are ideal. Pay attention to the outer fabric and inner lining since some suits are warmer than others. Your baby will also need a warm hat (babies can lose most of their body heat very quickly from an uncovered head), and several pairs of mittens and booties to keep hands and feet warm. Young babies can be relied upon to regularly lose these items by pulling or kicking them off. You will find countless varieties of shoes for babies in stores—all of them unnecessary and potentially harmful if they cramp your baby's toes. Your baby will only need shoes (properly fit in a children's shoe department) for later walks outside.

Most importantly, remember that you will be given many clothes as presents so try not to buy too much now. Your newborn can only use a few outfits per day even for the kind who throws up after every feeding. Before you can blink, clothes will no longer fit, which is why it is much better to stock up on missing items later on.

NATURAL FIBERS Clothes made of cotton and wool allow your baby's skin to breathe.

ESSENTIAL BABY CLOTHES

▶ Six cotton shirts with wide necks
▶ Six all-in-one suits
▶ Two cardigans (fleece-backed cotton or wool for winter, lighter cotton for warmer days)
▶ Two pairs of socks or soft cotton booties
▶ One shawl or cotton blanket
▶ One bonnet or sunhat, which shades eyes and neck
▶ One outdoor coat, including hood and booties, or one all-in-one outdoor suit, depending on season
▶ One pair of mittens, depending on season

CHOOSING A STROLLER

I cannot recommend a particular stroller (it would be out of date before you had finished reading this section), but I can offer a few points to keep in mind when you are choosing transportation for your newborn:

▶ For the first few months, your baby's spine needs proper support and he will need to lie totally flat. Any stroller that does not have a flat position should be rejected. If you buy a travel system, you can use the car seat clipped onto the stroller but only for short trips.

▶ Think about where you live. Some of the fanciest strollers have a very long chassis and big wheels—ideal for cross-country jogging but hard to maneuver up and down steps and in stores and on busy streets.

▶ For winter babies, look for a carriage that is well insulated and offers protection from the elements.

▶ Whatever time of year you give birth, you will need to have a rain hood but, unless you are having your baby in spring or summer, you will not need a sunshade for the time being.

▶ Make sure that the folded-up stroller or carriage fits inside your car trunk and the car seat fits in your car. Most baby equipment stores and department stores will let you try them out before you buy them.

Buying baby equipment

When you become pregnant for the first time, you discover a whole new world of products aimed at mothers, parents, and children. If you had little contact with babies until now, you might be surprised at the variety of what is available and even if this is not your first child, you may be surprised to see items in the stores this time round that were not around as recently as 18 months ago. Undoubtedly, many of these so-called "essential" products are conceived by the fertile imaginations of those who work in the baby and child-care industries, aided and abetted by all kinds of marketing wizardry.

Go to any baby section in a department store or leaf through any mother and baby mail-order catalog and you are suddenly confronted with everything from endless varieties of strollers and carriages—with horrifying price tags to match—to a plethora of little gadgets, which may or may not be of any use. After all, you ask yourself, how important is an in-car bottle and jar warmer? While there is no denying that many products are genuinely helpful and make our lives easier, it can be very difficult for a first-time parent to distinguish between what is a true must-have and what is a nice, but entirely optional, extra.

Strollers and carriages

The choice of stroller or carriage is probably the most important and costly item. Technically, a carriage is a crib on a chassis that can then be taken off and used for the baby to sleep in at night because it has sufficient depth to accommodate a mattress. A stroller may allow the baby to lie completely flat but it cannot be used for him to sleep in at night because it cannot be taken off the chassis and, crucially, it lacks the depth to accommodate a mattress.

Whether you go to a baby store or a department store, the sales help is likely to blind you with particulars as he or she demonstrates how the various models work. Before you know it, like a child's transformer toy, travel systems, strollers, and carriages are being assembled, folded, or taken apart to become forward and back facing strollers, car seats, and more. Some of the new generation of strollers and carriages now have four-wheel mechanisms and larger wheels allowing you to jog with your baby across the most rugged terrain.

As with all aspects of buying equipment, speak to as many people as you can who have recently had babies, since their advice will be unbiased and they can explain the pros and cons before you begin to look in stores. You will then have a clearer idea of what you are looking for and what your main criteria are.

If the price tags of strollers is alarming (and they often are) grandparents-to-be might like to contribute; alternatively, consider borrowing one from a friend or relative, or buying secondhand. Because of the high price and high turnover of nursery equipment, there is a thriving secondhand market in all these goods.

A baby front pack is a useful and inexpensive addition to having a stroller. It allows you to keep your baby close to you (with the bonus of a peaceful nap for your baby through any trips) and to have both hands free. It also saves the trouble of hauling your carriage in and out of stores and people's houses. Packs range from traditional hammocks that lie across the body to high-tech sport models with adjustable straps and back supports. Whichever one you choose make sure it supports your newborn baby's head. Try the pack on before you buy it to make sure that you can put the baby in it and fasten it by yourself.

Car seats

The other essential item you will need is a car seat: in fact, if you give birth in a hospital, you will not be allowed to leave unless you have one properly installed in your car. Some are part of a travel system—a suite of items that attach to a stroller base; or can be lifted out of the car and used as a babyseat. Many car seats for newborns are designed to last only around six months, so this is an item that you could consider borrowing from a friend if you know its history. If you buy a secondhand seat, check that it has not been involved in a serious accident, because there is a risk that it may no longer be safe as a result.

Cribs and beds

What will your baby sleep in at home? There are various options, from bassinets, cradles, and cribs to full-sized beds. Essentially, all options are fine from the start and, to be honest, many babies

BABY FRONT PACK The most comfortable packs have wide straps that support your back and shoulders and well-placed head support for your baby.

spend their first few weeks in simpler beds and are none the worse for it. The important thing is to buy a new mattress. A few years ago, there was a crib death scare that suggested a link between old mattresses and an increased risk of Sudden Infant Death Syndrome (SIDS). This theory has now been discredited but there is still a good reason for buying a new mattress: a secondhand one will have an indentation from the previous baby and will not give your newborn adequate spinal support. Some parents keep their babies in a cradle during the day, then transfer them to the crib at night, with the idea that the baby will gradually recognize that the latter is specifically for a long sleep at night. This is certainly worth a try, although it may make little difference before the baby is around three months old, by which time, he will have outgrown a cradle anyway. Babies move more freely and sleep better in a full-sized crib from the start, so there is really no need to invest heavily in the other options because you will soon be dispensing with them.

As far as bedding is concerned, cotton is always best, but it is important not to let your baby overheat by piling on layers of wool blankets. You can check to see if your baby feels cold by feeling the nape of his neck: if it feels warm, then he is fine. For each type of bed, you will need a minimum of two thermal blankets, two fitted sheets, and two top sheets. Babies should not have pillows, because their heads need to lie flat on the mattress. Even though crib bumpers, quilts, and comforters are widely available in stores, they are not recommended for small babies because they may move over the baby's head and restrict breathing.

Diaper-changing and feeding equipment

Regarding smaller items of baby equipment, you will need a plastic covered changing mat and a plastic bucket or bin (with a lid) in which to put disposable or reusable cloth diapers. Disposal

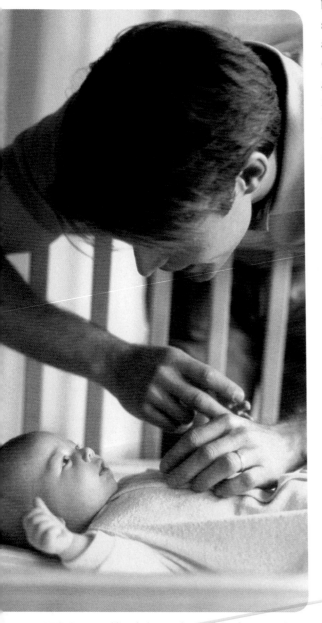

CRIB SAFETY Your baby is safest with his feet near the base of the crib so that he cannot slip under the blankets.

units store used disposable diapers in a sealed, fragranced container until you are ready to empty them. This is hardly an essential item but it reduces the amount of times you have to empty the trash and avoids unpleasant smells. A baby bath can make bath time easier on your back and nerves as you struggle with a small, slippery baby.

If you plan to use cloth reusable diapers, which are more environmentally friendly than disposable diapers, you will need to buy at least 30 of them to accomodate as many as 10 changes a day, together with plastic pants and diaper liners. They may appear to cost less than disposables but you need to take into account the fact that they need laundering either by you (make sure you have a good washing machine and dryer) or by a diaper laundering service (this latter option works out to roughly the same cost as using the disposables). The more modern-shaped reusable diapers are fully washable and tend to fit better than the old terry cloth styles because they come in different sizes, although all types require a liner (and plastic pants to protect clothes from soiling). You will find more information about the baby products you may want to buy closer to your delivery date on pages 224–27.

Feeding equipment

If you plan to bottle-feed from the start, you will need at least six bottles because you will be feeding your baby up to seven or eight meals per day (or at least making up that many feedings). If you plan to breast-feed, you should still buy two or three bottles so you will be prepared if or whenever you decide to start using bottles. Remember that babies can take your expressed breast milk, as well as formula, from a bottle. Buy slow-flowing

nipples; otherwise, your baby will struggle to swallow fast enough and may suffer from some stomach upset.

You will need to sterilize bottles until your baby is at least six months old using one of the following methods:

• A sterilizing tank uses water and sterilizing agent in tablet or liquid form. Bottles and nipples have to be submerged for several hours. There are some bottles that can be sterilized in different ways but you must carefully follow the directions on the label.

• The newer electric or microwave steamers use steam to sterilize bottles and nipples in a matter of minutes. Most kits are sold with bottles.

These are the essential items to have ready before your baby's birth but by all means go out and buy whatever seems useful afterward. The early weeks and months are about making life easier and, within reason, you should look for anything that helps you do just that.

ESSENTIAL EQUIPMENT

▶ Stroller or carriage that allows baby to lie flat
▶ Rain hood
▶ Crib or cradle with a new mattress
▶ Cotton bed sheets, including, for each size of bed:
  Two fitted sheets
  Two top sheets
  Two thermal blankets
▶ Rear-facing car seat
▶ Plastic changing mat and plastic diaper bucket (with lid)
▶ Bottles (six if bottle-feeding, two if breast-feeding) plus slow-flow nipples.
▶ Sterilizing equipment
▶ Baby carrier with good head support

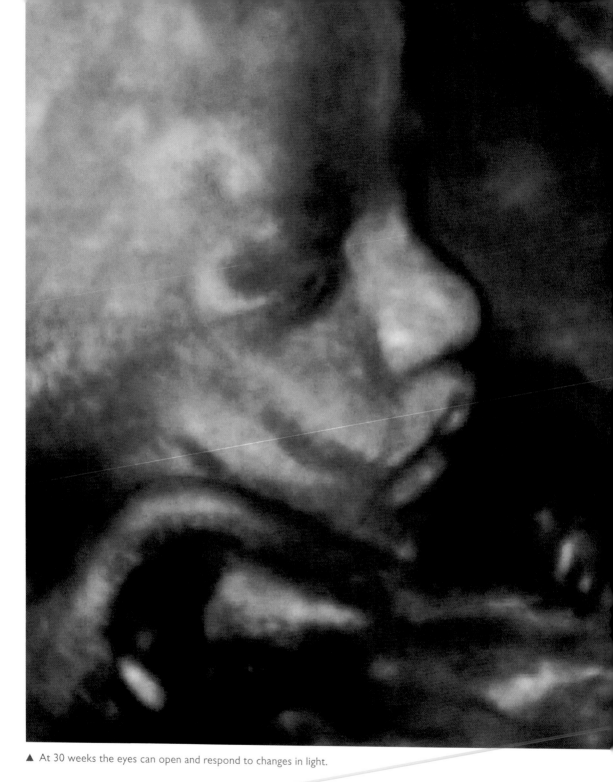

▲ At 30 weeks the eyes can open and respond to changes in light.

▶ **WEEKS 30–35**
The developing baby

YOUR BABY IS CONTINUING TO GROW IN LENGTH, but the really noticeable difference is your baby's weight. The layer of subcutaneous fat is increasing and the skin now looks pink and less wrinkled, especially in the face, which now looks smooth and chubby.

From 28 to 32 weeks the weekly weight gain is as much as 17oz (500g) and continues at a rate of about 9oz (250g) per week between 32 and 35 weeks, which means that the average baby weighs 5½bl (2.5kg) at 35 weeks. A baby born around this time will still look a little on the lean side but will no longer have the wrinkly, red, emaciated look of a few weeks ago. The surface coating of white waxy vernix will be very thick, but the lanugo hair is fast disappearing and will probably be present only in patches on the shoulders and back. If delivered now, your baby will be in less need of these protective mechanisms to combat the cold since its control of body temperature is becoming much more reliable.

Your baby's eyes are opening and closing now, blinking and learning to focus, because the pupils are able to contract and dilate in response to differences in light filtering through the wall of the uterus. The brain and nervous system are also fully developed, although some of the reflexes and limb movements will still be poorly coordinated if the baby is born now. The fingernails extend to the end of the fingertips, but the toenails will need a few more weeks to reach the end of the toes.

The sucking reflex becomes properly established at this stage and the baby will be repeatedly sucking on its thumbs and fingers. However, most babies born before 35 or 36 weeks still need a more practice, which can mean that breast-feeding is more difficult to establish. This is one of the reasons why the definition of a premature baby is still a baby born before 37 weeks. Although most babies born after 28 weeks have an excellent chance of survival, thanks to the special care they receive after the birth, there is no technical advance that can make a premature baby suck as effectively as a full-term baby. So if your baby is born at or before 35 weeks, you will probably need some help from breastfeeding counselors.

not life size

By 30 weeks, the fetus is about 11in (28cm) long and weighs about 2¼–3⅓lb (1–1.5kg). At 35 weeks its weight has increased to about 5½lb (2.5kg) and the fetus now measures 13in (32cm) from crown to rump and as much as 18in (45cm) from head to toe.

| 21 | 22 | 23 | 24 | 25 | 26 | 27 | 28 | 29 | **30** | **31** | **32** | **33** | **34** | **35** | 36 | 37 | 38 | 39 | 40 |

The lungs are maturing so fast between 30 and 35 weeks that every passing day reduces the time that your baby is likely to need assistance to breathe. In practical terms, a baby born at 34 weeks may need some help breathing, for days or even weeks, whereas one born at 36 weeks is almost always able to breathe unaided. In these next few weeks your baby crosses the divide; its lungs undergo the final steps of maturation that allow them to function independently.

The fetal adrenal glands, on the top of the kidneys, are pumping out cortisol to help stimulate the production of surfactant in the fetal lungs. They are working so hard that they are the same size adrenal glands as an adolescent and are producing 10 times the quantity of cortisol as those of an adult. Soon after your baby is born they will shrink down and only become active again at puberty.

Sex hormones

In boys and girls, the fetal adrenal glands continue to produce large quantities of an androgen-like hormone (DHEAS), which has to be processed by enzymes in the fetal liver before it can be passed on to the placenta for final conversion to estrogen. In boys, the fetal testes are producing testosterone and some is converted by special target cells in the genitals to DHT (dihydrotestosterone) essential for the development of the external genitals. It is quite common for these high levels of hormones to result in the external genitals of both boys and girls appearing large and swollen at birth. In the case of boys, the scrotal skin that surrounds the testes can be darkly pigmented. All of these changes disappear in the next few weeks as hormone production settles down.

3-D ULTRASOUND AT 30–35 WEEKS ···

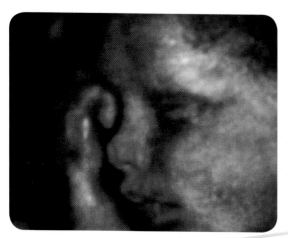

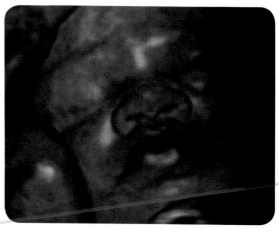

REFLECTIVE Images of a perfectly formed face begin to reveal a suggestion of the personality within.

SLEEP PERIODS Movement is now restricted by space and during quiet periods, the baby sleeps.

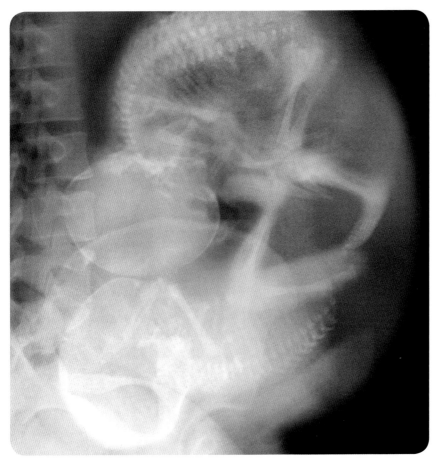

TWIN PREGNANCY An X-ray of a twin pregnancy reveals both babies lying transversely across the abdomen with both heads close to the mother's spine. Unless they change position, this will cause problems in normal labor and vaginal delivery. A cesarean section is the most likely option to deliver the babies safely.

Movements and lie

Your baby's movements will be strong but most probably slower than before for the simple reason that your baby can no longer move around as freely because space inside the uterus is now at a premium. However, if your baby's movement pattern changes from being very active to very quiet, or vice versa, seek advice immediately. Mothers-to-be are usually the best judge of whether problems are brewing in utero and you should never be worried that you are bothering people unnecessarily. It really does not matter how many false alarms there are.

Most babies are lying longitudinally (vertically) by week 35, but there are situations in which the baby may be transverse (horizontal) or oblique (diagonal). The risk of an abnormal lie is increased when the amniotic fluid level is high (see polyhydramnios, p.427); when the placenta is in the lower segment of the uterus (see placenta previa, p.240 and p.428); or when there is more than one baby.

Your baby's presentation is determined by which part of the baby is closest to your pelvis: head down is described as a cephalic presentation while bottom down (and head up) is called a breech presentation (see p.269). Cephalic is the most common presentation and by term, 95 percent of babies are in this position. At 32 weeks, as many as 25 percent of babies are breech, but this percentage has dropped to just 4 percent by 38 weeks. After 35–36 weeks the baby is much less likely to change presentation because the lack of space prevents major movements.

FLUID LEVELS

Levels of amniotic fluid increase rapidly from mid-pregnancy onward reaching a peak at 40 weeks. After 40 weeks the level needs to be checked regularly to make sure that the post-mature baby is not put at risk due to a decline in fluid.

The amniotic fluid

Your baby excretes about 16 fl oz (half a liter) of urine daily and at 35 weeks the amniotic fluid reaches a peak volume of 2 pints (1 liter). After this time the volume starts to decline and can be as little as 3.5–7fl oz (100–200ml) in a pregnancy that is post mature (overdue). Low levels of amniotic fluid (see oligohydramnios, p.426) can be a sign of a growth-restricted baby or a baby with kidney problems, while excessive amniotic fluid, or polyhydramnios, see p.426), may be seen in twin pregnancies and is sometimes associated with physical abnormalities in the baby or diabetes in the mother.

Your changing body

Now and for the next few weeks, the height of the fundus measured in centimeters will be about the same as the stage of your pregnancy in weeks. This changes slightly when the baby's head descends and causes the height of the fundus to drop slightly in most pregnancies.

Whatever the exact measurements at this stage, your uterus has expanded your abdomen so much that your belly button may have become inverted, giving it quite a prominent appearance. If this coincides with summertime and light clothing, it is often clearly visible through your clothes. Its change in appearance is not permanent; it will pop back into place once your baby is born.

Your blood volume will probably reach a peak of 10½ pints (5 liters), although some women will have another increase between 35 and 40 weeks. Most of the increase is due to the plasma or fluid content in your bloodstream, but the number of oxygen-carrying red cells does not increase at the same rate. This dilution of red cells by the increasing plasma fluid is a common cause of anemia in late pregnancy (known as dilutional anemia). Certainly, your blood count will be checked during this stage of pregnancy. However, rarely there is a serious problem as a result of lack of red blood cells, because the amount of hemoglobin (oxygen-carrying pigment) is now much higher than before you became pregnant. Rest assured, your baby is happily making use of the oxygen and nutrients it needs.

Varicose veins

If you are going to be troubled by varicose veins, this is the stage in pregnancy when you are most likely to notice them. Varicose veins are dilated veins just under the surface of the skin, most of which occur in the legs and anal region (see hemorrhoids, p.217). Wherever they pop up in pregnancy, they are caused by an unavoidable mechanical problem—the weight of your enlarging uterus pressing down on the main veins in your pelvis. These veins feed blood back to your heart and lungs, but they are now very dilated by your increased blood volume. When they meet a large obstacle in their path, such as your expanded uterus, the back pressure that develops forces your blood to pool in the dependent (smaller) veins in your legs, vulva, and anal region. Since your baby is going to get bigger over the next few weeks the discomfort from varicose veins is likely to get worse. Symptoms usually improve after the delivery of your baby but for some women varicose veins become a long-term problem.

Varicose veins in the vulval area are not as common, but they are often a cause of concern because they look unsightly and can become tender and uncomfortable. The symptoms are best treated in the same way as hemorrhoids. Although there is the potential for vulval varices to bleed heavily if damaged at the time of a vaginal delivery, this is only very rarely a problem. Furthermore, they usually completely disappear after the birth.

TIPS FOR DEALING WITH VARICOSE VEINS

▶ **Buy support pantyhose.** For maximum relief, put them on in the morning before you get out of bed.
▶ **Rest with your feet up** as high as possible whenever you can. This will help blood drain from the veins in your legs.

▶ **Walk briskly.** This will keep the muscles in your legs working and help to return blood to your heart.
▶ **If you have to stand** for any period of time, try to keep shifting your body weight from one leg to

another leg, instead of distributing your weight equally between both.
▶ **Keep a check on your weight gain.** Carrying extra pounds puts even more pressure on your legs and will make varicose veins more of a problem.

How you may feel physically

It is very common for women to feel uncomfortably large and unwieldy at this stage of pregnancy, particularly if it happens to coincide with the summer months when hot weather increases the risk of developing swollen hands, feet, and legs.

Even if you do not feel like the proverbial beached whale, you may well be moving around more slowly and laboriously than you normally do. Day-to-day tasks such as getting out of the car or putting on a pair of socks require a radical alteration of your usual technique. Although nobody likes to feel physically compromised and dependent on others for help, the best way to deal with these situations is to maintain your sense of humor and keep reminding yourself that this is temporary. If you are philosophical about your restricted movement now, it will also help to prepare you for the fact that young children will inevitably slow you down. After the birth, you will no longer be able to rush out of the house in three seconds flat, grabbing your car keys and bag as you go. You will just have to take things a bit slower. Try, though, to stay as active as possible during this last stage of your pregnancy— it will help you feel physically and mentally prepared for the challenges ahead of you, especially labor.

FINDING A POSITION
You may be most comfortable lying on your side with cushions supporting your belly and upper leg.

Hard to sleep

By now, you may not be sleeping well and this will affect how you physically feel during the day. It becomes increasingly difficult to find a comfortable position at night. Lying on your back should be avoided, because the weight of your uterus will press on the major veins returning blood to your heart, making you feel very faint and reducing the blood supply to your baby. The possibility of lying on your front disappeared some weeks ago and it is likely that your only practical option now is to lie on your side with your upper leg bent forward at the knee and, if necessary, supported by a pillow. However, you cannot stay in the same position all night,

MOVING SAFELY

Getting up from the floor or bed after relaxation or exercise can put strain on your abdominal muscles, which are already stretched to full capacity. Your altered center of gravity will also make large-scale movements difficult.

The following technique has been devised by yoga teachers to help you get to your feet safely. As with any strenuous maneuvers at this stage of pregnancy, move slowly and remember to breathe throughout.

Step one With your knees bent, roll onto your right side bringing your knee beneath you up to waist level. Keep your left hand aligned with your bent knee.

Step two Shift your weight onto your left hand and knee. Position your right knee under your right hip and your right hand under your shoulder and come up slowly on all fours.

and as you near term, turning over in bed becomes a major operation, involving shifting your increasing bulk, not to mention the supportive pillows around the bed. Added to this, your bladder is giving you regular wake-up calls and your baby may be continually kicking and wiggling.

The lack of sleep can make you extremely tired and irritable, which is why it is so important to try to find time to rest during the day. Even if you are still working full time, make sure that you reserve half an hour per day to sit down with your feet up. If you are at home, an hour's nap on your bed after lunch will really help make up for the poor quality sleep at night. If you establish a routine in which you break off from your tasks to rest during this prenatal period, it will be easier to continue after the birth when fatigue and lack of sleep are inevitable. You are going to need to train yourself to take advantage of the times when your newborn is sleeping to catch up on some rest or sleep for yourself and restore your energy and sanity.

Braxton Hicks' contractions

From now until the end of your pregnancy, the uterus starts to practice contracting mildly in preparation for labor. These painless tightenings, called Braxton Hicks' contractions, start at the top of the fundus and travel down the uterus causing it to harden for about 30 seconds. The 19th-century obstetrician John Braxton Hicks, from St. Mary's Hospital, London, was the first to describe

BREATHING Practicing deep, slow breathing can help you get rid of tension and regain control between contractions.

BEARING DOWN Exhaling with your knees wide apart and your head and elbows supported can help you prepare for second-stage contractions.

them. He realized that this painless activity toward the end of pregnancy gave the uterus the practice it needed to contract strongly enough to expel a baby through the birth canal and into the outside world. The contractions also help direct more blood into the placenta during the last few weeks of pregnancy.

Although some women are completely unaware of having Braxton Hicks' contractions, for others they can become quite strong and uncomfortable toward the end of pregnancy. If this is the case for you, try changing your position, getting up and walking around, or taking a warm bath, since these simple remedies can help relax the uterine muscles. Practicing some relaxation and breathing techniques you are preparing for labor are also likely to be of benefit, as will back massage.

If this is your first pregnancy, it may be difficult for you to know whether you are having strong Braxton Hicks' contractions or early labor pains so the rule here is that if you are not sure, go to your doctor for help. Similarly, it is essential to report any prolonged or painful uterine activity immediately, particularly if it is accompanied by lower backache, because you may be threatening to go into premature labor. As the mother of premature twins who spent a month in the neonatal intensive care unit (NICU) after delivery, I can assure you that this is a frightening and distressing experience for every parent, so make sure that you take every precaution to avoid delivering prematurely. Another possible cause of uterine pain and a low backache is placental abruption (see p.428), which requires urgent medical attention.

Your emotional response

At the top of the list of anxieties that patients share with me at this stage is the fear that their labor will be difficult, that it may go disastrously wrong, or that they will disgrace themselves during the delivery (see box opposite).

As always, the fear of the unknown is far more difficult to deal with than the reality. So once again, try to discard any horror stories that you have heard and hold on to the fact that most pregnant women are extremely healthy and their babies have a trouble-free entrance into this world.

Another common feature of late pregnancy is that women often find it hard to concentrate on specific tasks. Many of my patients tell me that their minds repeatedly wander off to baby-related issues, and for those of you who are still at work, this can be quite a problem. It may come as a bit of a shock to find yourself dreaming the days away and unable to focus on the job at hand. Tasks that used to demand high priority may no longer seem very important. I think the best way to manage this situation is to identify key tasks and ensure that they are finished, sideline nonessential jobs, and make sure that you do not take on anything new that is challenging or unlikely to be accomplished within a short time span. You should then be able to leave work comfortable in the knowledge that you have left things in a reasonable state.

Another feature of this last trimester is that sad or bad news tends to affect you more than usual. There is no doubt that pregnancy triggers intense emotional responses in many of us and makes us much more vulnerable to sad situations, particularly those involving children. Watching a TV program about any form of child deprivation or the loss of a child, for example, is likely to reduce you to a shower of tears, even if in the past you were able to take such things in stride. The only practical advice that I can offer is to try to limit your exposure to situations that are likely to make you feel distressed.

EMBARRASSING SITUATIONS

▶ **I hate feeling out of control. How can I avoid embarrassing myself during labor?**
Nothing you do during your labor and delivery will be considered embarrassing. This will be one of the few times in your life when you cannot be totally in control of your body, so instead of feeling embarrassed and agitated, just accept it. As for anyone in the delivery room being horrified or disgusted by anything that happens or upset by you groaning, swearing, or shouting at them, forget it; the labor nurses and doctors have seen and heard it all before. They would not be doing their chosen job if they did not

understand the practicalities of what happens when a 7½-lb (3.5-kg) baby is pushed through a woman's birth canal.

▶ **What if my water breaks when I am in a public place?**
It is unlikely that your water will break in the middle of the supermarket or other public place but even it does—so what? I have never heard anyone complain about having to help a pregnant woman whose membranes ruptured unexpectedly. But I have heard lots of people talk about how happy they were to be able to help when this entirely natural event occurred. The reality is that it is very rare for the amniotic fluid to gush out—it

is usually a trickle because most babies are head down, pressing on your cervix and preventing too much liquid from escaping.

▶ **I am worried that I might have a bowel movement during labor. Should I have an enema?**
Having a bowel movement during labor can happen because of the baby's descending head placing pressure on the rectum. However, it is unlikely that there will be much stool in front of your baby's descending head, so any problem is likely to be minimal. Compulsory enemas in early labor were routine practice years ago but are increasingly rare in modern-day maternity units.

Your prenatal care

Your doctor will want to monitor you more closely toward the end of your pregnancy. Take advantage of her expertise to ask questions about any procedures on symptoms that worry you.

The usual routine checks will be performed, but your doctor will be especially alert to signs of late pregnancy complications such as gestational diabetes (see p.427) or slow growth of your baby (see intrauterine growth restriction, p.429). Preeclampsia (see p.426) becomes increasingly common after week 30. Although it may develop without symptoms, there are usually some indicators. Any of the following should prompt you to see your doctor immediately to check for protein in your urine:

• Rings are too tight for your fingers and your feet too swollen for your shoes.

• Your face becomes puffy and swollen.

• Headaches have become constant or unbearable and you have flashes of light at the edge of your field of sight.

The height of your uterine fundus will be measured and if it is higher or lower than your dates suggest, you may be advised to have an ultrasound scan to check the size and health of your baby. If it shows that the baby is too small or too large for your dates, or if the amniotic fluid volume is increased or reduced, more testing (see pp.256–59) will be arranged. It may even be necessary to make plans to induce early delivery of your baby.

PLACENTA PREVIA

A placenta that remains low in the uterus causes problems if all or part is in front of the baby's head, overlapping or covering the internal part of the cervix (the cervical os). The first sign is often one or more episodes of painless bleeding, sometimes as early as 30 weeks, which must be assessed immediately in the hospital. If only the lower edge of the placenta approaches the cervix (marginal placenta previa), the baby's head may be able to pass through the dilated cervix and a vaginal delivery may be possible. If the placenta is lying centrally over the cervix (partial or complete placenta previa), there is a high risk of hemorrhage before or during labor. A cesarean section is the only safe option (see p.428).

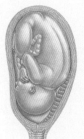

MARGINAL   **COMPLETE**

Palpating your abdomen will enable your doctor to determine the position in which your baby is lying. If your baby is breech (bottom down) it still has time to turn around and adopt the cephalic (head down) position, which is the best position for a normal vaginal delivery. However, if the baby's position stays breech, you may be advised to undergo external cephalic version (ECV), a manual procedure to turn the baby around in the uterus, which is usually performed after 37 weeks (see p.271). Remember that even the most skilled doctors can get the position of your baby wrong. Any obstetrician who says that he or she has never missed a breech presentation is either untruthful or has not been doing the job for long enough to have it happen!

   A complete blood count is usually performed at either 28 or 32 weeks to check for anemia (see p.423). The same blood sample will be used to check that you have not developed unusual red blood cell antibodies that might cause problems at a later stage, should you require a blood transfusion, for example, if you bleed severely during the delivery (see p.424). If you are Rhesus negative (see p.128 and p.424) you will be given a Rhogam injection and your blood will be checked to make sure no rhesus antibodies have been produced.

   Another ultrasound scan will be done at about 32–34 weeks if you were found to have a low-lying placenta earlier in your pregnancy. This is done to determine whether its position has changed. Even if the scan shows that your placenta is low, there are still several more weeks during which the lower segment of the uterus will continue to develop, so the chances of a low placenta causing problems at the time of delivery are reduced as the pregnancy becomes more advanced. The incidence of placenta previa (see left) is only 1 in 200 at term, whereas before 32 weeks it can be as high as 20 percent.

Remember that even the most skilled obstetricians can get the position of your baby wrong.

Diabetic pregnancies

All diabetic pregnancies will be closely watched from 35 weeks for late pregnancy problems (see p.408 and p.427). Poorly controlled glucose levels result in the baby being overweight (macrosomia), which increases the risk of shoulder dystocia (see p.429), birth injuries, and stillbirth. If your diabetes is well controlled and your baby's growth is normal, you may be able to wait for labor to begin and have a normal vaginal delivery. In uncomplicated diabetic pregnancies, there is no indication that an elective cesarean section improves the outcome.

   However, many hospitals have a policy of induction (see pp.294–97) for diabetic mothers anytime from 38 weeks, using continuous fetal monitoring and regular checks on the mother's blood sugar levels throughout labor and delivery. After 38 weeks, diabetic pregnancies are at increased risk of birth trauma, stillbirth, and neonatal complications. Induction tends to be earlier if you are required to take insulin during pregnancy. If fetal distress develops or there is poor progress in labor, an emergency cesarean section is sometimes required. After birth, babies born to diabetic mothers are thoroughly assessed because they can develop hypoglycemia (low blood sugar levels) during the first hours of life. They are also at increased risk of respiratory distress syndrome (see p.375), particularly if they are delivered prematurely.

Common concerns

Most of the physical ailments experienced in late pregnancy are related to your increasing size and are unlikely to get better until your baby is delivered. That said, if your baby's head engages in the pelvis before you go into labor, you may have some relief from them.

If you have been bothered by backache, it may get worse over the next few weeks.

If you are seriously troubled by breathlessness, try to cut down on unnecessary exertion while remaining reasonably active. Lying down flat often makes breathlessness worse, so you may find that you need to rest and sleep in a semi-propped-up position during the last trimester of pregnancy.

Palpitations

Missed heartbeats, a short run of fast heartbeats, or just being acutely aware of your heartbeat (often loosely referred to as palpitations) are common in late pregnancy. Normally they are nothing to worry about and are simply the result of changes in your blood circulation coupled with the mechanical disadvantages of having a large mass in your abdominal cavity. However, if you develop chest pain or severe breathlessness with palpitations or if they are occurring more and more frequently, you should consult your doctor.

Severe itching

It is very common to develop patches of dry, flaky, skin in late pregnancy but some women suffer from a severe form of itching on their abdomen and especially on the palms of their hands and soles of their feet, which does not respond to the usual moisturizers. Occasionally this is the first sign that you are developing cholestasis of pregnancy (see p.424), which is a rare condition in pregnancy caused by bile salts deposited under the skin. If it is severe, it can lead to maternal jaundice, liver failure, premature delivery, and even stillbirth so it is important to report severe, persistent itching to your doctor quickly.

Losing fluid

If you experience small gushes of fluid from your vagina when you make a sudden movement, it is most likely due to stress incontinence. However, at this stage of pregnancy, you should also consider the possibility that your membranes have ruptured (your water has broken) and you are leaking amniotic fluid. If you are unsure, seek advice from your midwife or doctor. They will usually perform a speculum examination to look for pooling of amniotic fluid in the vagina. Occasionally an ultrasound scan may provide additional information about the liquor volume, if the speculum examination is not conclusive. If your membranes have ruptured, both you and your baby are at risk of developing an infection, so if there are no signs of uterine contractions developing within 24 hours, most doctors will suggest that you are induced (see pp.294–97) if you are at least 34 weeks pregnant.

Localized backaches

If you have been bothered by backaches, it may get worse over the next few weeks. In addition to the generalized discomfort you have probably experienced up until now, you may find that mild, generalized lower backaches have been replaced by specific and clearly localized pain caused by a specific disorder such as sciatica that develops toward the end of pregnancy (see below). Although back pain is common during pregnancy, it is important to take severe lower back pain seriously and seek advice from your doctor.

Sciatica is characterized by a sharp, constant, or intermittent pain in the lower back or buttocks, which sometimes shoots down the back of one or both legs. The sciatic nerve is the largest nerve in the body and runs from the spinal cord, through the buttock and into the back of the leg. When crushed or compressed by your baby's head anywhere along its route, the sharp pain is frequently accompanied by numbness, tingling, weakness, and occasionally, a burning sensation. If the pain or weakness becomes severe, you will need to see your doctor to rule out the possibility of a slipped disk.

Gentle maneuvers to encourage the baby's head to change position and relieve the pressure on the sciatic nerve can be helpful, although it is sometimes easier said than done. Improving your posture and performing regular pelvic tilt exercises will bring relief (see p.219) as will yoga and stretching exercises, such as lying down flat on a firm mattress and trying to lengthen your spine by raising your head on pillows or books.

Coccygeal pain is pain in the very lowest part of your spine and tenderness when you press into the natal cleft of your buttocks. The coccyx is a sort of hinged appendage made up of four tiny bones at the end of your sacrum (the

LOCATING THE PAIN
Severe pain in your lower back is likely to have a specific cause and needs appropriate treatment.

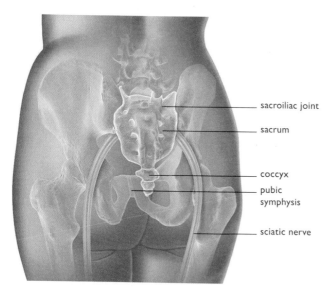

sacroiliac joint

sacrum

coccyx

pubic symphysis

sciatic nerve

TROUBLE SPOTS
Localized back problems are usually due to the softening of ligaments supporting the sacroiliac joint (the pubic symphysis joint between your pubic bones or the coccyx). Pain that radiates down one leg may be due to the baby's head pressing on a sciatic nerve.

large triangular bone at the base of the spine). Lax supporting ligaments can cause the coccyx to become displaced from the sacrum during late pregnancy and during delivery. A previous bruising impact injury to the area, such as a fall, often contributes to the problem. The pain can be excruciating, particularly when sitting; try applying local heat such as a compress and hot water bottles, or taking a hot bath. You can also take analgesics but try to limit them to times when nothing else brings relief.

Sacroiliac pain is usually experienced as a steady pain in the middle or lower back. At the lower end of the spine your sacrum interconnects with the left and right iliac bones at the sacroiliac joints (see picture), to help provide you with a stable pelvic girdle with which to walk and maintain an upright posture. Toward the end of pregnancy, hormones relax the ligaments to prepare for the passage of your baby through the pelvic canal and this, together with the increasing weight of your uterus, can make the sacroiliac joints unstable. This can result in severe pain, especially when walking, standing, or bending. You may need to see your doctor, physical therapist, or osteopath. Meanwhile, wear comfortable, low-heeled shoes and make sure you maintain good posture.

Pubic symphysis dysfunction is pain affecting the symphysis pubis, the narrowest point of your bony pelvic girdle lying just in front of your bladder. As the ligaments around the joint loosen in late pregnancy, the two pubic bones (pubic rami) may rub against each other uncomfortably when you walk and particularly when your legs rotate outward or your knees move wide apart. If you are suffering from this type of pain, avoid straddling movements by keeping your knees together and swinging your legs around from the hips when you get out of the car, bed, or bath. Cold packs over the painful area under your underwear for 10 minutes every three hours may reduce the swelling and pain. Knee squeezing and pelvic tilts (see p.219) or sitting on a birth ball can help ease the discomfort. If the two pubic bones actually separate from each other (a condition called diastasis of the symphysis pubis) the pain can be very severe. Although bed rest and local heat treatment can help, most women who develop this rare complication in late pregnancy are forced to limit their weight-bearing activities and often need to use crutches to move around.

Things to consider

By this time you will probably have started to develop views about how you would like your ideal labor and delivery conducted. You can talk to doctor about your preferences or consider writing a Birth Plan.

To make the process simpler, I have included a brief summary of the major childbirthing philosophies as well advice on compiling your own list of birth preferences, which can be incorporated in your handheld notes. However, before you make any decisions about the way you would like your labor and delivery handled, I suggest that you read the sections on pain relief, monitoring, labor, and birth to form a clear idea of what might be available and what to expect.

The change of title from the more usual "birth plan" is deliberate because I think that it is more useful to think of this document as a list of preferences. To me, a birth plan implies a rigidity of approach—as if it is a set of rules and commandments to which you and your medical caregivers must comply. As we all know, the most carefully laid plans can go awry

TOPICS FOR YOUR BIRTH PREFERENCE PLAN

Most maternity units make great efforts to help women fulfill their wishes during labor and delivery. Good communication with carers will help them work toward your requests wherever possible and prevent disappointments and unrealistic expectations.

Things to consider
▶ Who would you like to be with you during labor and birth—your partner, your mother, or friend? Usually this person has attended childbirth classes with you.
▶ What are your views on being cared for by student doctors (interns and residents)?

▶ Are you prepared to have your membranes ruptured and to receive medications to speed up your contractions (see pp.294–97)?
▶ What are your views about fetal monitoring (see pp.291–92)?
▶ How active or mobile do you want to be during labor?
▶ Would you want your partner to be with you during a cesarean delivery (see pp.360–69)?
▶ Do you have views on episiotomies or perineal tears (see p.330–31)?
▶ Do you want to hold your baby immediately following the birth or after first checks have been made?

▶ Who do you wish to cut the umbilical cord?

Questions to ask
▶ Will you be allowed to eat and drink normally in early labor?
▶ Is it OK to wear your own clothes?
▶ Will you have access to a bath, shower, or birthing pool?
▶ What types of pain relief are available and if there is a 24-hour epidural service (see pp.311–15)?
▶ Are different positions encouraged during delivery?
▶ Is there a time limit for the second stage of labor, even if progress is being made?

and labor can take an unpredictable turn, so the best way to avoid distress and disappointment about the final outcome is to be as flexible as possible.

At my hospital, we offer women a sheet of paper with a selection of issues that they may find helpful to consider prior to delivery. The points are arranged in boxes with space provided for women to make notes after they have considered the issues. We then encourage them to discuss their ideas with their doctor and their partner before putting a list of preferences together. There are some distinct advantages to this method:

• Your list sends a message to the team delivering your baby that you have thought about your labor and want to participate in the decision making.

• Putting together your list of preferences will help you feel more composed, because you will spend time thinking about your views on labor and delivery. If you find that you need more information about the many eventualities that could occur there is still time to seek out the missing facts.

• Early planning also gives your partner the chance to understand your preferences, to know what you expect of him during labor, and, if necessary, to speak on your behalf during labor and delivery.

UMBILICAL STEM CELL COLLECTION

Stem cell collection is a service that is available to parents. I mention it here simply to provide information rather than serving as an advocate for the procedure. At a cost of around $2,000 for collection and around $100 per month of storage, it offers parents the opportunity to have stem cells harvested from their baby's umbilical cord and stored as an insurance against potential illness in later life. Stem cells, which are found in embryonic tissue as well as umbilcal cord blood, have the potential to develop into different types of body cells. Although stem-cell treatment is still being developed, its most promising uses are in diabetes, in degenerative disorders such as Alzheimer's disease, and as a substitute for bone marrow in diseases such as leukemia.

You can find brochures advertizing umbilical cord sampling in some maternity wards, doctors' offices, and on the Internet. If you decide to do this, you will need to make arrangements several weeks before the birth. You will be sent a collection kit to take with you to the hospital and after the birth the sample will be sent to a laboratory to be frozen and stored for possible future use.

Testing cells Stem cells found in the umbilical cord blood can develop into white and red blood cells and platelets. White blood cells from a sample are being tested here.

Disadvantages of birth plans

The sheer length of your birth plan document can be a problem so keep in mind that three closely typed pages of detailed instructions will be harder for your doctor and labor nurses to absorb and follow than a single page of succinct bullet points. When you are putting together your wish list, try to take a positive approach and focus on what you would like to happen during the birth rather than making a negative list of things that you do not want to happen. Some doctors and labor nurses will even be put off by a lengthy list of "Don't do this and don't do that." However, there are other important reasons why birth plans can end up being counterproductive:

• There are so many different ways of experiencing labor that no birth plan can possibly anticipate them all. Indeed, I think that the more detailed the birth plan, the more likely it is for events to fall short of expectations.

• Some women who have spent a lot of time and effort constructing a natural birth plan feel desperately distressed and disappointed if their labor takes an unexpected turn and requires sudden medical intervention. I understand their disappointment but when they tell me that they feel like a failure because they were unable to deliver their baby naturally, it's my turn to feel distressed. No one who nurtures a baby in their uterus for the best part of a year and then delivers it safely into the outside world, by whatever route, can be considered anything other than extraordinarily successful. Similarly, when I hear comments such as, "control was wrested away from me by the doctors as if I did not matter," I do not feel defensive or angry. My immediate concern is that this woman may be at greater risk of developing postpartum depression because she feels negative about her birth experience and believes that she has been let down by her caregivers and by herself.

• Every person who is involved in the delivery of your baby (including you) has a common goal: the safe delivery of a healthy baby to a healthy mother. Although your prenatal doctor wants to help you experience the delivery of your dreams, there may be occasions when your wishes and expectations are just not compatible with protecting your safety and that of your baby. In such situations it is really important that you listen to the advice offered by the experts and understand why it may be necessary to override your birth plan.

My view is that ensuring that your medical team understands that you want to be involved in any decision-making is far more important than any written statement. Women give birth every minute of the day with no birth plan whatsoever—they simply use their voice to express their preferences.

Women give birth every minute of the day with no birth plan whatsoever —they simply use their voice to express their preferences.

APPROACHES TO CHILDBIRTH

A number of childbirth philosophers have had
a significant influence on the way pregnant women
and their maternity caregivers approach labor and
childbirth. Below is a brief summary of their ideas
and how they have been put into effect.

In the 1950s and 1960s, birth became highly medicalized in the Western world and the obstetrician's word was gospel. Hardly surprising, then, that in the decades that followed, advocates of a more natural approach to childbirth sprang up to question what had become the accepted way of giving birth. Collectively, their teaching and ideas have altered many aspects of prenatal and postpartum care, some of which we now take for granted because they have become such an integral part of obstetric care.

▶ **Dr. Grantley Dick-Read**, a British obstetrician, theorized in the 1950s that fear of childbirth was one of the major contributors to pain during labor. He introduced the idea of teaching breathing and relaxation techniques to help reduce fear and tension. He was also the first person to include fathers in prenatal education and encourage them to be present in the delivery room. Prenatal preparation is now considered essential to help women deal with the physical and emotional demands of labor.

▶ **Dr. Ferdinand Lamaze** developed a similar approach in France, using teaching about childbirth and relaxation techniques to counteract labor pains. Lamaze argued that women could be conditioned to deal positively with labor pains in the same way that the Russian scientist, Dr. Pavlov, had trained his dogs to respond to a learned stimulus. Both Dick-Read's and Lamaze's methods have had an enormous influence on the way women now prepare for and manage their labors. Indeed, Lamaze breathing and relaxation techniques are often used alongside preparation and education in many prenatal classes. It seems strange to us now to think that 50 years ago, women frequently went into labor in terror, ill informed, and reliant on anecdotal evidence!

▶ **Frederick Leboyer's** method of delivering babies is based on the theory that many problems in later life stem from trauma experienced at the time of birth. In his book *Birth*

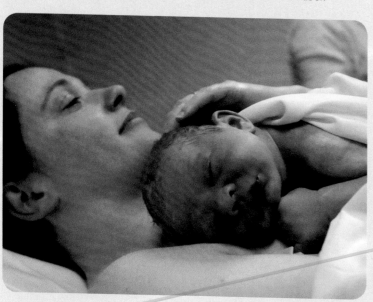

LEBOYER BIRTH The baby is delivered and put immediately into her mother's arms.

without Violence, Leboyer argued that babies needed to be born into calm, gentle, surroundings where noise and sudden movements are kept to a minimum. In a gentle Leboyer birth, the baby is placed immediately on the mother's skin and the umbilical cord is not cut until it has finished pulsating. He also advocates placing the baby in a warm bath immediately after delivery—a soothing replica of the watery world it has just left behind.

Soft lighting in the delivery room and facilities for water births are Leboyer influences, which are becoming common in many hospital maternity units and birthing centers. Although it may not always be appropriate to give birth in obscurity with only one labor nurse in attendance, thanks to Leboyer, newborn babies are no longer held head down by their feet and welcomed into the world with a slap on the bottom.

▶ **Sheila Kitzinger** emerged during the 1960s as one of the key figures in the natural childbirth movement. A founder member of the National Childbirth Trust in the UK (although she is no longer involved with them), Kitzinger argues that women should be allowed to reclaim some control over the way in which they give birth and participate actively in the process of birth. Having said that, she does not advocate natural birth techniques where they may endanger the mother or baby's welfare, but does campaign for the avoidance of unnecessary obstetric intervention. She believes that birth can be made a powerful, positive and personal experience for mothers

ACTIVE BIRTH CLASS Exercises strengthen the hips, pelvis, and thighs for delivery.

even when the labor becomes complicated, involves medicalized pain relief, or ends in delivery by cesarean section. As a result of Sheila Kitzinger's work, women are no longer shaved routinely or subjected to enemas when they go into labor, and episiotomies (see pp.330–31) are no longer routinely performed at delivery.

▶ **Michel Odent** is a French surgeon who uses active childbirth techniques in Pithiviers, which boasts the lowest rate of episiotomies, forceps, and cesarean deliveries in France. His belief is that women who are confined to bed with their legs in stirrups labor slowly and painfully because they face an uphill struggle to deliver their baby. His view is that women should be allowed to return to a primitive state (either upright or on all fours) in labor. Their instincts and loss of inhibitions help produce natural painkilling chemicals in the brain called endorphins, often eliminating the need for drugs to relieve pain.

▶ **Janet Balaskas** founded the Active Birth Movement in 1981 and from her Active Birth Centre in North London organizes a network of private classes teaching women yoga, massage, breathing techniques, and relaxation to help them prepare for labor. The Active Birth Centre, as well as the Healthy Mothers, Healthy Babies Coalition (see Useful Addresses, pp.436–38), emphasize the importance of postpartum support, focusing particularly on practical help with breastfeeding.

Of course, the reality is that many women now dip into some or all, of the above childbirth philosophies and extract the parts that they find the most helpful. There is nothing preventing you, for example, from learning about yoga, or massage, or breathing and relaxation techniques, and then choosing to have an epidural on your delivery day, if the pain become too difficult for you to bear.

BIRTH POOLS Spending your labor in water is now an option in some maternity units although it is somewhat rare and there may be only one birthing pool available.

Water births

Birthing pools and water-assisted births have become increasingly popular over the last 5 to 10 years, largely out of enthusiasm for gentle birth ideas of French obstetrician Dr. Leboyer, but also because they can be effective as a method of pain relief (see pp.319–20); so much so, that some birthing centers have installed birthing pools where you can spend some or all of your labor. My own hospital recently overcame the plumbing problems of an old Victorian building and have installed two on our labor ward and several in the new birthing unit, and they have proved to be very popular. Although the American College of Obstetricians and Gynecologists has not endorsed water births or issued guidelines, most obstetricians recommend that you not enter the pool until you have reached 5cm dilation for the most efficient labor, and leave the birthing pool for the third stage of labor because the relaxing effects of the water could theoretically increase bleeding after delivery or encourage retention of the placenta.

If you think that you might like this option, check in advance whether your hospital provides this facility and how likely it is that you will be able to use it (most birthing centers and an increasing number of hospitals have water-birthing rooms). Remember that a single pool will probably be allocated on a first-come-first-served basis, and it is difficult to predict exactly when you might go into labor. If you are hoping to rent a pool for a home birth, make sure the floor of the room that you are going to use will sustain the weight of a full birthing pool. The water should not be above body temperature (98.6° F/37° C); if it is hotter, you may become dehydrated or be at risk of high blood pressure. You will need to set it up close to where you wish to give birth and ensure that you have facilities both to fill it and empty it after the delivery.

The nesting instinct

With D- (delivery) day approaching, you may develop an urge to tidy up everything around the house, in preparation for the birth of your new baby. Although trying to conserve your energy in preparation for the birth would seem the best plan, many of you will be rushing around, trying to get the house ready. This strange compulsion, known as the nesting instinct, grips many women as the end of their pregnancy approaches. So don't be surprised if you find yourself steam-cleaning the carpet, cleaning out kitchen cabinets, attempting to dust the tallest bookshelves, or suddenly feeling inspired to repaint the living room.

I suspect that this nesting instinct is one of the means we use to help ourselves psychologically prepare for the birth. Many women tell me that they are only able to fully relax when they know that the house is completely ready for the new baby's arrival. Such is the enormous emotional relief and mental release from knowing that things are well prepared that some find they go into labor as soon as they feel their home is ready. Interestingly, women who deliver prematurely sometimes find the adjustment to the practicalities of motherhood more difficult and this may be because they have not had sufficient time to prepare for their baby.

This strange compulsion, known as the nesting instinct, grips many women as the end of their pregnancy approaches.

Air travel in late pregnancy

If you are planning a trip at this stage of your pregnancy you may have some concerns about the safety of air travel and at what point an airline may refuse to carry you. Generally speaking, most airlines do not accept pregnant women after the 36th week of pregnancy, but individual airlines vary. Although many people assume this is because the reduced cabin pressure in an aircraft can induce labor or harm the baby, there is no hard, scientific evidence to support this view. It is my belief that the ruling is based on the fact that some 10 percent of pregnancies are going to deliver prematurely, and airlines want to reduce the likelihood of having to deal with a woman in labor during their flight.

This is an ordeal that you probably do not want to willingly inflict on yourself either. If labor starts, there is a small chance you might give birth on the plane (on a long haul flight) and an even greater risk that you would be faced with the prospect of seeking help from doctors and nurses in a strange place.

If you do decide to fly in your last trimester, make sure that you have taken the precaution of finding out where the hospitals are located at your destination. You should also make sure that the airline is prepared to fly you not only on the outward trip, but on the return one as well.

Planning child care

Without a doubt, one of the biggest financial outlays for working parents is that of child care. Nurseries are not available in every area, and good ones can be heavily overbooked. As a result, many women have to find a child-care professional or nanny to take care of their children, particularly if their working life involves anything other than a strict nine-to-five day. Not surprisingly, many women are forced to conclude, after calculating the full costs of child care, that it is financially not worthwhile to work. This situation needs to change!

Although it may seem premature to be discussing child care before the baby has arrived, I can assure you that it is never too early to give thought to this crucial issue. If you are planning to go back to work, it is particularly important that you give thought to the type of child care that you would both prefer now, so that you have time to prepare while you are on your maternity leave. You will most likely be considering one of the following options:

Maternity nurses are usually employed to live in your home for a few weeks after the baby is born. They will help you with the practicalities of the 24-hour demands of your new baby, including feeding, diaper changing, doing the baby's laundry, and ensuring that you have regular rest periods and a good night's sleep. However, their most important role is to help teach you how to figure out a future routine for caring for your baby. Some women will welcome the organization and routine that a maternity nurse aims to provide, whereas others may find them intrusive, preferring to muddle through in the first few weeks, learning the ropes by themselves. There are plenty of private agencies offering maternity nurses, but a personal recommendation will be the best route to follow. If you decide to employ one, you will need to figure out exactly what you want her to do for you and your family. They are invariably expensive.

Nurseries or day-care centers are run either privately or by the state, and their hours and flexibility vary considerably. As mentioned above, they

> *Without a doubt, one of the biggest financial outlays for working parents is that of child care.*

may be hard to come by, especially good ones, and you will need to research what is available locally at the earliest opportunity. All nurseries adhere to strict legal requirements concerning safety, the ratio of caregivers to children, the suitability of the location, the space available, and the equipment they provide. All of the above points need to be considered carefully but friends' recommendations are also invaluable.

Family day care will take care of your child in their own home and are often paid per child and per hour, so although child care is usually cheaper than a nanny for just one child, if you are asking them to care for two or more children, they can become as expensive as a full-time nanny. Family day care tends to be less flexible because they often have other children to care for, including, though not always, their own. So if your child gets sick, they may not be able to take care of him—a situation that also applies to nurseries. In addition, if you work irregular hours or sometimes have to stay late, the day care may not be able to accommodate your individual needs, in which case you will have to pay for additional child care to fill in the gaps. Local referring agencies can help you find a licensed child-care provider. Licensing regulations for family day care vary from state to state.

A nanny will take care of your child in your own home. Nannies can either live in or live out, they can work for you alone or you can share them with another family. You will need to determine clearly the hours that they work during the day and whether you want them to take on other duties, which may include one or two evenings a week babysitting. Speaking from practical experience, the most important thing about employing a nanny is to make sure that you are completely up front about what you expect him or her to do for you. As with all employee/employer relationships, trust and good communication are crucial. No one can completely protect their family from the proverbial nanny from hell, but you can take some careful steps to make sure this is unlikely to happen to you.

In general, live-in nannies earn less money than live-out nannies because you also provide them with board and lodging. However, when you

BACK TO WORK Leaving your baby for the first time is a little easier when you are totally confident in your child-care provider.

add up the additional costs of having another adult living in your house, you will probably conclude that the overall outlay for a live-in nanny is higher than you originally thought. For some couples, the advantage of a live-in nanny is that they have someone in the house to call on in an emergency, whereas others find that another person in their home is an intrusion. Whatever the case, it is important to remember that if you are repeatedly late home from work, frequently expect them to help out on their weekend off, or fail to pay them for the extra hours they work, you will soon find that your nanny has become disgruntled and is looking for another job.

A live-out nanny is generally the most expensive child care option in terms of the salary you pay, but she goes home at the end of the day and you have your home to yourselves. When you are considering the financial implications of employing a nanny, it is important to remember that most nannies will expect you to deduct federal and local taxes from their salary. Setting up a nanny-share arrangement with another family is one way of reducing the expenditure but this requires a good deal of flexibility.

Start by buying an updated copy of *The Nanny Book: The Smart Parent's Guide to Hiring, Firing, and Every Sticky Situation in Between* by Susan Carlton and Coco Myers and follow up by getting advice from as many friends and acquaintances as possible, even those who are not currently using any child care. Above all, try to have clear ideas about what you are looking for before you start to search. Even if you decide to pay an agency to find a good nanny for you, you must make sure that you interview them and contact their references personally before you agree to employ them. Advertising and interviewing for a nanny is a time-consuming business, so plan ahead to make sure that he or she can start work on your preferred date. Resist the temptation to ask your new nanny to shorten the notice that she gives to a current employer and start sooner. If he or she does this for you, then they can just as easily leave you at short notice in the future.

Au pair are another possible source of child care and are usually much less expensive. In exchange for a room, board, and some pocket money they help you care for your baby or children and take on light housework. However, since they are usually young people from abroad with limited English skills and, possibly, no experience of caring for babies or young children, I think this

> ❝ …the most important thing about employing a nanny is to make sure that you are completely up front about what you expect him or her to do for you. ❞

option is much more suitable for school-age children rather than a newborn baby, particularly since many au pairs want to attend a language school for several hours every day. It is vital that you have total confidence in the person to whom you entrust your child or children and it is unlikely that a young au pair will fulfill your needs in that respect.

Younger grandparents with time on their hands are sometimes more than willing to care for a baby for one or two days a week so this may be an option, especially if you plan to work part-time. There are rich rewards in terms of the close bond that they will develop with their growing grandchild, but you need to be sure that you are not asking more of them than they can manage. Remember, too, that parenting styles differ: their views on, for example, feeding, sleep, crying, and snacks may conflict with your own and it is usually more difficult to address a problem with a relative than it would be with a professional caregiver. Much has changed over the years since your own parents or in-laws were bringing up a small baby and they may need a refresher course on the complexities of strollers and car seats and on important safety issues, especially if they plan to care for your baby in their own home.

If you only need a few hours to yourself for one or two days a week, reciprocal arrangements with a friend with children may be an option, but keep in mind that you will be caring for two or more children rather than just your own on your days off.

Whichever child-care option you choose, you will need to invest a significant amount of time and effort into finding the best affordable solution for you and your family. Start thinking about the options now, since you need to be very clear about what you require and the timeframe involved to find them when you actively start looking. This is usually two to three months before you plan to return to work if you decide on a nanny or day-care facility, but can be considerably longer if you are looking for a place in a popular day-care center.

GRANDPARENT CARE
When it works well, grandparent care can be a rewarding experience for everyone.

SPECIALIZED PRENATAL MONITORING

Most women and their babies are generally fit and healthy in late pregnancy and are unlikely to need any form of specialized monitoring. However, if your pregnancy goes past its term or you have developed, or are at risk of developing, a late-pregnancy complication, your doctor will arrange for you to have some specialized tests.

The kinds of problems that are likely to involve specialized monitoring include high blood pressure; a baby that is not growing well; reduced fetal movements; gestational diabetes that is poorly controlled; or a pregnancy that has gone past the due date (to mention just a few). Of course, the exact tests that you are given will be determined by the particular problem, but on most occasions you will have an ultrasound scan to assess your baby's growth together with a general assessment of the baby's well-being, called a biophysical profile. This will include a non-stress test (NST) and an electronic trace of the baby's heart rate. Many units will also perform a Doppler ultrasound scanning of the blood flow in the uterus, placenta, and in the baby's major blood vessels.

Most maternity units have facilities that provide detailed monitoring tests. The emphasis is on trying to keep the mother-to-be safe in an outpatient setting, albeit under careful scrutiny.

Fetal growth monitoring

If a problem with your baby's growth is suspected, you may be given ultrasound scans at 7–14 day intervals to establish the exact nature and cause. Your baby's head circumference and abdominal circumference will be measured as well as the length of the femur (thigh bone), another good indicator of growth.

There are several types of intra-uterine growth restriction (IUGR, see p.429) each with different causes and effects on fetal growth.

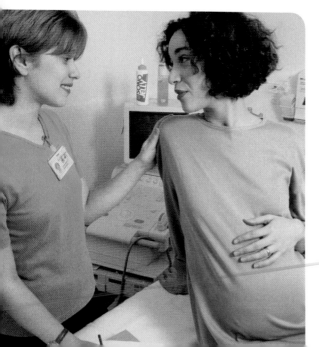

SPECIAL TESTS Usually these are performed in an outpatient setting avoiding the need for prolonged stays in the hospital before your delivery date.

DOPPLER ULTRASOUND SCANS

This highly sensitive form of scanning is performed like a normal ultrasound scan to assess the amount of blood flowing through the blood vessels in the uterus, placenta, umbilical cord, and in the baby's head.

▶ **When blood flow is being diverted to the brain and heart** and away from less vital body organs, the main blood vessels in the brain, particularly the middle cerebral artery, dilate (become less resistant) to accommodate the extra volume. This change can be detected by a scanner and gives a clear message that the baby is exposed to stresses such as low oxygen levels (hypoxia), and that action is needed in the near future.

▶ **Reduced blood flow in the umbilical artery** is a useful predictor of risk when a baby is growing slowly. On a normal Doppler scan blood pressure falls at the end of each heart pumping cycle but the supply to the baby is maintained. However, if the blood flow is interrupted at the end of each cycle, it is a sign that the baby is suffering from a lack of oxygen. If the scan shows blood flowing backward, immediate intervention is required.

high pressure at start of heart pumping cycle | low pressure at end of heart pumping cycle

high pressure at start of heart pumping cycle | absent flow at end of heart pumping cycle

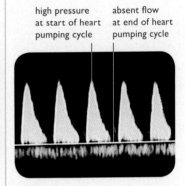

NORMAL Although the blood flow to the baby dips at the end of each heart pumping cycle it never stops. The supply is continuous.

ABNORMAL Small gaps between the peaks and troughs show an absence of blood flow to the baby at the end of each pumping cycle.

In symmetrical growth restriction, growth is restricted early in pregnancy and the baby's head and body are equally affected. Symmetrical growth restriction can be caused by congenital abnormalities, infections such as syphilis and rubella cytomegalovirus (see p.412 and p.415), and toxins such as alcohol, cigarette smoke, and heroin.

Asymmetrical growth restriction occurs after 20 weeks and is often referred to as placental insufficiency. It develops when a maternal or fetal problem affects placental function, and blood flow becomes insufficient to meet the needs of the growing baby. Examples include preeclampsia (see p.426), twin pregnancies and some fetal abnormalities (see pp.415–21) The baby responds by diverting blood to the brain and heart to protect the growth of these vital organs and the head becomes relatively larger than the abdomen as fats reserves in the liver and abdomen are used up. Subcutaneous fat is also absorbed and, as a result, the fetal limbs may become scrawny.

If your baby is growing too slowly with a head-sparing pattern, you may be offered a Doppler ultrasound scan (see above) to assess the severity of the situation. If there is no immediate danger, you may return for more growth scans at 7–14 day intervals. If a subsequent scan confirms that there has been no further growth or that your baby is growing very slowly, you will probably be advised to have an labor induced (see pp.294–97). In some cases, urgent delivery by cesarean section is needed.

Fetal heart recordings

Non-stress tests (NSTs) are pictorial printouts generated by an electronic machine that assess your baby's heart rate as well as the activity of the muscles of your uterus. NSTs are most commonly used during labor to assess how the baby is coping with contractions.

Two belts are strapped around your abdomen: one picks up any activity in your uterine muscles, while the other records your baby's heart rate. The combined recording shows whether the baby's heart rate pattern is normal or abnormal, and whether your uterus is actively contracting or quiet.

Computerized NSTs

This monitoring, for example, the Oxford NST, is used as a guide to the baby's general state of health and is usually reserved for specialized monitoring in labor and delivery. The Sonicaid System sets out a list of criteria that have to be met by the fetal heart rate pattern during a certain time interval. These might include a minimum baseline heart rate, episodes of high and low variability, presence of accelerations, lack of deep decelerations, and fetal movements. It is very reassuring when these criteria are met within a short space of time. The maximum recording time

INTERPRETING AN NST

Babies in utero normally have a baseline heart rate of 110 to 160 beats per minute. This varies constantly by 5–15 beats, except during periods of sleep that last about 30 minutes. This variability is an important sign of well-being: if there is a lack of variability for more than 30 minutes, the baby may be experiencing stress.

Healthy babies also have frequent accelerations or increases in their heart rate (defined as an increase of more than 15 beats per minute for more than 15 seconds), which are usually associated with fetal movements and are sometimes provoked by external stimuli, such as firmly prodding your abdomen, and uterine contractions.

Decelerations in the baby's heart rate may occur following fetal movements or uterine contractions. However, repeated decelerations of more than 15 beats per minute for more than 15 seconds are another sign of possible fetal distress, particularly when they have not been provoked by contractions. That said, there are often variations in the way individuals interpret NSTs, which is one of the reasons for developing computerized analysis of the recordings.

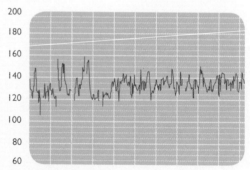

GOOD VARIABILITY The peaks and troughs on this NST show a healthy pattern of accelerations and decelerations in the fetal heartbeat over a short period of time.

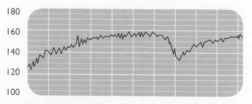

POOR VARIABILITY An NST showing little variation in the fetal heartbeat over a period of more than 30 minutes suggests that the fetus may be suffering from stress.

is 60 minutes, but the computer starts analyzing the NST after 10 minutes. If the criteria have been met, the analysis stops and it is unlikely that another NST will be necessary.

If the criteria are not met within 10 minutes, the computer will continue analyzing the signals every two minutes, until criteria are met. If this does not occur within 60 minutes of recording, this will cause concern about the baby's well-being.

As with all forms of testing, the computerized analysis may produce a false positive result, suggesting that there may be a problem when none exists. Although this invariably causes alarm, I think it's much safer to use a test that eliminates the risk of doctors failing to identify a baby in distress needing immediate help—even if it does throw out the occasional false positive.

Contraction stress test

If the NST shows no change in the fetal heart rate when the fetus moves, the American College of Obstetricians and Gynecologists recommends a contraction stress test. This test measures the reaction of the fetal heart rate to contractions of the uterus. When the uterus contracts, the blood flow to the placenta decreases for a short time. Under normal conditions the fetal heart rate is not affected, but the contraction can decrease the oxygen flow and cause the fetal heart rate to drop if the baby has a problem or there is something wrong with the placenta.

To make a woman's uterus contract mildly, she is given oxytocin. The response of the fetal heart rate to the contractions is measured by Doppler ultrasound. For results to be obtained, three contractions must take place over about 10 minutes, and each must last about 40 seconds. If the response is abnormal, more testing or treatment is needed.

Amniotic fluid volume

The volume of your amniotic fluid is usually assessed by measuring the depth of the pools around the baby using ultrasound scanning. When the total fluid volume (measured by adding the depths of the pools in four separate areas) is reduced, intervention and prompt delivery is usually advised.

Exactly why fluid volume surrounding a baby near to term is so important in determining the outcome of pregnancy has been difficult to establish scientifically. The logical explanation is that increased or reduced fluid volume are indications that the fetal kidneys and metabolism are not working optimally, but testing these important functions while the baby is still in utero is nearly impossible. Nonetheless, my personal experience is that evidence of reduced fluid volume should always be taken seriously. When a pregnancy is near term or past its due date, I almost always make the decision to deliver the baby when the fluid volume is very low.

Biophysical profile

This used to be the first test to recognize the importance of using a combination of factors to estimate fetal well-being. It uses a scoring system to assess fetal breathing movements, body movements, musular tone and posture, the volume of amniotic fluid, and the results of an NST. Today, a reduced volume of amniotic fluid and sub-optimal NST analysis are considered the most important indicators that a baby may require prompt action. If, for example, your fluid volume is low but the NST is fine, your doctor will automatically look at additional parameters to help them decide whether they should watch and wait or if they should intervene.

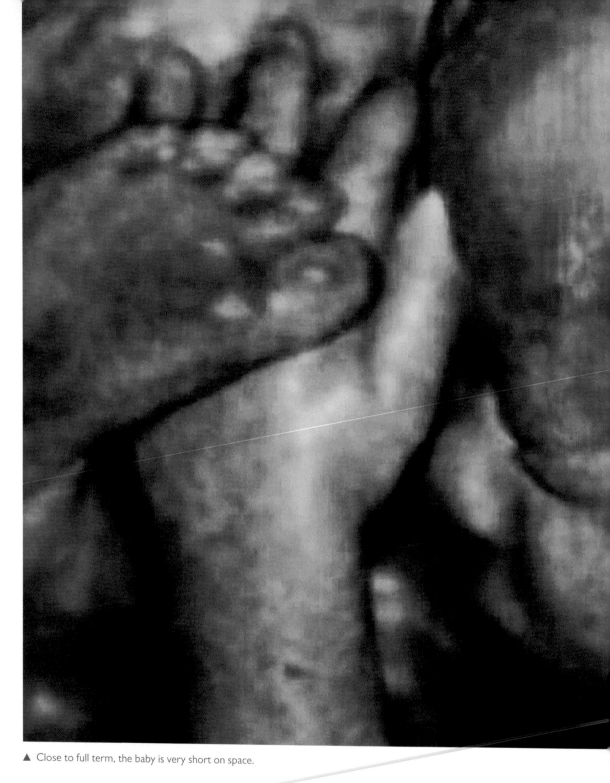

▲ Close to full term, the baby is very short on space.

1	2	3	4	5	6	7	8	9	10	11	12	13	14	15	16	17	18	19	20

▶ **WEEKS 0–6**   ▶ **WEEKS 6–10**   ▶ **WEEKS 10–13**   ▶ **WEEKS 13–17**   ▶ **WEEKS 17–21**

▶ **FIRST TRIMESTER**   ▶ **SECOND TRIMESTER**

▶ **WEEKS 35–40**

The developing baby

YOUR BABY IS QUITE A SNUG FIT INSIDE YOUR UTERUS NOW. THE BABY IS USUALLY CURLED UP TIGHTLY, HEAD POINTING DOWNWARD, WAITING FOR LABOR TO START. MOVEMENTS ARE MORE LIMITED BUT YOU WILL PROBABLY NOTICE REGULAR CHANGES IN THE CONTOURS OF YOUR BELLY AS THE BABY SHIFTS POSITION.

Your baby continues to gain weight steadily during this final stage of pregnancy, almost entirely due to more fat being laid down under the skin, around the muscles, and around some of the abdominal organs. The average full term baby will have a plump rounded appearance and will weigh 6½–9lb (3–4kg); it is common for boys to weigh a little more than girls. Although your baby is too cramped to move freely now, you should still be able to feel movement and you may experience the occasional sharp twinge as your baby throws a punch and indents the uterine wall. Remember that any sudden change in the pattern of your baby's movements needs to be seen to urgently.

Most of the lanugo hair has disappeared, although some slippery vernix is still present to help the passage of the baby through the birth canal. Post-mature babies commonly have cracked and peeling skin because they have been without their protective coating of vernix for longer periods of time; some even have scratch marks on their faces from their long fingernails. The amount of hair babies have at birth is variable, ranging from completely bald, to downy patches to a complete head of hair. Most of it is lost in the first weeks but this will hardly be noticeable because it is replaced simultaneously with more hair.

Ready for birth

The lungs are now fully mature, and the baby continues to produce large quantities of cortisol to provide plenty of surfactant in the lungs and make sure that the transition to breathing air in the outside world goes smoothly. The heart is beating at a rate of 120–160 beats per minute. Dramatic changes will occur in the heart and circulatory system at the time of delivery when the baby takes the first breath (see pp.378–79).

not life size

At 38 to 40 weeks your baby will weigh between 6½ and 9lb (3–4kg) and will measure as much as 20in (50cm) from the crown of the head to the tip of the toes.

| 21 | 22 | 23 | 24 | 25 | 26 | 27 | 28 | 29 | 30 | 31 | 32 | 33 | 34 | **35** | **36** | **37** | **38** | **39** | **40** |

▶ WEEKS 21–26   ▶ WEEKS 26–30   ▶ WEEKS 30–35   ▶ WEEKS 35–40

▶ **THIRD TRIMESTER**

The digestive system is now ready to accept liquid foods. The intestines become filled with a dark green sticky substance called meconium, which is made up of dead skin cells, remnants of lanugo hair, and secretions from the baby's bowel, liver, and gall bladder. This meconium plug will normally be passed in the first few days of life, but if your baby becomes distressed or frightened before delivery, it may have a bowel action into the amniotic fluid. If meconium is seen in the amniotic fluid after the water has broken, it is evidence that the baby has already been stressed and may need to be monitored closely during labor (see pp.291–92). For boys, the testes descend into the scrotum during this period, which explains why premature babies are often born with undescended testes.

Your baby's immune system is now capable of protecting against a variety of infections, but this is mainly due to the transfer of antibodies from your own blood. After birth, babies continue to receive antibodies from breast milk. One of the main reasons for trying to establish successful breastfeeding is that you can continue to give your baby protection against infection in the first few months of life, before he or she is capable of producing antibodies.

How the head is adapted

Your baby's head is relatively much smaller than it was earlier in the pregnancy, but the circumference is still as big as its abdomen. By full term, the head remains one of the largest parts of the baby's body, so delivering it safely through the birth canal during labor is an important consideration. This is one of the reasons why the fetal skull bones do not fuse together until much later in neonatal life. Although the baby's brain needs to be protected by bone, these are quite soft compared to an adult skull and can slide over each other and overlap. This allows the head to adapt to the shape of the mother's pelvis and greatly eases the passage through the birth canal and vagina.

In a normal pregnancy, the fetal head will move down into the pelvic brim and become engaged in preparation for the start of labor. In first pregnancies, head descent can start as early as 36 weeks, whereas in second or third pregnancies, the fetal head may not engage in the pelvis until immediately before labor starts.

The placenta at term

Your placenta now looks like a discus and measures about 8–10in (20–25cm) in diameter and is about 1in (2–3cm) thick. This large surface area facilitates the transfer of oxygen and nutrients to your baby and the passage of waste products from baby back to mother. At term, the placenta will weigh around

1½lb (700g), just less than one-sixth of the fetal weight. Although nearly 45 percent of pregnancies are undelivered at 40 weeks, most doctors will advise that the pregnancy not continue after 41 weeks. At this stage, they will probably suggest an induction of labor (see pp.294–97) because the placenta will no longer be functioning effectively. Its reserves are now pretty well exhausted, which is why the risk of having a stillborn baby is increased in post-mature pregnancies. After 42 weeks, your baby can be more reliably cared for in the outside world.

Your changing body

If your baby's head has started to engage, or settle into the pelvis, your belly will appear to be lying lower in your abdominal cavity. There is sometimes quite a noticeable change in your body shape and you may hear people remarking that you have "started to drop."

This does not mean that you are about to go into labor and literally drop your baby; you may still have several weeks to go. It is merely an indication that your uterus and your baby are both getting prepared for labor. As I mentioned earlier, if this is your first baby, engagement is likely to occur sooner rather than later. This is because the muscles of your uterus are tight since they have not been stretched by a previous labor and are able to exert more pressure on the baby's head. Also, the arrangement of the pelvic bones is slightly altered after a previous vaginal delivery and this may delay engagement.

If your baby's head has started to engage, your breathing may be easier now and the decrease in pressure under your diaphragm and ribs may make it possible to eat a complete meal. This is why engagement is sometimes referred to as lightening (of the abdominal pressure). The downside to this change is that the baby's head (if head down) is pressing directly on your bladder. You will need to urinate frequently, and at night your increased bulk will make getting in and out of bed to use the toilet a major undertaking.

Further loosening of your pelvic ligaments and joints will occur in preparation for the birth and can result in a variety of aches and pains in the pelvic area and an increasingly sore lower

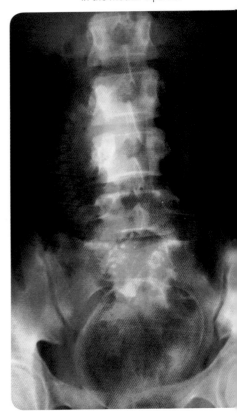

READY FOR BIRTH
A color-enhanced X-ray shows a full term baby with its head down and engaged in the mother's pelvis.

abdomen. The problem is made worse because your posture changes again as your baby moves into your pelvis. Your own weight gain usually slows down and may even stop during the final few weeks of pregnancy, although the baby can put on as much as 2¼lb (1kg). But if you suddenly feel swollen and puffy, see your doctor immediately to check that you are not developing preeclampsia (see p.426).

Hormone effects

The pregnancy hormones produced by the placenta will bring about more changes to your body. Your breasts will swell even more and may fill up with milk, even squirting small quantities at unexpected times. However, not every woman experiences this symptom and you may never see a drop of milk or colostrum leaking from your breasts until after the baby has been born and you start to breast-feed. Many women notice an increase in their vaginal discharge, which may look slightly brown or pink, particularly if you have had sex recently. This is usually nothing to worry about and is merely another sign that your cervix has become softer because of the increased blood supply it is receiving. As a result it can become bruised and bleed slightly even with light contact. However, any bright red vaginal bleeding, particularly if it is accompanied by pain, should be reported urgently.

...this is not the time to worry that you cannot bother your doctor again because you have already called her three times this week...

How you may feel physically

You have now reached your maximum size and find that you bump into things and feel quite clumsy. You need to take care when going up and down stairs because your center of gravity has altered significantly and you can no longer see your feet.

During these last few weeks of your pregnancy, Braxton Hicks' contractions (see pp.237–38) will be a constant reminder of the fact that labor could start at any time. True labor contractions are much stronger and more painful, but if you doubt what you are experiencing, always seek advice. Your doctor will encourage you to come into the maternity unit and be checked over, rather than stay at home feeling anxious. This is not the time to be ignoring abdominal pain and hoping that it will go away. Nor is it the time to worry that you cannot bother your doctor again because you have already called her three times this week with the same symptoms. It does not matter how many false alarms there are, it is essential that uterine pain is always investigated promptly and carefully.

No matter how much you are trying to rest at this stage, you will probably still be feeling tired because you are unlikely to be getting enough of the continuous, uninterrupted sleep that you need to restore your mental and physical energy. Quality sleep involves cycles of four different stages from light sleep to deep sleep followed by REM (Rapid Eye Movement) sleep, which is when you dream. If you wake up during any of these stages, the sleep cycle goes back to stage one. As a result, you miss out on important deep sleep and REM stages and wake feeling fatigued. Even if you manage to doze and sleep for long stretches of time, the repeated lack of quality rest will mean that you become progressively weary and exhausted.

Impatience and frustration

The most common emotions women experience at this late stage are impatience that they are still waiting for D–day to arrive and frustration because there is no way of knowing when it will occur. By now, many women want to reach the end of their pregnancy, however enjoyable it has been.

MAGNIFICENT By the end of your pregnancy, your belly can be a source of amusement coupled with sheer amazement.

If you are feeling like this, remind yourself that the end is in sight. Even if you are overdue, there is only a set number of days left to go. Nevertheless, one of the other pressures you will have to deal with at this point is that the closer you get to your due date, the more you will have to field questions about when you are due, and telephone calls asking if you have had the baby yet. Your friends and family obviously mean well when they show such interest, but many of you will feel even more irritated by being constantly reminded that your long-awaited baby has still not arrived.

Labor represents an enormous emotional and physical challenge and I suspect that many of you will see it with a mixture of excitement and apprehension, because it is almost impossible to predict exactly how it will proceed and how your body will respond. As several women have commented to me recently, it is easier to train for a marathon than it is for labor. If you still feel seriously apprehensive, have a detailed discussion with your doctor so that they can understand and address your fears.

Your prenatal care

During the last weeks of pregnancy, you will have checkups every week. Be sure to tell your doctor about any new or unusual symptom, any issues that are worrying you, or anything that does not feel quite right with the baby, even if you cannot put your finger on the exact problem.

The usual prenatal tests will be performed and you may have another blood count if you have been feeling very tired, or have been taking an iron supplement for anemia that has been previously diagnosed. Your doctors will look for obvious signs of severe fluid retention (edema) and if you have noticed any sudden swelling in your fingers, ankles, or face and there is any suspicion that you might be developing preeclampsia, they will arrange for blood tests and more frequent checks of your blood pressure. Women with late pregnancy complications might be referred to a specialist who can provide more detailed monitoring and expertise (see pp.256–59).

Your doctor will palpate your abdomen carefully at each of your prenatal visits and record the findings in your medical reports. At this stage of your pregnancy, it becomes important to assess the lie and presentation of your baby and whether the presenting part has started to engage in the pelvis. These findings influence the plans that are made for your labor, may determine the type of delivery that is best for both you and your baby, and will help your doctor assess your progress when labor is underway.

...a deeply engaged head is usually a good sign that labor will be swift and un-complicated.

Is the head engaged?

Pregnant women are often unsure about what the term "engagement" means, so a brief explanation of engagement and the terms used to describe it in your medical records may be useful. Strictly speaking, the baby's head is not properly engaged until more than half of it (three-fifths of the head) has passed through the pelvic brim in the mother's abdomen. The best way to assess engagement is with an abdominal palpation.

• (High/Free): if your doctor can feel all of the head in the mother's abdomen, the baby's head is high or free.

• (NE): when she can feel more than half (three-fifths or four-fifths) of the head above the pubic bone, she will determine that the baby is not engaged in the pelvis or state exactly how much of the head is palpable abdominally, for example, three-fifths or four-fifths.

• (E): when she can feel less than half the head (just two-fifths of it) above the pubic bone, the baby's head is engaged. If there is only one-fifth or no fifths left to feel, the head is deeply engaged.

The other way to assess engagement is by performing a vaginal examination. and these are also performed regularly during labor to monitor the progress of your baby's downward descent through the pelvis. There are occasions when a prenatal vaginal examination is helpful. For example, it can be difficult to assess the height of the head abdominally in very overweight women near full term. Similarly, when the baby's head is very deeply engaged and the shoulder is positioned just above the pelvic brim, it can be difficult to decide which part of the baby is being palpated. Accuracy is important here because, if you go into labor with your baby's head high and free, there is the potential for serious complications such as cord prolapse (see p.430). On the other hand, a very deeply engaged head before labor is usually a good sign that the labor will be swift and uncomplicated.

Occasionally, an ultrasound scan may reveal an obstruction to your baby's head that stops it from engaging, such as a low-lying placenta or a uterine fibroid. If the baby's head cannot travel past the obstruction, delivery usually has to be by cesarean section. Occasionally, the high head may be because your pelvis is too small to allow the fetal head to engage—the medical term for this problem is cephalopelvic disproportion (CPD). This is always a relative diagnosis: a woman with an average-sized pelvis may develop CPD if her baby is very large, but in her next pregnancy may have a smaller baby and not

ENGAGEMENT ··

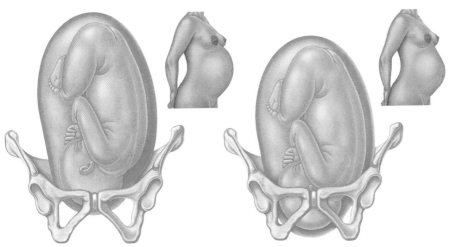

NOT ENGAGED
The baby's head is still at the brim of your pelvis and the uterus is at its maximum height.

ENGAGED The baby has dropped into the pelvis, producing a sudden change in the outline of your belly.

NOT ENGAGED

ENGAGED

experience any disproportion. True CPD—where the mother's pelvis is too narrow to allow even the smallest of babies to engage—is rare.

Having said that, high heads can settle down and engage in the pelvis right up until the last minute of pregnancy. Adopting a wait-and-see attitude is usually the best policy when it comes to engagement.

Presentation and lie

Your prenatal records will probably contain written entries by the time you reach term and the notes may include own abbreviations to describe the lie, presentation, and position of your baby, you may be feeling somewhat confused at this stage. The following information should help to give you a clearer picture of your baby's position.

• As I have mentioned previously, the lie of your baby is either longitudinal (in a vertical position in your uterus), transverse (lying horizontally), or oblique (in a diagonal position) (see p.213).

• The presenting part of your baby is the part that lies closest to the cervix and the one that will therefore present itself to the world first. In a longitudinal lie, this can be either cephalic (head down), or breech (bottom downward). When the lie is transverse or oblique, there is no presenting part. By 35–36 weeks most babies are cephalic presentations and by term 95 percent will be head down, four percent will be breech presentations, and one percent will be transverse or oblique.

• The position of the baby refers to the relationship between the baby's spine and the back of its head (occiput) and the inner wall of the uterine cavity. Hence the baby's postion can be anterior (in front), lateral (to the side), posterior (at the back), and facing either to the right or left. An anterior or lateral position at the onset of labor is considered normal.

• The attitude of the baby describes the relationship between the head and the rest of the baby's body. The normal attitude is fully flexed or curled up with the limbs and head tucked into the body. If the baby's head and neck are extended backward, this will result in an abnormal brow presentation (see p.429).

Posterior presentation

If the baby takes up a posterior position in the pelvis, the occiput rotates toward the mother's spine and the baby is effectively facing forward, labor is likely to be longer and more difficult. The baby's head simply does not fit well into the pelvis in a posterior position and the normal mechanics of labor are interfered with. When I was training I was taught to think of posterior positions as similar to trying to fit the right shoe on the left foot—possible, but clumsy.

> *When I was training I was taught to think of posterior positions similar to trying to fit the right shoe on the left foot...*

Fortunately, only about 13 percent of babies (and they are usually first babies) start labor in the posterior position, and about 65 percent of these turn during labor and can be delivered normally. Occasionally a baby is born spontaneously positioned face up.

A breech presentation

If your baby is breech at 35–36 weeks, it is still possible that it may turn spontaneously. If your baby is breech at term, which occurs in about four percent of pregnancies, a vaginal delivery may be possible (see p.357).

During the first stage of labor, a breech baby will not dilate the cervix as effectively as a cephalic presentation and, as a result, labor is more likely to be prolonged and the baby more prone to develop distress, requiring emergency interventions. If your water breaks and your baby is still in a breech position, you are at risk of a cord prolapse and need to go to your hsopital immediately. This is because a breech baby does not fit the pelvis as snugly as the head would, so the umbilical cord can slip past the baby's bottom or legs and fall through the cervix, which is a potentially life-threatening situation for your baby. Another major concern occurs during the second stage of labor, because there is no way of knowing whether your pelvis can accommodate the largest part of the baby's body—the head—before the limbs and trunk have been delivered. There are three main breech positions:

• In a frank breech (or extended breech), the legs are flexed at the hip and the knees are extended straight up in front of the baby—this is the best position for a vaginal delivery.

• In a complete breech (or flexed breech), the legs are flexed and folded tightly in front of the baby — a vaginal delivery is sometimes possible.

• In a footling breech, the legs are extended below the baby and one or both feet are presenting first—a vaginal delivery is inadvisable.

You may be advised to deliver by caesarean section, as recent research suggests that this may be the safest mode of delivery for a breech baby, in terms of both labor and delivery complications and possibly the long-term neurological development of the baby. If you are anxious to have a vaginal delivery, it is worth having your doctor try turning the baby around manually, a procedure called external cephalic version (ECV). The procedure is not suitable in every case and is advised against if you have had complications in a previous or current pregnancy.

ECV done in the labor ward or specialized outpatient care at about 37 weeks and should only be attempted by an experienced practitioner. There should be immediate access to emergency delivery if complications arise. You

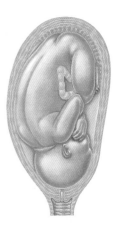

ANTERIOR PRESENTATION

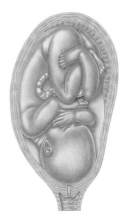

POSTERIOR PRESENTATION

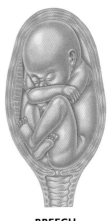

BREECH PRESENTATION

YOUR BABY'S POSITION

Your baby's position is determined by where its occiput and spine are lying in the uterus as the baby passes into the pelvic brim. The six most common positions are shown here along with their abbreviations and percentages indicating how often they occur. Direct anterior (OA) and direct posterior (OP) positions, in which the baby is facing directly toward or away from your spine, are rare. Breech presentations are defined by the position of the baby's bottom (sacrum). The most common is right sacro anterior (RSA) with the baby's spine toward the front of the uterus.

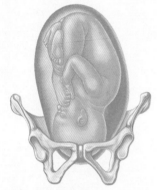

LOT: LEFT OCCIPITO-TRANSVERSE (40%) The baby's back and occiput are positioned on the left side of the uterus at right angles to your spine.

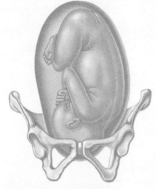

LOA: LEFT OCCIPITO-ANTERIOR (12%) In this position the baby's back and occiput are closer to the front of your uterus on the left.

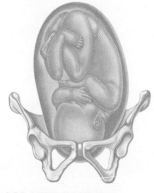

LOP: LEFT OCCIPITO-POSTERIOR (3%) The baby's back and occiput are toward your spine on the left side of your uterus.

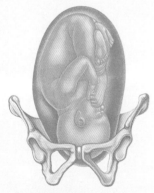

ROT: RIGHT OCCIPITO-TRANSVERSE (25%) The baby's back and occiput are at right angles to your spine on the right-hand side of your uterus.

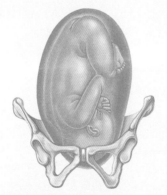

ROA: RIGHT OCCIPITO-ANTERIOR (10%) The baby's back and occiput face the front of your uterus on the right-hand side.

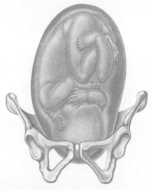

ROP: RIGHT OCCIPITO-POSTERIOR (10%) In this position, the baby's spine and occiput face your spine on the right-hand side of the uterus.

will have an ultrasound scan and a NST of the baby's heart and be asked to empty your bladder before the obstetrician attempts to rotate your baby inside the uterus by gentle sustained pressure, keeping the baby's head flexed. Raising the end of the bed may help disengage the breech from the pelvis and you may be given a subcutaneous injection of terbutaline to relax the uterine muscles. The scan and fetal heart trace are usually repeated after the procedure and if your blood group is Rhesus negative, you will be given a Rhogam injection. About 50–70 percent of ECVs are successful and the baby remains head down.

When your baby is overdue

If your due date arrives and you have not yet gone into labor, your pregnancy is referred to as post-mature or overdue. about 45 percent of women are still pregnant at 40 weeks, but the majority deliver during the next week and only 15 percent go beyond 41 weeks.

POST-MATURE Your doctor will assess the position and engagement of your baby's head.

What happens when your baby is overdue depends on the type of birth you are hoping for and the policy on induction of labor (see pp.294–97) at your maternity unit. The sequence below is what is offered in my own hospital and although timings may vary, the basic procedures will be the same in all prenatal clinics.

• The first thing that your doctor will do is to check the accuracy of your expected delivery date, using a combination of last menstrual period (LMP) dates and early scan measurements—wherever available. Many doctors advise you to continue until the 41-week checkup, either at home or in the hospital, after which time they will assess the position and engagement of the fetal head and may offer an internal examination to assess how ripe the cervix is.

• If the head is down and the cervix sufficiently soft and dilated, your doctor will usually offer to "sweep" the membranes around the baby at the top of the cervix to release chemicals called prostaglandins, which may help start uterine contractions. If this is not an option, your doctor will discuss the pros and cons of induction (starting labor artificially) versus waiting a little longer.

• If you decide to wait, you will probably be asked to visit the maternity outpatient unit for a post-maturity assessment (see below) around or just after your due date. If the assessment is satisfactory and no problems are found, you may want to continue your wait until 40 weeks plus 10 days, when you will most likely induce labor. In my experience, it is unusual for women to choose to continue beyond 42 weeks, but if they do, regular fetal heart traces and biophysical profile scoring will be advised.

• If at any stage, your post-maturity assessment is abnormal in any way, your obstetrician will discuss the need to induce labor and deliver your baby. Although research shows that induction is likely to result in a higher incidence of prolonged labor and instrumental deliveries, there are conclusive studies showing that it does not increase cesarean-section rate. On the other hand, leaving the pregnancy until the end of 42 weeks may increase the risk of fetal distress or unexplained stillbirth if the placenta stops functioning well. Even when the placenta is working well and the baby is continuing to grow after 41 weeks, there is also the risk of a difficult or obstructed labor if the pregnancy is left for too long because the baby becomes increasingly large.

It is unusual for the assessment findings to be so dramatic that they prompt an immediate decision to deliver the baby but, if this is the case for you, you may be advised that cesarean section is the safest course of action. That said, everyone involved in taking care you will do his or her very best to achieve an induction of labor and vaginal delivery if this is what you prefer.

• If you need to have a post-maturity assessment, your doctor will start by checking the exact size of your baby (see p.256–57) and the amount of amniotic fluid in the uterine sac (see p.259). The results may prompt them to perform a detailed Doppler blood flow assessment (see p.257). The scanners will also perform a biophysical profile of the baby (see p.259) looking carefully at limb movements, muscular tone, breathing movements, and heart rate pattern in order to help assess the baby's general well-being. Most hospitals will also offer a computerized fetal heart trace (see pp.258–59), a pictoral printout generated by an electronic machine that assesses your baby's heart rate to see whether the criteria for a healthy baby are met within a certain period of time. The obstetrician will also look at the placenta and grade its appearance and texture, which will give a rough indication of how well it is working. All of these tests are prone to error and should be considered as indicators, rather than as definite diagnosis.

Having sex is another option that, theoretically, should encourage labor to start…

COMMON QUESTIONS IN LATE PREGNANCY

▶ How can I prepare for the pain of labor?

You are likely to have plenty of conversations with your doctor and birthing instructors about the different pain relief options available to you during early labor (see pp. 308-23). In early labor, many women find TENS machines, breathing exercises, massage, and having a long soak in a warm bath especially soothing and relaxing. If you have been feeling frightened about the birth, discuss your anxiety with your doctor so they can understand your fears and respond to them.

▶ My baby seems less active than before. How can I be sure that everything is OK?

Many babies change their pattern of movement during the last few weeks of pregnancy, usually because there is no longer sufficient space in your uterus for them to move around and kick as freely as before. If you have not felt the baby moving during the last few hours, try lying down in a quiet place on your left side or have something cold to drink. If this does not work, then you need to seek advice immediately. You will most likely be advised to go to the hospital for a fetal heart trace (see pp.258–59), to make sure that all is well. You may also be given a fetal movement or kick chart to fill in over the next couple of days but, as I have mentioned before, I have mixed feelings about these because I firmly believe that babies develop their own individual patterns of movement late in pregnancy, and it is a change in this pattern that needs to be reported to the doctor and investigated promptly, not the actual number of movements or kicks. If you are at 37 weeks or more, your doctors may advise an induction, as there is emerging evidence that there may be an increased risk of "term" stillbirth with reduced fetal movement.

▶ What can I do to help start my labor?

Although the following ideas are by no means proven, some are worth a try. Eating spicy food is traditionally thought to help initiate action, presumably because it may encourage a bowel movement. Taking castor oil is a much less enjoyable version of the same idea.

If you have not felt like moving around much for the last few weeks, some exercise may help move the baby down a bit inside your pelvis. The more pressure exerted on your cervix, the more likely that labor will start. Try taking a long walk to see if this helps start labor.

Having sex is another option that, theoretically, should encourage labor to start, because semen contains prostaglandins—chemicals similar to those in the suppository used to induce labor. So if you're not too exhausted, give it a try. Nipple stimulation is often cited as a way of inducing labor because it releases oxytocin, which stimulates the uterus. That said, you or your partner would need to stimulate your nipples for around one hour, three times a day, to have any significant effect, so I suspect that this is not likely to be the most helpful option!

▶ How can I distinguish between vaginal bleeding and a show?

The simple answer is that you cannot know until you have been fully checked out, so speak to your doctor immediately. A typical bloody show includes the loss of a mucuslike plug, mixed with fresh red and old brown blood, but it is always best to make sure that there is no other cause of fresh bleeding, particularly if it is accompanied by sudden abdominal pain. (See placenta previa, placental abruption, and prepartum hemorrhage, p.428).

> ❝ *Many babies change their pattern of movement during the last few weeks of pregnancy...* ❞

PREPARING FOR A HOME BIRTH

If you are planning a home birth, make sure that
the practical arrangements are not left to the last minute. You
don't want to find yourself hunting for towels at the bottom of
the laundry basket in the middle of a contraction.

The certified nurse-midwife who will attend to your delivery will bring all the necessary medical equipment with her (see list, below) but make sure you have discussed with her, several weeks before the expected date, exactly what she expects you to provide. She will also be able to give you some useful tips about additional, nonessential items that you may find helpful during your labor and delivery.

DOCTOR'S DELIVERY TOOLS

EQUIPMENT
▶ blood pressure monitor
▶ thermometer
▶ fetal Doppler
▶ gloves
▶ oxygen cylinder
▶ baby resuscitation
   equipment
▶ antiseptic solutions
▶ intravenous set
▶ urine test sticks
▶ scissors
▶ stitching equipment

PAIN RELIEF
▶ opiate drugs such as
   Demerol
▶ local anesthetic

PACK A MATERNITY BAG
You might think this is a strange idea since you are not actually going to the hospital, but packing a maternity bag (see p.277) is a good way to be sure that your essential personal items are gathered together in one place, ready for when the time comes. This will also be useful if things do not go according to plan and you end up in the hospital for any reason.

ORGANIZE YOUR BABY EQUIPMENT
Assembling a bag or assigning a drawer or cabinet for all the basic items you will need for the new baby is a very sensible idea. You may be planning to deliver your baby at home, but very soon after that you will be going out with your new baby, so don't forget to include a portable crib or cradle and car seat.

PLAN WHERE YOU WANT TO DELIVER THE BABY
The essential requirements are comfort, warmth, and cleanliness. Make sure that you have plenty of plastic sheeting to protect bedding, mattresses, chairs, and the floor and large plastic bags to clean up any trash. You will also need lots of towels, hot water, soap, bowls, and sponges. If you are going to use your bed, it should be easily accessible to your midwife from both sides. Several changes of bed sheets will come in handy, together with extra pillows or cushions.

EXTRA COMFORTS
You may want to have a birthing ball, bean bag, or large floor cushions available to you during labor. There is no reason why you should not use a birthing pool at home, but this will have to be arranged well in advance. Although you may prefer to have dim lighting to help relax during labor, your certified nurse-midwife will need a good source of light to see what she is doing, especially after the delivery when you may need stitches, so have a portable, directable lamp close at hand.

Things to consider

Most of the considerations at this stage of pregnancy tend to be practical, concerning preparations for the birth and the period afterward. That said, your most pressing concern is likely to be knowing for sure when your labor has really begun.

There are lots of possible indicators but no hard and fast rules so interpreting exactly what is happening to you when you are in labor, especially if it is the first time, is not easy. I've included the key signs and symptoms at the beginning of the Labor and Birth chapter (see pp.283–86) so I suggest you turn to those pages now. As always, seek advice and reassurance whenever you feel unsure—no one will accuse you of wasting time if your symptoms turn out to be a false alarm.

Clothes for comfort

In the last few weeks of the third trimester, you may find it useful to buy some special maternity underpants in order to accommodate your vastly enlarged abdomen. These garments will never win any fashion prizes, but they can make a world of difference for your comfort, particularly if you have been suffering with underwear that is continually slipping down or riding up in uncomfortable positions. Some women also find that wearing a light girdle, positioned under their belly (not constricting it), helps support their abdomen and reduce symptoms of backaches and weariness. Both of these items can be found in maternity stores and online.

If you are planning to breast-feed, you will need a couple of good breast-feeding bras so that you are comfortable in the hospital and when you return home. These need to be properly fitted by a trained fitter who knows how much room to allow since your breasts will expand when your milk comes in. You will also need to find nightgown with a low button opening on the front that will allow you to breast-feed. This may sound overly complicated, but breast-feeding really does become easier if you feel comfortable and at ease. If every meal involves struggling in and out of your clothing, you will become increasingly irritated—especially when you are embarking on your third meal of the night at 5am.

This is a good time to choose clothes to wear into the hospital—something comfortable that you don't mind getting dirty—a loose-fitting dress, or a baggy T-shirt with sweatpants. You will also need a fresh set of pajamas for after the delivery and a bathrobe and slippers for the postpartum ward.

> *...maternity underpants won't win any fashion prizes, but they can make a world of difference for your comfort.*

Getting ready for your baby

If you are going home from the hospital by car, the law requires that you have a special baby seat in which to take your baby home. Your partner can bring it in when you are ready to leave, together with a hat, outdoor clothes, and shawl or light blanket in which to wrap your new baby. You do not need a cradle, carriage, or stroller until you arrive home.

You may already have arranged your baby's room in your home or may choose to have your baby in your bedroom for the first few weeks, when he or she will need several meals each night. You will need a crib, or a cot, with a waterproof mattress, cotton sheets, and warm cotton thermal blankets. Don't use a pillow for your baby and if you choose to have padding inside the crib to prevent your baby hitting from the edges, it should be free of ribbons, tassels, and bows so that there is no risk of your baby putting them in his or her mouth or becoming tangled up.

If you plan to breastfeed, you will probably do this in bed at night, so make sure you have plenty of soft washcloths on hand to cope with the inevitable regurgitated milk. You may be planning to move your baby into a separate room in a while, so make sure there is a comfortable chair to sit on for nighttime feedings. If you have decided to bottle-feed, then you will need a sterilizing unit and some of bottles and nipples available when you return home (see p.229). To reduce the risk of gastrointestinal problems it is best to make the bottles up as and when you need them rather than storing them for any length of time.

Whichever sort of diapers you have chosen to buy (see p.229), make sure you have a good supply and also cotton balls and baby wipes for cleaning up your baby. Your newborn baby has very sensitive skin and the best way to avoid getting diaper rash is to use cotton balls and cooled, boiled (sterile) water at first. Wipes are useful when you are traveling or out of the house, but opt for the gentlest (hypoallergenic) type for your newborn baby. Some baby products are bulky to carry home and it is worth considering online shopping—many supermarkets and other retailers can supply a wide range of baby products directly to your front door.

Stock your cabinet and freezer with easy-to-prepare food or prepared meals to get you through the first few days when you return home. This is not

66 *Don't forget that grandparents can make an enormously valuable contribution in the first few days after the birth.* **99**

the time to feel guilty about the lack of homemade food. Also make sure that you have enough coffee, tea, milk, and snacks for visitors.

If you have other children at home, of any age, make some plans in advance of your delivery date to entertain them when you first return home from the hospital with the new baby, so that they do not feel left out or away from the limelight. You will find that other mothers of small children are more than happy to rally around you and invite your children out for a play date, a day out, or even a sleepover, depending on the age of the child, given a signal from you that offers of practical help are welcome. Aunts, uncles, and other relatives might want to pitch in with your older children. Also, don't forget that grandparents can make an enormously valuable contribution during the first few days after the birth of your baby. There is nothing they will like more than making your toddler feel special and nothing your toddler wants more than to feel special when a new baby comes on to the scene.

YOUR MATERNITY BAG

Try to avoid packing a heavy suitcase big enough for a two-week, long-haul vacation. Remember that your partner, family, and friends can always bring any extra or missing items. Unless you find yourself giving birth in an emergency situation and have not had a chance to pack a bag, most maternity units will expect you to supply most of the items that you will need during labor and your stay after the birth.

ESSENTIALS FOR YOUR MATERNITY BAG
▶ Nightgown or large T-shirt
▶ Personal toiletries
▶ Sanitary napkins
▶ Changes of underwear
▶ Camera
▶ Loose change/phone card
▶ Cell phone with the numbers of people you want to call. Be sensitive to those around you and to your own needs when using a smart phone. Too much technology may distract you from the task at hand.

OPTIONAL ITEMS
▶ Face spray, sponge, lip balm, massage oil
▶ Mp3-player, music, magazines, books
▶ Change of clothes for your partner
▶ Food and drink for your partner

THE BABY'S BAG
▶ A pack of newborn-size diapers
▶ Zinc barrier/diaper cream
▶ Cotton balls
▶ Two sleepsuits
▶ Two shirts

FOR AFTER THE BIRTH
▶ One nightgown (front-opening if breast-feeding)
▶ Disposable panties or several of your oldest pairs
▶ Extra-absorbent sanitary napkins (bulky but a must)
▶ Breast pads
▶ Toiletries
▶ Towel
▶ Slippers

▶ Bathrobe
▶ Favorite snacks, high-energy foods, drinks
▶ Ice pack and/or a heated pad

OPTIONAL ITEMS
▶ Earplugs and eye mask (to help block out noise and light)
▶ Reading material including a good book about practical child care
▶ Pillow (many maternity units are unable to provide more than one)

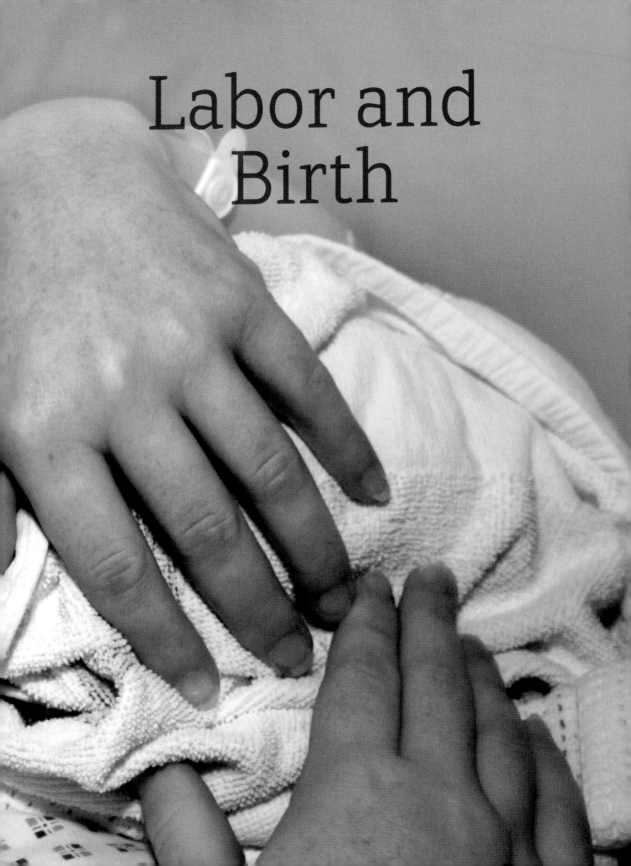

Labor and Birth

The stages of labor

The only predictable thing about pregnancy is that it always comes to an end—in the majority of cases with the delivery of a healthy baby to a healthy mother after normal labor and a vaginal delivery. I regularly reflect on the fact that this is an extraordinary achievement, since decades of research have failed to establish exactly what triggers labor. Indeed, if we better understood this process, we would be better able to predict when it will happen and take steps to prevent it from occurring prematurely.

CONTENTS

The first stage

GOING INTO LABOR IS ALWAYS AN EXCITING TIME as your pregnancy nears its end and you embark on the next steps of this amazing journey. During the first stage, your uterus will contract repeatedly, causing your cervix to thin, shorten, and dilate. Only when your cervix is fully dilated can your baby's head pass into the birth canal for delivery.

There is no right or wrong way to go into labor. Every woman does it differently and no two labors are the same. Interestingly, toward the end of pregnancy, women are often more concerned about the symptoms and signs that they are likely to experience before labor and during early phases of giving birth than in the three stages of established labor. The questions they ask are not simple to answer, because the pre-labor stage can last for days or you may skip it completely and become dilated 5cm before you know it.

Symptoms and signs of early labor

There are various signs and symptoms that indicate the end of pregnancy and that the start of labor is not too far away.

Engagement of the baby's head is one of the signs that labor is likely to occur soon for first-time mothers. During the last few weeks of pregnancy your doctor will be assessing the position and descent of your baby's head into the pelvis. When it engages (see p.267 and p.302) you will probably notice that your breathing becomes easier and that your indigestion and heartburn have disappeared now that the pressure on your abdomen has been somewhat relieved. Instead, you will be feeling a new pressure in your pelvis and it is likely that you will need to urinate frequently. However, if this is not your first baby, engagement of your baby's head may not occur until shortly before, or even after, labor starts.

Braxton Hicks' contractions (see p.237) are likely to become stronger and more frequent in the prelabor stage. They are usually painless, although some women may find them uncomfortable. It is easy to mistake them for real labor contractions, particularly if they are strong and this is your first baby. However, Braxton Hicks' contractions occur irregularly, rarely more than two per hour, and then fade away, whereas labor pains start slowly but gradually build up in strength and frequency.

> 66
> *...every woman goes into labor differently and no two labors are the same.*
> 99

Your cervical canal has contained a plug of mucus throughout pregnancy, designed to prevent infection from ascending from the vagina into the uterus. As the cervix starts to soften, shorten, and dilate, this plug becomes dislodged and the thick discharge that results is referred to as the "bloody show." It is frequently tinged with small quantities of blood, since the mucus plug was attached to your cervical canal by small blood vessels (capillaries). The appearance of the show is often interpreted as a sign that labor is imminent. The reality is that you can have a show and still remain pregnant many days later. Nonetheless, loss of the cervical plug is a demonstration that your cervix is changing and a sign that the end of pregnancy is near.

If your vaginal discharge becomes watery or you pass a gush of clear fluid, your water may have broken (ruptured membranes) or you may have merely leaked some urine. It is important to find out which has occurred so put on a sanitary pad to soak up any further fluid and contact your doctor immediately. One of them will examine you to decide whether you are leaking amniotic fluid or urine (see p.286).

The prelabor emotional symptoms experienced by pregnant women vary. Some women find that their nesting instinct goes into overdrive and they rush around finishing as many tasks as possible. Others prefer to avoid venturing too far from home in case anything happens. This is undoubtedly a strange and unpredictable period of time, during which you may feel as if you are in limbo. This can give rise to feelings of anticipation, excitement, impatience, anxiety, and fear. The fear is usually focused on what might happen during labor and, in particular, on how you will deal with the pain. Reading about the available methods of pain relief (see pp.308–23) will help you feel more relaxed, confident, and in control of the exciting developments ahead.

Your antenatal carers are there to give you advice and practical help 24 hours a day.

Recognizing labor

This is one of the biggest concerns for many women. Thinking that labor has started only to be sent home by the hospital is disappointing. But remember, prenatal caregivers deal with these uncertainties on a daily basis and it does not matter how many false alarms there are, as long as you and your baby are safe. Most pregnant women experience symptoms and signs that suggest they have begun labor (see box opposite) and need to seek advice. However, these signs come in no specific order and you may not experience all of them. It is also important to remember that early labor does not always progress at a steady rate. You might experience one or several symptoms followed by hours of no activity at all, only for events suddenly to pick up speed again. Overall, there will be a tendency for your contractions to become gradually stronger and more painful, but this increase in strength does not necessarily occur steadily. It is quite common to have a period of painful contractions followed by a series that are less intense in strength.

Most first-time mothers are very aware when they are in labor because the contractions take longer to establish since the uterus has never attempted to expel a baby before. In second and subsequent births, labor can be quick and if the contractions have been bearable throughout, there are occasions when the mother will not realize she has reached full dilation until she experiences an overwhelming urge to push the baby out. However, it is unusual for women not to make it to the hospital in time, or to find themselves delivering on their own at home before the doctor arrives (see p.289).

SIGNS OF TRUE LABOUR

▶ Your contractions are occurring regularly every 15 minutes or so (time them).
▶ Your contractions are getting longer, stronger and closer together.
▶ Walking around or changing positions does not make your contractions go away.
▶ You have pain in your lower back, as opposed to your lower abdomen.
▶ You feel the need to empty your bowels.
▶ You are passing fluid which you don't think is urine (ruptured membranes, see p.286).
▶ Your cervix is undergoing changes (these will be apparent from a midwife's examination).

Contacting your doctor

Any time you are worried or unsure about what is happening or what you should do, contact your doctor or maternity unit. Your caregivers are there to give you advice and practical help and are available 24 hours a day. Whether or not you are advised to go to the hospital will depend on many factors:

• whether this is a first, second, or subsequent birth
• the strength and frequency of your contractions
• how you are dealing with contractions at home
• if you have had any vaginal bleeding (more than a bloody show)
• how far you live from the hospital
• whether your water has broken
• whether your baby's movements have changed significantly.

Broadly speaking, if you have not had any complications during pregnancy and this is your first baby, your doctor will probably advise that you stay home until your contractions are regular. There are no hard and fast rules but if the contractions are occurring every 15 minutes, lasting for about one minute (time them), and are so uncomfortable that you are forced to stop what you are doing, you should be thinking about going to the hospital. If you live far away from the maternity unit or are likely to meet difficulties or delays on thetrip, you should allow plenty of time to get to the hospital.

Another important consideration is how you are coping with the contractions and whether you will need pain relief. Many women feel more comfortable, both physically and emotionally, with the knowledge that a variety of pain relief is close at hand once they reach the hospital (see pp.308–23).

Waters breaking

...if your water breaks before your contractions start, this invariably jump-starts labor.

If your water breaks (membranes rupture) before the onset of regular or irregular uterine contractions, it is a good idea to seek advice from your doctor. If you are near to term and you and your doctor know that the baby's head is deeply engaged, you may not need to be checked over and can safely wait at home for several hours to see what happens. However, now that the protective amniotic seal around your baby has been broken, you should not take a bath (shower instead) and make sure that you clean yourself carefully after passing a stool to reduce the risk of infection developing in the uterus.

If, on the other hand, your water breaks before 37 weeks or if the amniotic fluid is not pale yellow, but is tinged green or black, you need to contact the hospital immediately. This is a sign that your baby is or has been excreting meconium into the amniotic fluid, which invariably means that he has been exposed to stress and delivery should be sooner rather than later. Meconium is a thick sticky substance present in the baby's digestive system during pregnancy. If the baby is distressed, the response of the nervous system affects the digestive system, causing meconium to be expelled from the gut into the amniotic fluid.

In fact, in only 15 percent of pregnancies does the water break before contractions have started, and when this does happen it invariably jump-starts labor: 60 percent of women at term will go into labor within 24 hours of their membranes rupturing. Keep in mind, however, that once the membranes have ruptured, there is a greater chance of an infection developing and affecting the baby in the uterus. As a general rule, most hospitals advise women who have reached 35 weeks or more to have labor induced if uterine contractions have not started within 24 hours of the water breaking (see pp.294–97).

Finally, if you see a lot of blood mixed with the fluid that you are passing, or have bright red, fresh bleeding that continues after the water has broken, you must treat this as a potential emergency. Call your doctor and go to the maternity unit immediately.

Going to the hospital

The first thing to do once your doctor has advised you to go to the hospital is to pick up your maternity bag and have your partner or someone else take you to the hospital.

As for the trip to the hospital, if you are planning to travel by car, make sure you and your partner have figured out the route well in advance and know how long the trip will take at any given time of day. It goes without saying that you should not drive yourself to the hospital, except in very unusual circumstances. Strong uterine contractions are distracting and totally incompatible with the concentration necessary for safe driving.

It is a good idea to have also checked out the parking facilities at the hospital or maternity unit and to have some loose change in your maternity bag for the parking garage or street meters. If you arrive in a hurry or in an emergency situation, it may not be possible to park the car as you had planned. In this case, your driver should leave a note behind the windshield and let the hospital security desk know that you need to get to the labor ward urgently and somebody will attend to the car as soon as possible.

If you don't have access to a car, you will need to call either a taxi or an ambulance to transport you to the hospital. Make sure that you give the dispatcher clear instructions to find your home and provide a telephone number to avoid unnecessary delays. Ambulance crews are well trained at dealing with the practicalities of labor and ensuring that you reach the hospital safely—indeed, they can even be relied upon to deliver your baby at home or in the ambulance if the need arises.

Make certain that you know which entrance to the hospital you should use when you get there and the exact directions to the maternity unit—many hospitals may use a different entrance to the maternity unit at night for security reasons.

...it goes without saying that you should not be driving yourself to the labour ward...

Admission to the hospital

Call your doctor to let her know that you are on your way so that they can prepare for your arrival. Your doctor, if not attending another delivery, will meet you at the maternity unit. A labor nurse or doctor will check your temperature, pulse, blood pressure, and urine (for protein and glucose) and will palpate your abdomen to establish the presentation of your baby and listen to the fetal heart. She will also be assessing your contractions and she will

ARRIVING AT THE HOSPITAL Let the hospital know in advance if you think you are in labor so that they can prepare for your arrival.

ask you questions about the pattern of activity you have noticed in your uterus thus far, whether your water has broken, and if you need pain relief at this time. The answers to these questions and the results of the abdominal examination may suggest that you need to have an internal examination in order to assess the dilation of your cervix. If you have prepared a written birth plan, this will also be discussed with you at this time.

The findings will be recorded in your medical records and, if you have started to dilate (more than 4cm) and are contracting regularly, you are considered to be "in active labor." Most maternity records include a partogram or labor curve—a section with a graphical record of how your labor is progressing over time (see p.303).

What happens next will depend on your labor nurse or doctor's assessment and how your maternity unit is arranged. Some units have early labor rooms in which you will stay through much of your labor, until you are transferred to the main delivery unit just before your baby is born. Other maternity units prefer to admit you directly to a room where you will stay for the duration of labor and delivery. Whatever happens, your partner will be welcome to stay with you if this is what you want.

If you are not in labor

If your contractions are weak and irregular and your membranes are still intact, an internal examination may not be necessary although you and your doctor will probably conclude that a careful assessment of the cervix will help decide the next step. If you are not in labor and your doctor is certain that both you and your baby are well, they will want to discuss with you whether you would prefer to remain in the hospital for observation, or whether you can safely return home to wait for your labor to become established enough to return to the hospital. The decision will depend on many factors, including how your pregnancy has been so far, your past obstetric history, how anxious or apprehensive you are feeling, and how close you live to the hospital. There is no need to feel embarrassed if your arrival at the hospital ends up being a false alarm—these are very common, particularly with first babies, and you cannot possibly be expected to know when labor has really started.

MANAGING A SUDDEN DELIVERY

It is rare to have labor that is so unexpected and speedy that you have to deliver your own baby. However, here is some practical advice in case you do find yourself in this situation.

Try to remain calm. This is always easier said than done, but panic will only make the situation more difficult to handle. Hopefully, your birth partner will be with you, but if this is not the case, then try to contact a neighbor or friend who is nearby and can provide you with immediate practical help and support.

Call emergency services (911) and request an ambulance. Explain what is happening to you and also ask them to contact your doctor. Ambulance dispatch centers and crews are well trained in talking women through labor and can assist you. Make sure you keep your cell phone with you at all times.

Wash your hands and your vaginal area with soap and water if possible. Boil some water and collect plenty of towels or sheets. If you have time, cover the bed or the floor with plastic sheeting, blankets, sheets, newspapers, or clean towels and find a spare bowl that you can use to catch the amniotic fluid and blood. Then either get onto the bed or lie down on the floor.

If you feel the overwhelming urge to lie down, relax and start inhaling and blowing in short, controlled breaths. You may have practiced this in your childbirth classes and it will help prevent the baby's head from suddenly delivering.

If, despite your breathing, the baby's head starts to deliver before help arrives, reach down and place your hands on the baby's head at your vulva, applying gentle pressure to be sure the head emerges gradually. When the baby's head has delivered, check with your fingers that there are no loops of umbilical cord around the neck and if there are, hook them under your finger and carefully over the baby's head.

Gently stroke the sides of the baby's nose downward and the neck and chin upward to help expel mucus and amniotic fluid from the nose and mouth.

Help has usually arrived by now, and the delivery of the baby's body can be aided by the ambulance crew. If you have to deal with this on your own, place your hands around the baby's head and apply firm pressure downward (never pull or jerk) to help deliver the first shoulder. Then gently sweep the head and first shoulder upward toward your pubic bone. This will allow the second shoulder to deliver and the rest of the body will then slip out easily and can be placed onto your abdomen. Wrap the baby in towels or blankets immediately to keep him warm.

Do not pull on the umbilical cord, but if the placenta delivers spontaneously, elevate it so that the blood drains into the baby. You can wait until professional help arrives to cut the cord, unless help is delayed.

The most important thing is for you and your baby to keep warm until professional help arrives.

❝ Try to remain calm. Panicking will only make the situation more difficult to handle. ❞

The delivery room

It is a good idea to take advantage of maternity ward tours offered from your doctor. These provide an opportunity to see the delivery rooms and ask questions about the procedures and equipment you are likely to come across during labor and delivery.

The delivery bed will be higher than your own bed for practical reasons. It can be raised and lowered electronically. The end of the bed will be detachable and there will be places for stirrups or IV poles to be attached.

The sphygomanometer is an instrument attached to a wide band of inflatable material that is placed around your upper arm to measure your blood pressure. Some are portable hand-held units, others are wall-mounted and the most modern are automatic machines on small trolleys.

Piped gas outlets and tubing These are installed in hopsitals in the UK and parts of Canada to the walls above delivery beds. They deliver oxygen for women to breathe, through a mask or mouth piece for comfort during labor.

The baby resuscitator is a high, movable crib with a platform covered by a mattress for your baby to lie on. The resuscitator is equipped with an overhead heater to keep the baby warm, a piped oxygen supply, and drawers containing equipment that the pediatrician may need. In addition, there will be a simple comfortable crib in the room for the baby to lie in after delivery.

Stirrups can be attached to special slots at either side of the delivery bed. These allow your legs to be raised and supported so you can be examined more thoroughly should you need to have forceps or vacuum delivery or if you require stitches after the birth. They are no longer routinely used for internal examinations during labor and are usually stored under the bed.

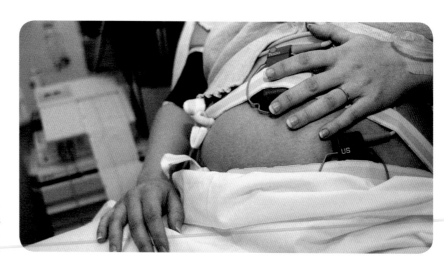

EXTERNAL ELECTRONIC FETAL MONITOR
This noninvasive device is used to measure the baby's heart rate and the strength of the uterine contractions.

An IV pole will be attached to the delivery bed or it will be on wheels nearby. There are several situations when it will be essential for you to have an intravenous line during labor:

• if you decide to have an epidural (see pp.311–15)

• if you are having labor induced (see pp.294–97)

• if you require drugs to strengthen your contractions (see p.304)

• if you are bleeding and the doctors need to have immediate access to your veins, to make sure they can restore your blood pressure if it drops.

• if you require IV antibiotics.

  Urinary catheters and bed pans are required in some circumstances during labor when you are unable to get to a bathroom.

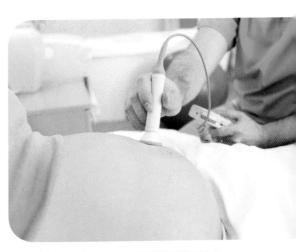

THE HANDHELD DOPPLER This battery-operated device is used by doctors to listen to the baby's heartbeat during normal hospital labors or home births.

Fetal monitoring equipment

As a medical student I remember an eminent obstetrician telling me that "traveling through the birth canal is the most dangerous journey a human being ever embarks upon." It is not surprising, therefore, that we spend a lot of time trying to monitor the progress of labor and the baby's ability to cope with it.

  The simplest method of monitoring the fetus during labor and delivery is with a handheld, battery-operated Doppler sonicaid machine that can be used at regular intervals to listen to the baby's heartbeat.

  Electronic fetal monitoring (EFM) measures the baby's heart rate continuously and also the frequency of the uterine contractions and displays the information on a paper printout or tracing. A healthy baby has a baseline heart rate of 120–160 beats per minute, which is continuously changing by 5 to 15 beats—this is described as "good variability." Lack of variability between contractions will alert your doctor to the possibility that the baby is not coping well with the stress of labor. Similarly, a low baseline heart rate of 100 beats per minute or less or a high heart rate over 160 beats per minute are more signs of possible distress. However, the most useful information gained from the fetal heart tracing is how the baby's heart rate responds to the stress of the contracting uterus.

  There are two types of electronic fetal monitoring: external and internal. External EFM is completely noninvasive. Two small devices are strapped to your abdomen with soft belts—one picks up the fetal heart beat; the other measures the intensity and duration of each contraction. They are linked by wires to the electronic fetal monitor and their findings are also displayed as flashing numbers on the front of the machine. In most labors, external

> *In most labors... there is no need for the woman to be confined to her bed.*

monitoring is required only intermittently, and there is no need for the woman to be confined to her bed.

Internal EFM is used when the tracing of the baby's heart rate is difficult to pick up or when there is clear evidence of fetal distress or slow progress is being made and the medical team need to know exactly what is happening. This is called a fetal scalp electrode (FSE). The fetal heart tracing obtained is more accurate than the one produced by an external monitor. A small electrode is clipped to the baby's head and linked up to the electronic fetal monitor. This clip can be applied only if the cervix has dilated to 2.5cm or more and your water has broken or been artificially ruptured. A belt around your abdomen holds the pressure transducer that measures your uterine contractions. During the first stage of labor the nurse or doctor will listen for one minute after every contraction every 15 minutes; during the second stage she will listen every five minutes and for one minute after each contraction. Alternatively, a fluid-filled catheter is inserted into your uterus to gauge the pressure of the contractions.

Electronic fetal monitoring was adopted widely and enthusiastically when it was first introduced in the 1970s and was used by some hospitals, either intermittently or continuously, during childbirth for almost all women. However, recent studies have shown that its routine use in labor significantly increases the chance of unnecessary interventions for supposed "fetal distress" and does not actually improve the outcome for the baby during a normal uncomplicated labor.

Fetal blood sampling

If the EFM print-out shows signs of fetal distress, your doctor may suggest that a fetal blood sample (FBS) be taken. A sample of blood is taken from the baby's scalp and the pH (acidity/alkalinity) is measured in a special machine. The more acidic the reading, the more likely it is that the baby lacks oxygen and, therefore, that intervention is needed. To obtain the blood sample, you will need to be examined internally either with your legs raised in lithotomy poles (stirrups) or while lying on your left side (left lateral position).

If measurements confirm that the baby is distressed, the best course of action will depend on how far advanced you are in labor. Obviously, if you are only dilated a few centimeters, it will be necessary to deliver you by emergency cesarean section to avoid delay. However, if you are dilated nearly 10cm or in the second stage of labor, it is often possible to achieve a vaginal delivery more quickly than an abdominal one, albeit with the help of forceps or vacuum extraction in some cases. If the baby is not distressed, it is likely that you can continue safely with a vaginal delivery.

Your birth partner's role

It serves no purpose to be judgmental as to whether or not the father should be present at the birth of his child. Watching someone you care about go through the pain and physical trauma of labor and birth is difficult and some men decide that they don't want to be there, however much they love their partner. That said, it is undoubtedly true that the man who attends the birth of his child will witness an extraordinary event that he will remember for the rest of his life. In addition, he will be able to share it with the mother-to-be and that common experience will often strengthen the bond between them. Encourage your partner to do what feels right for him.

The most important things a birth partner can do are, first, know what the mother-to-be wants and, second, know what is going on during the labor and delivery. The first can only be achieved if the two of you have had detailed discussions, and the second is probably best achieved by your partner attending the partner's prenatal preparation session (there is usually at least one class) and reading the labor chapter in this book.

Laboring women have different needs—some want physical reassurance such as massaging, hand-holding, and wiping sweat from their faces. Others need emotional reassurance and verbal encouragement. Your partner will have to be prepared for all eventualities and accept that you might want physical closeness one minute and not want him anywhere near you the next. It is also important for your partner to try not to let his fear or anxiety show.

Gathering information

A major part of your partner's role will involve liaising with the medical staff. There may be instances when something occurs that you do not fully understand. This is where your partner can really help, by asking for clarification to make sure you understand the medical reasons for a particular procedure. He needs to be prepared to ask questions about what is happening in a nonconfrontational way and to gather information so you are both kept fully informed and can participate in any medical decisions that need to be made.

YOUR PARTNER The most important thing your partner can do is make sure he is fully informed before the birth so you can rely on him for support as your labor progresses.

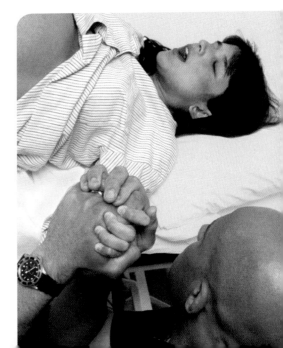

Inducing labor

When labor is induced it is brought forward artificially, before it has started of its own accord. Induction may be necessary if the mother or her baby need to be delivered promptly and it is unlikely to happen spontaneously.

...there is no such thing as a textbook induction and it is difficult to predict how things will go.

Induction may be used for women who have never shown any signs of going into labor and for a few whose water has broken but who fail to start contractions within 24 hours. It is important to understand that having labor induced does not usually involve a single intervention being performed at a specific time. Being induced is a process and is likely to necessitate a complex set of interventions and interactions, the exact nature of which will be dictated by the events that occur during labor. Hence there is no such thing as a textbook induction and it is often difficult to predict how things will go. This is why most doctors are hesitant to recommend induction unless there is a really good reason to do so.

Induction rates vary considerably between different countries, different maternity units, and even between individual obstetricians in the same unit. The rate is influenced by many factors, but particularly by the complexity of the pregnancy and the individual obstetrician's perception of what circumstances may place the mother or the fetus at risk.

Overall, nearly 70–80 percent of inductions result in a vaginal delivery, but the procedure increases the risk of an assisted delivery with either forceps or vacuum extraction (see pp.352–55). The best chances of a successful induction are when the mother has had a previous vaginal delivery, her cervix is ripe, the baby is of average size, and the head is engaged in a normally sized pelvis.

Indications for induction

The absolute indications for induction are that the baby would be better cared for in the outside world than in the uterus, or that the mother's health demands that the pregnancy end promptly. All other indications are relative and will frequently include a mixture of fetal and maternal considerations.

There are many reasons why an induction may be considered:

• **Fetal**—when prenatal monitoring suggests that the baby's growth has slowed down or stopped or there are signs that the baby is distressed in utero. This may include a reduction in fetal movements or volume of amniotic fluid and is usually because the placenta is no longer functioning properly.

Other fetal indications for induction arise if the baby has been affected by maternal rhesus isoimmunization (see p.128 and p.425) or the mother has diabetes mellitus or has had several small bleeds throughout the pregnancy, which places the baby at increased risk in the last few weeks (see p.408). Similarly, if the baby is known to have an abnormality that will require surgery immediately after delivery, it is safer for the baby to be delivered at a time when the necessary expertise is readily available. The decision to induce depends on balancing the risks of delivering a baby prematurely with those of leaving it in utero. However, with modern neonatal expertise, most babies delivered at or after 28 weeks survive without serious problems.

• **Maternal**—severe preeclampsia, poorly controlled diabetes, preexisting kidney, liver, or heart disease, and autoimmune disorders may require induction.

• **A combination of fetal and maternal indications**—many of the fetal and maternal indications listed above can lead to the decision that induction is the best way of ensuring the welfare of both mother and baby. However, preeclampsia and maternal diabetes are the most frequent combined indications for induction, together with premature rupture of the membranes.

• **Post-maturity**—most units offer induction to women whose pregnancies are prolonged past 41 weeks because it reduces the risks of unexplained stillbirth and other late-pregnancy complications, without increasing the cesarean section rate.

BISHOP'S SCORES

If an induction is being considered, most maternity units use a Bishop's score table to make an objective assessment as to whether or not the cervix is favorable for induction. Your doctor will perform a vaginal examination and cervical dilation, measuring length, consistency, position, and the station that the fetal head has reached in the pelvis (see p.302), which are given a score of 0 to 3. Total scores of 5 or more are considered favorable for induction because the cervix is ripe.

SCORE	CERVICAL STATE				
	Dilation (cm)	Length (cm)	Consistency	Position	Station of head
0	Closed	3	Firm	Posterior	-3
1	1–2	2	Medium	Middle	-2
2	3–4	1	Soft	Anterior	-1
3	5+	0			0

METHODS OF INDUCTION

When the decision is made to induce labor, the method used will depend on whether this is a first or subsequent pregnancy, the presence of a previous uterine scar, whether or not the membranes are intact, and the state of the cervix. It is unusual to plan an induction if the baby is not in a cephalic presentation (see p.268) and the head is not well, or nearly, engaged.

You will be admitted to the labor unit and your doctor will examine you and monitor the baby's heart rate with an EFM recording to make sure there are no signs of fetal distress (see p.291). The examination will start with one of them palpating your abdomen, confirming the longitudinal lie and cephalic presentation, and assessing the degree of engagement of the head (see p.302). Your doctor will then perform a gentle vaginal examination and use the Bishop's scoring system (see p.295) to assess the state of your cervix. A favorable cervix is critical if the induction is to be successful.

PROSTAGLANDIN GEL OR PILLS

The hormone prostaglandin occurs naturally in the uterine lining and stimulates the uterus to start contracting. If your cervix is unfavorable, your doctor will suggest putting some synthetic prostaglandin tablets/gel or a slow-release preparation in the form of a tampon into the vagina to help ripen it. A second vaginal dose of prostaglandins may be needed about six hours after the first and some women require more doses to ripen the cervix sufficiently. From a practical point of view this is often best done overnight, in the hope that the woman will wake up refreshed the next morning and will either start to go into labor or be better prepared to undergo the next part of the induction process. In pregnancies that have been straightforward, this can be carried out on the prenatal ward, but if recognized risk factors are present, the woman will be moved to the main delivery unit for this procedure.

After each insertion the baby will be monitored electronically for 30 minutes, but as long as the fetal heart tracing is normal, the mother will be encouraged to be mobile. As soon as contractions start, another period of electronic monitoring will be advised.

AMNIOTOMY (AROM)

When the cervix is dilated (2.5–5cm) it is usually possible to artificially rupture the membranes easily by inserting a long, thin plastic hook into the vagina and through the cervix to break the delicate membranes and allow the amniotic

STRIPPING THE MEMBRANES

Before deciding on a date for a formal labor induction, it is likely that your doctor will suggest attempting to strip the membranes. The doctor will gently insert one or two fingers through the cervix and literally sweep them around its circumference. This often helps start the contractions because it releases prostaglandins from the cervix. It can be a little uncomfortable and often results in light bleeding or a "bloody show." However, it is completely safe and is often an effective way of stimulating early labor contractions which is why you will be offered a strip at 41 weeks.

fluid to start draining away. This releases more prostaglandins, and these to help establish regular uterine contractions. In some inductions, no more interventions are required after an AROM (artificial rupture of membranes) has been performed because the uterine contractions become well established in a relatively short period of time. However, when this is not the case you will need to move to the next part of the induction process—an oxytocin IV to establish uterine contractions.

OXYTOCIN

Oxytocin is a hormone that is produced by the pituitary gland in the brain and causes the uterine muscle to contract. Oxytocin is given via an IV line that is inserted into a vein in your lower arm. The oxytocin drug is injected into a sterile bag of fluid (usually a mixture of salt and sugar solutions), which is then attached to an IV stand. The starting dose is always small and is gradually increased until effective uterine contractions have been established, usually defined as three moderate to strong contractions occurring every 10 minutes. The quantity of oxytocin that you will receive will be carefully measured by a special machine or infusion pump that is attached to the IV. This allows for the dose of oxytocin to be adjusted up or down, depending on the progress of your labor and, most importantly, how the baby responds to the uterine contractions.

Because uterine contractions that are induced by oxytocin can be strong and start suddenly before the baby has been exposed to more gentle uterine activity, there is a greater risk of the baby becoming distressed. Of course, this potential problem is frequently aggravated by the underlying reasons for which the induction is being performed. For example, if the baby needs to be delivered because of poor growth, then its reserves may already be lower than a baby who is normally grown and going through spontaneous labor. For this reason, once the oxytocin infusion has started, you will need to be monitored continuously with an electronic monitor (see pp.291–92).

You may find it useful at this point to read the section on the latent phase of labor (see p.298), where I explain the prolonged period of uterine activity usually needed before normal contractions become strong enough and regular enough to dilate the cervix effectively.

Once you have understood these points, I think it becomes much easier to appreciate why an induction delivery that requires an oxytocin intravenous line often results in an apparently lengthy and more painful labor. The labor is not really longer but there is a lot of catching up to do to reach the point when the cervix starts to dilate. Furthermore, the induced

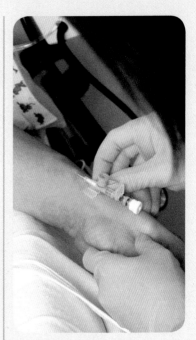

OXYTOCIN The drug is administered via an intravenous line inserted into a vein in your hand or forearm.

contractions may appear to be more painful because you have not experienced the gradual increase in uterine activity that occurs in the latent phase of spontaneous labor. This is one of the reasons why most maternity units in hospitals and birthing centers suggest that you have epidural anesthesia if you are induced and why the epidural is often inserted before the oxytocin IV is turned on.

> ❝ *A favorable cervix is critical if the induction is to be successful.* ❞

Established first stage

Theoretically, the first stage of labor starts when regular, painful uterine contractions become established and ends when the cervix is dilated 10cm. This first stage can further be divided into three phases: the latent phase, an established first stage, and a passive second stage.

The latent phase

During the latent phase of the first stage of labor, uterine activity starts with usually mild and irregular contractions. For many women, these contractions feel like menstrual period cramps or a backache and are usually not seriously distressing. This early activity changes the cervix from being thick and barrel-shaped, about 1in (2cm) in length, into a much thinner, softer, shorter structure.

Although you may not be aware of the mild contractions spreading down your uterus, they are thinning the lower part of the uterus and cervix, drawing them up over the baby's head (if that is the presenting part) like a glove. This process is called effacement and needs to occur before the cervix can open up or dilate. The latent phase can last for eight hours (sometimes more for first-time mothers), but if you have had several children it is usually much shorter—indeed you may not even realize that something has started to happen.

The hormones secreted during the last few weeks of pregnancy help soften the cervix in preparation for labor, but dilation in the first stage can only be achieved by the onset of progressively stronger uterine contractions. Hence, in the latent phase, the mild contractions usually occur every 15 to 20 minutes, lasting for no more than 30 to 60 seconds.

FIRST STAGE PHASES

As the first stage of labor unfolds, the cervix effaces and dilates. It changes from being a structure that is firmly closed to one that is dilated 10cm and able to allow the baby's head, or other presenting part, to pass through

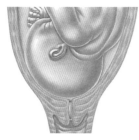 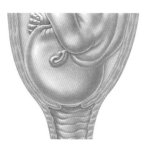 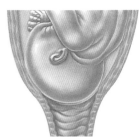

THE LATENT PHASE The cervix thins (effaces) and begins to stretch and open.

ESTABLISHED FIRST STAGE Contractions are stronger as the cervix dilates.

THE PASSIVE SECOND STAGE The cervix is fully dilated and ready for baby to descend.

Rest assured that if you find the pain you are experiencing in the latent phase distressing, you will be offered relief. However, until active labor has started, your doctor will probably recommend an injection of morphine or demerol rather than an epidural (see pp.308–23). This is because they will want to encourage you to stay upright and active for as long as possible, rather than confining you to a bed, in order to allow gravity to play its part in helping you reach the active phase of labor.

Established first stage

The established first stage is said to have started when the cervix has dilated to about 4cm and uterine contractions have become more regular and rhythmical. Your doctor will be able to judge if you have reached the established first stage based on what your uterine contractions are like, rather than what has happened to your cervix.

As the first stage progresses to the active phase, the contractions will become more noticeable around the center of your belly and will be accompanied by a hardening and tightening of the uterine muscles that you can feel with your hand. Uterine contractions are painful because the uterus is a massive muscular organ and requires enormous amounts of energy to work efficiently. During a contraction, the blood vessels in the walls of the uterus are compressed and the muscle becomes short of oxygen, resulting in the release of chemical substances that mediate pain. These are washed away in the recovery period between contractions.

The other thing to remember about contractions is that they slightly reduce the oxygen supply to your baby, because blood vessels in the uterus that are supplying the placenta are also compressed. As a result, your baby's heart rate may slow down at the peak of a contraction. This will be carefully monitored as your labor progresses (see pp.291–92) to make sure your baby is not getting too tired or distressed.

…the uterus is a massive muscular organ and requires enormous amounts of energy to work efficiently.

Changing contractions

As you enter the established first stage of labor, your contractions will change in nature. First, they become stronger, regular, and more painful. Second, instead of being concentrated in the lower part of the uterus, they now begin at the top and spread downward through the entire uterus. This allows the baby's head (or other presenting part) to be pushed against the cervix, since the priority now is to get the cervix opened to a 10cm diameter. Your contractions will occur every 10 to 15 minutes, then every five minutes, then every two minutes, the timing being calculated from the start of one contraction to the start of the next.

❝ *…the baby's head and shoulders must descend deep into the pelvic cavity before the second stage of labor starts.* ❞

By the end of the established first stage, each contraction will last about 60 to 90 seconds, so there will be little time for rest in between. As they get stronger, you will feel as if you are being gripped around the abdomen by a tight band as the uterine muscles harden and tighten. During each contraction, the pain will usually start slowly, rise to a peak lasting about 30 seconds and then subside.

From 4–10cm the cervix usually goes through the most rapid phase of dilation, after which there may be a deceleration. Cervical dilation is not the only marker of progress, however; it is important for the baby's head and shoulders to descend deep into the pelvic cavity before the second stage of labor.

The total length of the active phase of the first stage varies because it is determined by whether or not this is your first baby. In a first labor, you are likely to dilate at about 0.5–1cm per hour, whereas if you have already had one or more babies you may dilate considerably faster.

Rupture of the membranes

In 15 percent of term pregnancies, the membranes rupture spontaneously (SROM) before labor starts and, in the majority of these cases, contractions and progressive cervical dilation follow within 24 hours. This leaves 85 percent of pregnancies in which the water is still intact when labor commences. In most cases, the membranes will rupture spontaneously as labor progresses but occasionally, and usually in very fast or precipitate labors, the baby will be delivered still surrounded by the amniotic sac.

If your labor is progressing normally, there is no benefit in having an AROM performed routinely, particularly if you feel strongly about not wanting any interventions. Also, AROM is associated with an increased need for analgesics.

If your labor has been induced (see pp.294–97) or needs to be augmented because the progress is slow (see p.304), you may be advised to have an AROM. The AROM will first help labor progress. It will also facilitate the monitoring of your progress because labors that are not completely straightforward need to be watched closely to make sure the baby does not become distressed. If the FHT (fetal heart tracing) has any unusual results, your doctor will probably suggest that the amniotic sac be ruptured in order to attach an electronic monitoring clip to the baby's head (see p.292). One of them will also inspect the amniotic fluid and check that there is no meconium staining present, which would suggest that the baby has already been exposed to stress.

AROM is usually a painless procedure (see pp.296–97) if you are already partially dilated. There are a few situations where AROM is not advisable, for example, if you go into premature labor. In such a case, it is always best to leave the amniotic fluid intact for as long as possible, since it protects and cushions the fragile premature baby during labor and delivery (see p.341).

The passive second stage

This is the name given to the interlude that sometimes, but not always, occurs during the final phase of the first stage of labor. This phase is marked by the cervix being fully dilated, and, characteristic of the second stage of labor, you experience the intense urge to push. At this point, the second stage of labor is fully underway.

The passive second stage can last a few minutes or continue for an hour or more. For some women it can be the most difficult part of their labor because they already feel exhausted by the hours of contractions they have experienced. Some women report that they feel nauseous during this stage. The contractions will be intense, coming every 30 to 90 seconds and lasting for 60 to 90 seconds. Hence there is very little time between them and many women feel frightened that they have completely lost control of what is happening. In some ways this is true, since the labor has developed a momentum of its own and there is no stopping until the baby is delivered. The good news is that this is not far off, so try to think of the passive second stage as a very positive sign that the end of your labor is in sight.

Wanting to push

Some women experience an intense urge to push during the passive second stage before the cervix is fully dilated. If you start pushing when the cervix is 8 or 9cm dilated, it will develop into a thick, swollen ring around the top of the baby's head, instead of the paper-thin pliable membrane that will allow the head to descend and slip past it. If you do develop the urge to push before you reach full dilation, your labor nurse will show you how to pant or take short shallow breaths when you are having contractions, to try to prevent you from breathing too deeply and bearing down. She may also suggest that you change to a position that takes the pressure of the baby's head off your cervix (such as on hands and knees with your bottom raised), since an upright position will encourage the urge to bear down. Alternatively, if you have an epidural in place, it may be a good idea to have a small increase in dosage at this point so

> 66
> ...think of the passive second stage as a positive sign that the end of your labor is in sight.
> 99

that full dilatation and further head descent can be achieved before you start to do your pushing.

Examinations in labor

The onset of labor depends on the cervix starting to dilate, but progress in labor requires both continued cervical dilation and descent of the baby through the pelvic birth canal. There is no absolute size of baby or pelvis that can guarantee good progress during labor, nor is there a magical number of strong contractions that can predict a swift and smooth delivery. This is why it is so important for your doctor to perform an abdominal examination as well as a vaginal/internal examination whenever she is assessing your progress.

How often you are examined during labor will depend on many different factors, not least of which is how long or short your labor lasts. At each examination, the doctor will palpate your abdomen and confirm the lie and presentation of the baby (see p.268). In the vast majority of pregnancies this will be longitudinal and cephalic, so the rest of this section on progress in labor assumes this is the case. Details on how breech presentations and transverse or oblique lie labors are managed can be found on pp.356–59 and p.430.

ENGAGEMENT AND STATIONS

Now that you are in labor, it is important to assess how far the baby's head has descended into the pelvis. By definition, engagement means that the largest diameter of the head has entered the pelvic brim, which means that no more than two fifths of the head will be palpable in the abdomen. The less head that can be felt on examination, the better the progress of labor is likely to be. You will have a vaginal examination to identify how many centimeters your cervix is dilated and the level that the leading part of the head has descended within the pelvis. The levels are referred to as stations, which are best thought of as imaginary horizontal lines drawn through the pelvis at centimeter intervals. When the head first enters the pelvic brim it is said to be at –5 station. When the tip of the head reaches the middle of the pelvic cavity it is at zero station. When it reaches the vaginal opening it is at +5 station.

THE DESCENT The position of the head is described according to where it is in cm above (-) or below (+) the ischial spines (the narrow part of the pelvis).

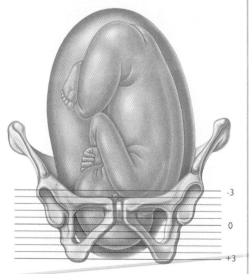

-3

0

+3

LABOR CURVES

The best way of monitoring the progress of labour is to use a partogram. This is a large chart consisting of various graphs onto which all of the observations made in labour are entered.

Your personal details will be noted at the top of the labor curve along with any special instructions or reminders to the labor nurses and doctors. Underneath this are several graphs recording the fetal heart rate, the number of contractions you are having in each 10-minute interval, your temperature, blood pressure, pulse, and the results of any urine tests. If you need pain relief or an oxytocin drip, the dosages and the timing of these drugs will also be recorded on the chart.

The most useful part of the labor curve is the graph of your cervical dilation and the station that the fetal head has reached, which the labor nurse or doctor will plot at each assessment. This shows at a glance how labor is progressing and will quickly identify any delays. Indeed, most labor curve graphs have bold black lines already drawn on them showing the anticipated curved pattern of cervical dilation from 0 to 10 cm. The ideal curve for first-time mothers moves along and up the graph toward the 10cm mark more gradually, whereas the curve for mothers who have already had a baby is shorter and sharper. The ideal curve for head descent will move along and down the graph. Not every labor will follow these guidelines exactly, but if progress is slow (the entries are appearing significantly to the right of the ideal progress line), then interventions are likely to be needed. The earlier this is recognized, the better for both you and your baby.

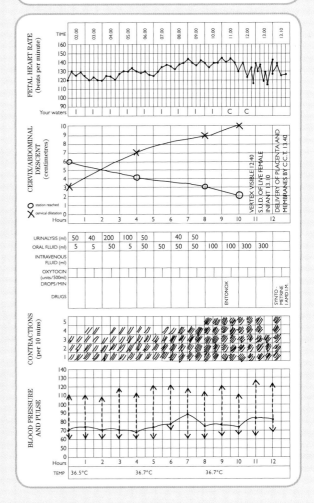

PROGRESS OF THE FIRST STAGE

Good progress is achieved by a combination of three important P factors:

▶ **The power**—strong uterine contractions that effectively dilate the cervix.

▶ **The passenger**—a baby that can fit through the mother's pelvis and is in a good position for an easy exit.

▶ **The passage**—a pelvis that is sufficiently roomy to allow the baby to pass through it.

All three of these P factors are relative to each other, and the progress of the first stage of your labor will depend on how they interact with each other.

...in the vast majority of pregnancies, the lie will be longitudinal and cephalic

Augmentation of labor

If the progress of a spontaneous labor begins to slow down, it may become necessary to accelerate or augment it. Using a labor curve is the best way of deciding when augmentation should be started. If the contractions are satisfactory but the membranes are intact, then an artificial rupturing of the membranes (see p.300) may be all that is needed to speed things up. If the contractions are weak, infrequent or irregular, and your water has already broken, then an oxytocin infusion will be started to strengthen and regularize the contractions. As with an induction (see pp.294–97), the dose is small to begin with and then gradually increased, aiming for about three to four moderately strong contractions every 10 minutes.

Continuous monitoring will be needed to make sure that the baby does not become distressed and another examination will be done about two hours after regular contractions. Usually this shows that there has been some progress in the cervical dilation or descent of the baby's head, in which case the oxytocin infusion will be continued until the next examination in four hours' time. Observations will be added to the labor curve (see p.303), and the curve of cervical dilation should appear normal. Occasionally, however, if there is no progress after several hours, doctors will reassess you and may advise delivery by cesarean section.

Prolonged labor

Prolonged labor is a labor that takes longer than expected by the known labor curves. It complicates five to eight percent of all labors and is much more common in first-time mothers than those who have had a baby before.

Labor becomes prolonged when the cervix is slow to dilate or the baby is unable to descend through the birth canal or rotate into the optimal position for a straightforward delivery or the uterine contractions are suboptimal. In reality it is often a combination of factors, since dilation, descent, and rotation are all interdependent. Cephalopelvic disproportion, fetal or maternal obstruction, inefficient uterine activity, and occipito-posterior presentations are usual causes of prolonged labor, and they are often closely interlinked

Cephalopelvic disproportion

Cephalopelvic disproportion (CPD) means that the baby's head is too large for the mother's pelvis. It is a relative term, since a different-sized baby might slip through the same pelvis easily. For first-time mothers, CPD may be suspected if the head has not entered the pelvic brim at term. Further clues may be obtained by looking at the mother's height—if she is less than 5ft (1.5m) tall, she may have a small pelvis that will make a vaginal delivery difficult or impossible.

In cases where there is a strong suspicion of cephalopelvic disportion prenatally, doctors will often advise a cesarean section to avoid exposing the fetus to a prolonged and difficult labor. On the other hand, if the head has entered the pelvic brim and the mother wants to deliver vaginally, she may choose to have a trial of labor. If this is attempted, progress will be carefully monitored using a labor curve (see p.303). If progress is poor, then plans will be made for an abdominal delivery.

In second or subsequent pregnancies, the head does not always enter the pelvis until the onset of labor, so predicting CPD may be more difficult. A careful review of previous labors and the birthweight of the babies may provide useful clues. The reality is that however restricted or capacious the bony measurements may be, the only way of knowing whether or not the baby's head will pass through the pelvis is by monitoring labor as it progresses.

> " The only way of knowing whether or not the baby's head will pass through the pelvis is by monitoring labor as it progresses. "

Obstructed labor

This is usually the end result of a poorly managed or neglected labor in which CPD or a malpresentation, such as a shoulder or transverse lie (see p.430), has been missed. Other causes include pelvic masses in the mother, such as a uterine fibroid (see p.423), or a congenital abnormality in the baby, such as hydrocephalus (see p.420). Happily, I can tell you that obstructed labor is a rare event these days, since most causes will be identified before you go into labor.

In a first labor, the uterus contracts strongly to try to overcome the obstruction and then may become inactive. However, if an obstruction occurs in a second labor, the uterus more commonly continues to contract strongly, leading to the development of a contraction ring in the uterus that is known as a Bandl's ring. The upper part of the uterus becomes thick and short and the lower part becomes progressively stretched and thinner. In this situation, urgent intervention with delivery by cesarean section is needed in order to prevent the uterus from rupturing (see p.428).

Inefficient contractions

Labor will progress normally only if your uterine contractions are efficient and move downward through the entire uterus. When they fail to do so, labor usually becomes prolonged. The inefficient uterine activity can be under- (hypo) or over- (hyper) active and develops in 5 percent of first labors and about one percent of all labors. Underactivity—called uterine inertia—often responds to stimulation with oxytocin (see p.297) unless disproportion or another type of obstruction is present. Overactivity—also called incoordinate uterine activity—is when the various parts of the uterus contract independently. These contractions fail to dilate the cervix efficiently and are often very painful. Now that epidural anesthesia is used so widely, the severity and exact location of the mother's pain in labor can be difficult to monitor. For this reason, a labor curve is useful in identifying the different types of inefficient uterine activity. If slow cervical dilation persists and the cervix fails to dilate by more than 2cm over a period of four hours, accompanied by poor head descent, then plans will need to be made for delivery by cesarean section.

> *Labor progresses normally only when your contractions are efficient, moving down through the entire uterus.*

Occipito-posterior presentation

Labor and delivery are more likely to be swift and uncomplicated when the baby is in the occipito-anterior position (OA, see p.270)—the back of the baby's head (the crown or occiput) is toward the mother's front (anterior). When in the occipito-posterior (OP) position, its face is pointing forward and the back of its head is pressing against the mother's back (sacrum) with the baby's spine lying on top of the mother's spine. In this position it is more difficult for the baby to flex its neck and chin because the baby's bony crown is pushed up against the bony maternal sacrum, and, consequently, a larger than normal proportion of the head is presenting. This may prolong the labor and it often results in the mother experiencing more pain and discomfort in her lower back. This is why occipito-posterior positions are sometimes referred to as backache labors.

To take the pressure off your lower back, try getting down on your hands and knees, sitting with your legs crossed in front of you and leaning forward, or rocking your pelvis. Alternatively you may find that a vigorous massage of the painful area is helpful. One of the best ways of encouraging rotation of the baby's head is to stay upright and mobile for as long as possible. If you need to lie down, then adopt a position that will encourage internal rotation (your doctor will advise you). Always avoid lying flat on your back because in this position the complete weight of your baby will be pressing down on your spine and lower back, and this may make you feel faint.

Rotating the baby

Just over 10 percent of babies are in a posterior position at the start of labor (see p.270), so this is not an uncommon problem. In the majority of these cases the babies tend to rotate themselves into an anterior position before the end of the first stage, but labor invariably gets off to a slower start and the progress is often more painful and protracted, which is very tiring for the mother.

When the baby is in the OP position both epidural anesthesia and labor augmentation are often required during labor and, not surprisingly, maternal exhaustion and fetal distress are frequent complications. If the baby fails to move into the anterior position by the beginning of the second stage, the doctor will probably suggest increasing the dosage on your epidural (to prevent any urges to push), changing your position to encourage further rotation, and letting you rest for an hour or so. If the baby is still OP, a vaginal delivery is still possible but the baby is born facing upward which may carry a greater risk of perineal damage. If the baby is in an occipito-lateral position then it may be necessary to rotate the baby into the occipito-anterior position with rotational forceps or vacuum extraction to achieve a smooth delivery. When this proves to be unsuccessful, the only remaining option for delivery is an emergency cesarean section.

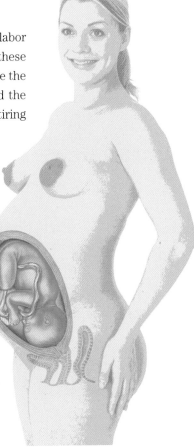

OCCIPITO-POSTERIOR
In this position the back of your baby's head is toward your back, with its spine pressing against yours.

PAIN RELIEF IN LABOR

Pain relief is an important issue, since nobody wants to be exposed to pain in normal circumstances. However, uterine contractions that are capable of pushing your baby through your birth canal will be painful, although there are effective ways to help alleviate the discomfort.

The issue of pain relief in labor is a thorny one which has been fiercely debated over the years by pregnant women and health-care professionals. As with all heated debates, there is a tendency for views to become polarized. Some have the view that childbirth is an entirely natural process, and since they believe that pain is an integral part of this process, will refuse to consider any form of medicalized pain relief. At the other end of the scale are women who will want to have an epidural at the first sign of discomfort, and may even request an elective cesarean section in order to bypass the process of labor altogether.

I have no real problem with either of these attitudes as long as the woman has been given sound advice about all the options available to her before she decides on her approach. There is no way for a first-time mother to know how the pain of labor will affect her, since it is probably the first time in her life that she has been exposed to significant pain. Women who have already experienced labor may have a better idea of their own pain threshold but, since every labor is different, they cannot be expected to anticipate accurately how they will cope with pain this time around. I cannot accept that a woman who is capable of delivering a healthy baby into this world should ever feel that she has failed because she needed pain relief to help her.

My personal opinion is that adopting an open-minded attitude and waiting to see what feels right for you at the time is usually the best approach. I urge you to be as informed as possible; there is no doubt that the more you know about different methods of pain relief available, the better you will be able to make the best choice for you and for your baby on the day.

MEDICALIZED PAIN RELIEF

THERE ARE THREE BROAD GROUPS:
Regional anesthesias, which create a localized block of pain sensation (numbness), including:
▶ epidurals
▶ spinal blocks
▶ pudendal blocks

General anesthesias, which produce a loss of consciousness and therefore no pain is felt.

Analgesics, which relieve pain or the perception of pain. These are:
▶ inhalational analgesics (Not used in the US)
▶ systemic analgesics such as morphine or meperidine

ANALGESIC DRUGS

Analgesic drugs work on receptors in the brain, dulling the pain messages sent by the nervous system. Morphine and meperidine (Demerol) are the most common types of analgesics used for labor in the US.

Inhalation analgesia

While not used in the US, inhalation analgesia is a popular form of analgesia in the UK. Nitrous oxide is an analgesic gas, which, when used as a 50:50 mix with air, numbs pain centers in the brain and dulls the pain messages being sent to the brain without causing loss of consciousness or significant sedation. Its popularity is partly due to the fact that the mother is in control of how much gas she receives and is reassured that she cannot overdose nor will her baby suffer any side effects.

How gas and air is used in the UK

Women wait until the start of a contraction, then inhale slowly and deeply through your mouth using a mouth piece. After five or six breaths the gas will have reached an analgesic level in the brain and women will notice the pain relief and probably also a floating feeling of well-being. Inhaling and exhaling continue until the contraction has subsided.

Usage stops between contractions, since continued use does nothing to help reduce the pain of the next contraction, but it will increase your chances of feeling sick and disorientated. Some women find that the smell of the rubber face mask and/or the gas makes them feel sick, which is why mouth pieces are better to use. Also, having something in your mouth that you

PARTNER PARTICIPATION Partners are taught to recognize the woman's pain and to coach her in abdominal breathing and other methods of deep relaxation.

can bite on hard can be very helpful in the middle of a strong contraction.

Nitrous oxide may be most helpful in the early stages of labor. Some women manage using only nitrous oxide throughout, whereas others find they need an additional form of analgesia to cope with strong contractions or during the second stage of labor. I am sure that part of the benefit of using gas and air during the first stage is that it forces women to concentrate on breathing technique, which helps women feel more in control of the situation and can also provide a degree of pain relief.

Although nitrous oxide readily crosses the placenta, it is eliminated both from your body as well as from your baby's body very rapidly, causing no adverse effects. The baby will not be affected.

Systemic analgesia

Systemic pain relief affects the whole body. Opioids, especially morphine, Nubain, and Statdol, are generally not used that much in labor these days but they are still used in some cases. They are members of the family of narcotic (opiate) drugs, which literally means that they induce drowsiness as well as reducing pain. Morphine relieves pain by stimulating specific receptors in the brain and spinal cord so that the pain messages that are relayed to the brain by the nervous system are dulled. These are the same receptors that are used by endorphins—our body's naturally occurring analgesics.

Morphine is administered quickly and easily by injection, usually into a muscle either in your upper thigh or buttocks or, alternatively, intravenously, and you will feel its effects within 15 to 20 minutes. The pain relief will wear off after three to four hours, when you will need to have a repeated dose if that is what your pain relief method is.

Problems with morphine

Morphine has a poor track record as a labor analgesic, however, because the dosages that are required for effective pain relief can also result in the mother becoming overly sedated. This can occasionally lead to breathing difficulties and low oxygen levels in the mother between contractions. Morphine can also be accompanied by additional side effects, such as nausea, vomiting, indigestion, and delayed gastric emptying and many mothers tell me that they dislike morphine because it makes them feel out of touch with what is happening during labor.

The other major disadvantage of morphine is that it rapidly crosses the placenta and goes into the baby, in whom it can cause drowsiness. As the baby becomes sedated, there is often a reduction in the baseline variability on the fetal heart tracing that is displayed on the electronic fetal monitoring machine (see p.291), which makes interpretation of the fetal heart trace difficult. In addition, when morphine is administered close to the delivery, the baby tends to be born with lower Apgar scores (see p.375) and, consequently, is more likely to require an injection of naloxone in order to help reverse the effect of the morphine in the baby's body. Morphine is only usually given late in labor and usually not in preterm babies but it depends on how premature the baby is and this will be something to talk to your doctor about prior to your due date.

REGIONAL ANESTHESIA

A variety of regional anesthesias can be used to block the pain associated with labor, delivery, and the completion of the third stage of labor. All of these depend on blocking nerve fibers by injecting them with local anesthesia drugs.

The type of regional anesthesia used to block pain depends on the procedure. For example, epidurals can be used throughout labor and all types of delivery, whereas spinal blocks are usually reserved for cesarean sections or manual removal of the placenta. Similarly, if used alone, a pudendal or cervical block will only provide sufficient analgesia for simple forceps or vacuum extraction deliveries.

Epidural blocks

A good epidural block will numb all sensation in your abdomen and stop you from feeling painful uterine contractions. This method of pain relief involves significant medical resources and can be offered routinely in hospitals that provide 24-hour anesthesia coverage in the labor ward. When you are admitted, tell your obstetrician or nurse that you might want to have an epidural so she can alert the anesthesiologist that he or she may be needed at a later stage.

How epidurals work

Your spinal cord is covered by a thick membrane or sheath, which is called the dura, and is further protected by the bony vertebral column that surrounds it (see diagram, p.313). The epidural space lies between the bony vertebral column and spinal cord.

The nerve fibers that control your contraction pains pass out of the spinal cord, pierce the dural sheath, and cross through the epidural space before passing between the vertebrae to enter your abdomen. The anesthesia drugs injected into this epidural space penetrate the nerve fibers and block your pain pathway.

If the dosage of anesthesia drug is high, some of the motor nerve fibers that control your legs

WHEN ARE EPIDURALS USEFUL?

▶ At your request: to help with painful uterine contractions in first or second stage of labor
▶ Multiple pregnancy
▶ Premature labor
▶ Prolonged labor: posterior positions; inefficient/irregular

uterine contractions; suspected cephalopelvic disproportion following induction or augmentation of labor (see p.304)
▶ Instrumental delivery: vacuum extraction or forceps
▶ Breech delivery

▶ Elective cesarean section— unless there are specific contraindications (see p.362)
▶ Emergency cesarean section— unless there are contraindications or insufficient time
▶ Perineal repairs/episiotomies—

EPIDURALS

▶ **Will the procedure be painful?**

Since the skin of your back is numbed with local anesthetic before the epidural needle is inserted, it is unusual for you to feel anything other than mild discomfort.

▶ **What happens if the epidural does not work?**

If the local anesthesia solution spreads unevenly in the epidural space, you may find that an area of your abdomen or a patch of your thigh is not effectively blocked. Occasionally, the block is only on one side of your body. However, these problems can be quickly resolved by the anesthesiologist, who will adjust the position of the catheter and may ask you to change your position to ensure that the drugs reach all of the nerve fibers evenly. Rarely, the epidural does not "take" and in this situation your anesthesiologist will probably decide to reinsert it.

▶ **I have had a previous back injury. Will I be able to have an epidural?**

The answer to this will depend on the nature and severity of the previous back injury. But generally speaking, it is unusual for an existing back injury to make an epidural out of the question. The most sensible thing to do is to arrange to meet with one of the obstetric anesthesiologists during the prenatal period so that you can be offered advice regarding the best form of analgesia.

▶ **Can the epidural catheter damage my spine?**

It is extremely rare for the catheter to move inside your spine. If this did happen your caregivers would notice it quickly, since the area of numbness would move to a higher or lower level, which they will fix immediately. It is virtually impossible for an epidural to damage your spinal cord or paralyze you.

▶ **Will I be able to push my baby out in the second stage?**

The answer to this is yes, you should be able to, but it will be more difficult with an epidural. This is because you cannot feel your contractions and will not experience the strong urge to push that accompanies second-stage contractions. You will also be less able to judge where to direct your pushing efforts. However, your nurse or obstetrician will help by telling you when the contractions are occurring and between the two of you, you can find a way of using them effectively. The other way of dealing with the second stage is to let the epidural wear off slightly, so that when you start pushing, you are aware of your contractions and more importantly, you have a better idea of where in your perineum you should be aiming for.

▶ **Am I more likely to have an operative delivery?**

There is good evidence that epidurals increase the likelihood of an operative vaginal delivery such as vacuum or forceps because the woman is less likely to be able to push in the second stage. My view is that an experienced nurse or doctor, working with a motivated mother who really wants to deliver vaginally, can usually manage any uncomplicated delays in labor and achieve a vaginal delivery.

▶ **Will my baby be affected?**

Some modern epidural solutions contain opiods that cross the placenta and in large doses may cause short-term respiratory depression in the baby. The epidural may occasionally cause a fall in your blood pressure, which, if sudden or sustained, could result in fetal distress. This is why CTG monitoring is recommended for at least 30 minutes after insertion of an epidural and after each top off.

> **66** *Your nurse will help by telling you when the contractions are occurring.* **99**

and bladder will also be blocked, which will make your legs feel quite heavy and difficult to move, and you may no longer be able to tell when your bladder is full.

Preparing for an epidural

If you choose to have an epidural, your nurse or doctor will explain the procedure to you and the anesthesiologist will answer any questions you may have and will obtain your verbal or written consent to proceed.

You will be asked either to lie on a bed on your left side with your legs bent forward or to sit up on the bed and lean forward, stabilizing yourself by leaning your arms on a table. Lying down is usually more comfortable if you are already contracting strongly, whereas the upright position may be the best choice if you are having the anesthesia inserted before an elective cesarean section. Lying on your left side prevents the weight of your uterus from pressing on the major veins in your pelvis, which could make you feel light-headed and reduce the blood supply to your baby during the 20 to 40 minutes it will take to set up the epidural.

You will need to lie or sit very still during this time, but the anesthesiologist will interrupt the procedure each time you have a contraction and wait until it is over. If you have not needed an intravenous fluid line before now, one will be inserted in your non-writing arm before the procedure starts and an infusion of fluid, usually saline, will be started. This ensures that your blood pressure does not drop suddenly when the epidural block starts to work (see p. 315).

Your lower back (lumbar region) will be cleaned with antiseptic, sterile covers placed over the rest of your back and legs to reduce the risk of infection, and local anesthetic injected over the

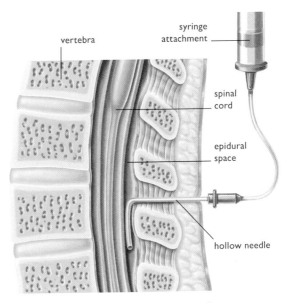

INSERTING AN EPIDURAL A hollow needle is inserted into the epidural space, leaving the spinal cord and its protective dural sheath untouched.

site chosen to insert the epidural to numb your skin and minimize your discomfort. It is unusual for the procedure itself to be painful.

The procedure

The anesthesiologist will carefully insert a fine hollow needle between two vertebral bones in the lumbar region of your lower back and from there, into the epidural space (see above). To check that the needle is in the right place, a small quantity of anesthesia will be injected. If this numbs your abdomen satisfactorily, a thin hollow plastic catheter will be threaded through the hollow needle and positioned into the epidural space. The needle is then removed and a protective bacterial filter is attached to the end of the catheter.

The long length of catheter outside your body will be led up your back and over your shoulder and taped onto your skin. This prevents the catheter from becoming dislodged and allows you to move around and receive small doses of pain relief throughout labor. The first dose of anesthesia will then be injected through the catheter and you will experience a cold sensation like an ice cube running down your lower spine as the drugs reach the spot.

The anesthesiologist will check your blood pressure immediately, then every 10 minutes for the next 30 minutes, and at regular intervals afterward. An electronic record of the baby's heart (FHT) will also be done during this period. In most maternity units, continuous fetal monitoring is advised after an epidural has been inserted, but in an otherwise normal labor intermittent EFM monitoring may be possible. Your abdomen will quickly go numb, but epidural anesthesia that is deep enough to allow a painless operative delivery, such as cesarean section or forceps, takes longer to develop—usually 20 to 30 minutes.

Conventional epidurals also block the nerve fibers that control your bladder. As a result, you will not be aware of developing a full bladder and will find it very difficult to urinate on your own. In order to overcome this, a urinary catheter will be gently inserted into your urethra to drain your bladder continuously or intermittently. If mobile epidurals are available in your maternity unit, you will probably be able to urinate without the need of a catheter.

Once your epidural is working effectively, it can be increased regularly to maintain your pain relief. This is usually needed every three to four hours, but the exact timing will depend on each individual's needs. For women who reach full dilation with an effective epidural, the decision to delay active pushing has been shown to reduce the risk of an instrumental vaginal delivery.

Mobile epidurals

Most hospitals have introduced mobile epidurals which use lower doses of drug to block the pain fibers but leave the motor fibers controlling your leg movement relatively unaffected. As a result the woman is usually less numb from her knees down and can be more mobile and benefit from gravity that may help the progress of her labor. She is also less likely to need a urinary catheter to empty her bladder. Another advantage of a mobile epidural is that, since the low-dose increases are usually needed every hour, the analgesia can be tailored more

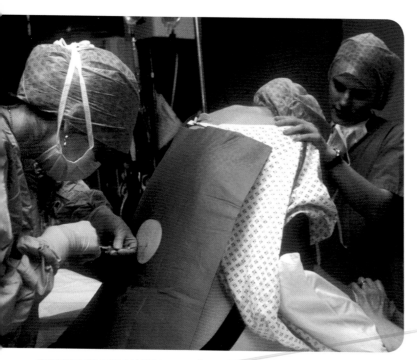

PREPARING FOR AN EPIDURAL Sterile covers are placed over your back and then a local anesthesia is injected.

accurately to suit the stage of labor. However, mobile epidurals are not available everywhere.

When an epidural may not be possible

There are few absolute contraindications to having an epidural. They can be summarized as situations where the introduction of the needle could result in a blood clot (hematoma) or abscess (collection of pus) forming and placing pressure on the spinal cord. These are potentially very serious complications, which could lead to paralysis. In reality, they are uncommon, but both inherited and acquired bleeding disorders can put you at risk. Similarly, if you have been taking large doses of anticoagulant drugs (for example, because you have suffered a thrombosis during pregnancy—see p.424) an epidural may not be recommended for you.

It is very rare for a pregnant women to have an infection in her lower back, but occasionally it may result from chronic tuberculosis or osteomyelitis (a serious infection of the bone pulp). In these situations, the risk of introducing an infection that could cause an abscess in the

MEDICAL COMPLICATIONS OF EPIDURALS

Hypotension (low blood pressure)
A fall in your blood pressure is a common side effect of epidurals, usually most marked with the first dose. This is because in addition to blocking pain fibers, the anesthesia drugs also block some of the nerve fibers that control the size of the blood vessels in your pelvis and legs.
▶ As a result, these blood vessels dilate and the blood will tend to pool in them, reducing the volume of blood returning to your heart and head.
▶ This can cause a reduction in the blood flow through the placenta, which in turn reduces the oxygen supply to your baby.
▶ This is why the anesthesiologist will always take the precautions of inserting an intravenous drip before inserting the epidural and checking your blood pressure regularly after the first and subsequent doses of drug are injected. Meanwhile, your doctor or nurse will be monitoring your baby on the EFM machine.

Headaches
Headaches are a well-documented side effect of epidurals, but in reality they affect only a small percentage of women.
▶ Severe post-epidural headache is usually the result of accidental puncture of the membrane covering the spinal cord as the needle is being inserted into the epidural space.
▶ The pain is caused by the leakage of small quantities of spinal fluid, which results in traction or pulling of the membranes around the brain. It tends to be relieved by lying down.
▶ Similarly, some mothers report a slight tingling or numbness in one of their limbs and some also suffer from back pain.
▶ I do want to emphasize that, although this all sounds frightening, symptoms are only temporary and nothing to worry about.
▶ All these side effects will disappear, most within hours of the birth, some up to a few weeks later.

Backaches
Recent studies have concluded that epidurals are not a genuine cause of long-term backaches. Postpartum backaches are much more likely to be related to preexisting backaches, perhaps caused by poor posture, rather than the use of epidural anesthesias during labor.
▶ Poor posture and strain on your sacroiliac joints is almost inevitable in the latter stages of pregnancy and during delivery, but many mothers forget the discomfort they experienced prenatally and blame their postpartum backache on the epidural they had during labor.
▶ Interestingly, recent reports suggest that the use of low-dose mobile epidurals, which allow the mother to remain upright during labor, may reduce the chances of her experiencing postpartum backaches.

epidural or dural space will mean that epidurals and spinal anesthesia are not options.

Spinal block

Much of the general information that I have already offered about epidural blocks also applies to spinal regional anesthesias. The principles in pain relief are the same—namely to block nerve fibers that conduct pain from the pelvic organs. The difference with a spinal block is that instead of trying to avoid piercing the dural sheath (in the case of an epidural), the anesthesiologist deliberately passes the needle through the epidural space and punctures the dural membrane to inject anesthetic drugs into the fluid surrounding the spinal cord.

In recent years, spinal blocks have become increasingly popular for cesarean sections and emergency obstetric procedures because they are very quick to take effect—the pain relief is almost instantaneous, whereas an epidural will require 20–30 minutes. However, the spinal block is a one-shot technique and lasts for about an hour, possibly two, so it is not a useful method of pain relief throughout labor. Many anesthesiologists favor a combined spinal and epidural approach (CSE) for cesarean sections. The spinal block offers instant pain relief and the insertion of an additional epidural catheter allows the delivery of small doses of analgesic drugs in the post-operative period. CSE blocks are also useful during labor when rapid analgesia is needed in severe fetal distress in the late first stage or second stage of labor.

Pudendal block

This is a injection of local anesthetic into the vaginal tissues surrounding the left- and right-sided pudendal nerves (these supply sensation to the lower half of the vagina). A good pudendal block will markedly reduce pain in the vagina and perineum during the second stage of labor, but it has no effect on the pain of uterine contractions. Therefore, a pudenal block is usually reserved for low, uncomplicated lift-out forceps or vacuum extraction deliveries when the mother has been given no other method of pain relief. The pudendal block will last long enough to perform the delivery of the baby and repair the episiotomy or any vaginal or perineal tears that may have occurred.

Since this injection is deep within the vagina, the pudendal needle can be quite long. A pudendal anesthesic will have no effect on the baby and can be used in combination with morphine or nitrous oxide. Although it is usually inserted by a doctor, there is no need for an anesthesiologist to be present. Therefore, pudendal blocks are more commonly used in low-risk delivery units that do not have a 24-hour anesthesia service available.

Paracervical block

This local anesthetic is injected into the sides of the cervix toward the end of the first stage of labor to block the nerve fibers from transmitting the discomfort of cervical dilation. Since a paracervical block takes effect immediately, it is occasionally useful for an emergency forceps delivery.

> 66 *In recent years, spinal blocks have become increasingly popular for cesarean sections and emergency obstetric procedures because they are very quick to take effect.* 99

GENERAL ANESTHESIA

Although over the past 20 years epidurals have become increasingly popular for deliveries by cesarean section, to the extent that they have virtually replaced general anesthesia, general anesthesia is still sometimes used for abdominal delivery.

Having general anesthesia

All the technical preparations for the operation will be done while you are wide awake in the operating room. Your birth partner will be asked to leave as you are about to become unconscious. You will be asked to breathe deeply from an oxygen mask for several minutes to increase your oxygen levels, and to lie on the operating table on your left side, which further improves the oxygen supply reaching the placenta.

Only when everything is absolutely ready will the anesthesiologist ask you to inhale the drugs to put you to sleep. This will be followed immediately by the insertion of an endotracheal tube (airway) into your mouth and down your throat, to ensure that oxygen can be delivered to your lungs and to prevent you from regurgitating food or fluid from your stomach. Other drugs to relax your abdominal muscles will be given intravenously and the surgeon is then able to perform the operation very quickly, delivering the baby in a few minutes, before significant amounts of the anesthetic drugs have crossed the placenta.

Overall, you will be asleep for about 45 to 60 minutes. It takes a lot longer to stitch the tissue layers back into place and make sure that the bleeding is under control than it does to open up the uterus and deliver the baby.

REASONS FOR USING GENERAL ANESTHESIA

Maternal request
Acute fear of needles, backaches, the operative procedure (if a cesarean section is planned), or a previous traumatic delivery are all valid reasons for a woman to request general anesthesia.

Obstetric indications
Extreme emergencies, such as severe placental abruption or cord prolapse, are situations where the baby's life is at risk unless delivery is performed immediately. Severe bleeding may prompt the anesthesiologist to use general anesthesia, or even convert your regional block to one, because he or she will be better able to stabilize your cardiovascular system. Many favor general anesthesia for women with an anterior placenta previa (see p.428).

Maternal indications
Women with heart disease may be best delivered with the help of general anesthesia. Similarly, there are pregnant women with such severe abnormalities of their spine (such as curvature of the spine or spina bifida) that it is technically too difficult to insert a regional anesthesia.

Coagulation problems following maternal infection, hemorrhage, or preeclampsia can make it unsafe to insert a regional block due to risk of bleeding into the epidural subarachnoid space (see diagram, p.313).

NONPHARMACOLOGICAL PAIN RELIEF

There are various ways in which pain can be reduced without using any drugs, although you should understand that their effectiveness varies and what might help one woman could prove totally ineffective for another.

Broadly speaking these methods of pain relief can be divided into two main groups:
- methods involving the use of equipment or expert practitioners including acupuncture, hypnotherapy, reflexology, transcutaneous electrical nerve stimulation, and water births
- natural methods for use by yourself or with a birth partner including breathing and relaxation, massage, aromatherapy, and homeopathy.

Transcutaneous electrical nerve stimulation (TENS)

The TENS machine is a battery-operated device, connected by wires to small electrodes that are attached to both sides of your lower back by four adhesive pads. It works by conducting a small electric current through your skin to stimulate the production of endorphins (your body's own natural analgesics), which in turn help block the pain impulses carried by your nerves to your brain. The advantages of TENS are that you remain free to walk around, carrying the handheld control box, and you are in control of the frequency and strength of the electrical current. The TENS will not affect your baby in any way. It cannot be used while you are in water, so you will need to take it off while in a bath, or if you have chosen to have a water birth. Check with your maternity unit to see if TENS machines are available; if not, you may be able to rent one from another source. TENS is not a good method of pain relief in established labor.

Childbirth education classes

The three most common types of childbirth education classes—Lamaze, Bradley, and Dick-Read—vary greatly but the philosophy behind all three is the same: that the pain of childbirth is increased by fear and tension. All three classes teach the mother breathing exercises and

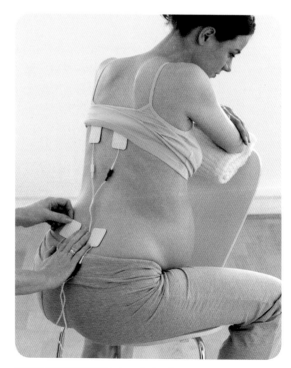

TENS MACHINE This is a choice for pain relief in early labor since the mother remains free to walk around and is in complete control of the level of pain relief she receives.

other physical conditioning and encourage the participation of the father in the birthing process, acting as coach. A new approach combines the techniques of Dr. Dick-Read with hypnotherapy.

Acupuncture

Acupuncture is another method of stimulating your body to produce endorphins. Instead of electrical currents, firm finger pressure or fine needles are applied to specific points on your body by a trained acupuncturist.

The theory behind acupuncture is based on the age-old Chinese belief that a life force called "chi" flows through the body and that medical disorders occur when there are imbalances in this life force. The balance is restored when the "chi" is unblocked by the insertion of fine needles into key areas of the body. Many women find acupuncture helpful during pregnancy to treat symptoms such as morning sickness, headaches, allergies, indigestion, backache and emotional disturbances including depression. It can be a very effective method of pain relief both before and during labor.

If you want to explore this option, you will need to find an acupuncturist who has experience in treating women in labor and is prepared to attend your labor at home or come into the hospital with you.

Hypnotherapy

Hypnotherapy works by suggestion. To relieve the pain of labor this means that you will need to be hypnotized into believing that you can control the pain of your contractions and, as a result, will be less troubled by them. There are several ways to achieve this, but they will all require careful preparation and plenty of practice before you go into labor.

You can hire a hypnotherapist to be with you during your labor, or you could arrange to have your partner trained to help hypnotize you. Self-hypnosis is a further possibility, but I suspect that for most women, with all that is going on during labor, this may be more difficult to achieve than anticipated. Nonetheless, hypnosis has been shown to be an effective method of pain relief in labor, may reduce the length of labor, and can be a useful tool with which to improve your emotional state both during and after delivery.

Reflexology

In reflexology, gentle pressure or massage is applied to specific points on the feet. This stimulates nerve endings, which helps relieve problems in other parts of the body. During pregnancy, reflexology can be used to help ease backache, general aches and pains and, in combination with more orthodox medical remedies, treat problems such as high blood pressure and gestational diabetes. Some therapists believe that attending regular reflexology sessions during pregnancy allows you the option of using this as a method of pain relief during labor. It is also claimed that reflexology can help the progress of labor by making the uterus contract more efficiently and the cervix dilate faster. If you decide to explore this option, you will need to spend some time with a therapist during pregnancy and make sure that both you and your partner learn exactly where and how to massage the trigger points on your feet during labor.

Water and water births

There is no doubt that immersion in water can relieve pain, particularly in the early stages of labor. The warmth of the water helps relax your muscles and the buoyancy of the water also supports your body, thereby relieving some of the pressure of the baby's head pressing down in your pelvis. Provided that your membranes have not ruptured, you can enjoy sitting in a warm bath either at home or in the maternity unit, for as long as you like during the early stages of labor.

I should mention here that most water-assisted births do not actually take place under water in the birthing pool. Even if you spend most of the first and second stage of labor in the water, some certified nurse midwives and doctors will still prefer you to be on "dry land" when the baby is actually born. This is quite simply because they will want to ensure that they have maximum access to you and your baby in the final critical stages of the delivery and also because not all certified nurse midwives are experienced in water births.

In the past, fears were expressed that delivering a baby under water might lead to problems with him inhaling water into the lungs as the first breath is taken. This is unlikely to happen if the baby is brought to the surface quickly, since the umbilical cord continues to deliver a good supply of oxygen for several minutes after the birth, providing it is not cut.

The other concern is that the mother's body temperature may rise if she remains immersed in warm water for a long period of time. This will result in the baby developing a high temperature and increased heart rate, which could lead to hypoxia (lowered oxygen levels). This is why your nurse will be checking your temperature regularly and will advise you to leave the birthing pool if your temperature increases by more than 33.8°F (1°C) during labor or delivery.

Breathing and relaxation

There is no doubt that we all feel pain more acutely when we are tense or frightened. By learning how to relax and breathe properly, you will feel much calmer and as a result, will be in a much better position to cope with the labor. The breathing and relaxation techniques that you have been taught in your childbirth classes will be put to good use during the last few weeks of pregnancy and particularly when you are in labor. Try to spend a few minutes each day practicing inhaling deeply and exhaling slowly. It is a good idea to enlist your partner's help here, so that you can rely on someone else to remind

WATER BIRTH Immersion in water is undoubtedly a valuable method of pain relief, particularly in the early stages of labour.

BIRTH STORY

Muriel, 31, has one daughter, two years and one month old
Second baby Killian, born 39 weeks + 5 days, weight 7lb (3.1kg)
Length of labor from first contraction: approximately 12 hours

My first baby, Maela, had been born at home—on dry land. For our second baby, we were planning to have another home birth but this time with the added benefit of a birthing pool and a certified nurse midwife. About three days before my due date, I felt a few contractions in the evening. They were quite strong but didn't last very long. The next day, when I took Maela to her day care, I told the staff there that I would have the baby that night. I had a strong feeling that labor would start for good that evening—which it did.

Right from the start, the contractions felt strong, although manageable. Because I was still breast-feeding Maela in the evenings, I fed her hoping she would sleep soundly through the night and this increased the strength of the contractions. They were coming every four minutes at that stage and I needed to concentrate on breathing and relaxing. My husband, Steve, started to get the birthing pool ready but we decided not to fill it until Margaret, my certified nurse midwife, said it was OK for me to get in.

At about 11pm, I phoned to tell Margaret that my labor had started and she came over to my house to see how I was doing. I was reallly happy to see her because I wanted to get into the pool, so I asked her to do an internal examination and it turned out that I had not dilated at all but my cervix was fully thinned. Margaret stayed for a little while and then went home. She had told me not to get into the birthing pool, so I just took a bath instead and felt it did help a bit. We were taking the contractions one at a time and trying to accept them or even welcome them (not very successfully). At about 3:30am, I did a self-examination since I am a childbirth instructor. I thought I had reached about 4cm (about 1½in) dilation so Steve phoned Margaret again and she arrived at our house at about 4am. This time she let me get into the pool and I finally felt I was where I should be to give birth to my baby.

The contractions were coming thick and fast and by 7am and I really wanted to push. I was in a half squatting position in the pool. By now, the contractions had changed and I could now feel the baby's head coming down. Margaret told me to look at her and pant. Our little baby was coming.

I grabbed the tiny body, brought it to the surface and held our little boy's head out of the water. We got out of the pool and I felt the placenta being delivered. Margaret asked me to cut the umbilical cord, which I did. It felt wonderful to at last be holding my newborn baby boy.

> 66 *I grabbed the tiny body, brought it to the surface, and held our little boy's head out of the water.* 99

you of these simple but important principles when you need them most.

In the early stages of labor, you will need to concentrate on breathing in and out slowly as each contraction starts. The secret is to close your eyes and inhale calmly through your nose, imagining that the breath is filling every part of your body. Concentrate on relaxing all of your muscles as you do so. Then, exhale slowly through your mouth, this time imagining that you are drawing out and exhaling the pain of your contraction. You may have been taught visualization techniques in your childbirth classes. The principle here is to focus on an image or visit a place in your mind that you find calming and soothing in order to take your mind off the pain. This is a sort of hypnosis and works better for some women than others.

As the contractions become more intense, you will probably find that you need to take shorter breaths, in groups of two or three, since strong contractions make it difficult to keep up the slow breathing rhythm. At this stage you need to remember that breathing out is the most important part, since the breathing in will take care of itself if you are breathing out properly. My best tip here is to imagine that you are blowing wind through trees and that every breath needs to reach a spot about 12in (30cm) away. I often find that if I sit on the bed beside a woman in strong labor and ask her to ensure that every outward breath hits my nose, that she gets the hang of the new rhythm very quickly. Once achieved, it is easy to get her partner to take up the same position and encourage her to continue breathing out in the same way.

Massage

A good massage is always relaxing and when you are in labor, it can really help relieve a backache. If your baby is lying occiput-posterior with his

MASSAGE This provides a great opportunity for your partner to become physically involved in helping you cope with your labor. Encourage him to master the art of giving a good massage while you are pregnant.

spine closest to your spine and sacral bones, you will be particularly glad to have someone massage your lower back, preferably in slow, firm circular movements just above the cleft of your buttocks.

In addition to the physical comfort this provides, I think that it is also a source of emotional comfort. The fact that someone else is with you and helping you deal with the discomfort or pain helps reduce the feelings of isolation and fear that being in an unknown situation always engenders.

Asking your birth partner to extend the massage to your shoulders, neck, face, forehead, and temples will further help relieve tension and anxiety, enabling you to feel more relaxed during labor. Make sure that the person doing the massage warms his or her hands and removes any jewelry. Using scented oils or creams will help the hands slide easily over the skin.

Aromatherapy

In aromatherapy, essential oils are used to soothe and relax the body. It is thought that the scented oils trigger the nervous system to produce natural endorphins and that this helps reduce tension and alleviate some of the pain.

During early labor, the diluted oils can be absorbed through your skin when combined with massaging or by adding them to your bath or birthing pool. Alternatively, you can inhale them from a burner or vaporizer, which gently warms the oils producing a soothing, scented atmosphere.

If you want to use aromatherapy during labor, you will need to take the oils and the vaporizer with you to the maternity unit. Check beforehand that all of the oils you are planning to use are suitable for use in pregnancy.

Homeopathy and herbs

There are a wide variety of homeopathic and herbal remedies that can be used during labor to help relieve stress and discomfort. However, it is important to ensure that you obtain specialized advice about the types of herbs that are suitable and the doses that should be used. Remember, too, that you need to consult with your nurse or doctors about anything you are taking when you are in labor.

MY PRACTICAL TIPS TO HELP WITH LABOR PAINS

Here is a selection of practical tips and thoughts that I hope will come in handy on the big day. I was given them by a close friend and have since shared them with many of my patients.

▶ Labor is like walking a tightrope. The goal is to stay balanced and on the rope.

▶ Therefore, one step at a time, one contraction at a time.

▶ Don't think about how much farther you have to walk. No one can tell you how much longer it will be before you give birth.

▶ Concentrate on getting through the next contraction with your breathing techniques.

▶ Don't think about how much pain you may be in three contractions'

time, just think about getting through your next "step" on the tightrope.

▶ Levels of pain vary along the tightrope and are unpredictable. What you are going through now could be better or worse in half an hour's time, so there is no point in worrying about it.

▶ Each contraction gets you one step closer to the end of the tightrope—the birth of your baby—so there is a very good reason to keep going.

▶ You should try to stay fed and hydrated in order to keep your energy levels up.

▶ Try to find ways of distracting yourself in order to help take your mind off the pain.

▶ Find different positions that help you relieve your pain.

▶ Above all, try your best to stay as relaxed as possible. Tension increases pain.

▶ If you want pain relief, ask for it sooner rather than later and do not hesitate to request a stronger dose of morphine or an increase to your epidural, if you need it.

You may find it helpful to write out a list of your own reminders or practical coping tips for labor and delivery, which you can then pack in your hospital bag with your birthing plan to be sure that they are close by when you need them, or just in case your mind goes totally blank in the middle of your labor.

The second and third stages

THE SECOND STAGE OF LABOR begins when your cervix is fully dilated and ends with the birth of your baby. This is followed by the third stage, when the placenta and membranes are delivered. You will probably feel tired as you enter these phases, but knowing that the end is in sight will, I hope, spur you on through these final stages.

During the second stage of labor your baby is forced through your birth canal by uterine contractions that are now stronger and more frequent, occurring every two to four minutes, and lasting for 60–90 seconds. At this stage, you will probably feel as if you are contracting continuously and that labor has taken on a momentum of its own, which it has. There is now no stopping normal labor until the baby has been expelled from the birth canal. In first labors, the second stage may last for two to three hours, although the average length of time is one hour. In second or subsequent labors, the second stage usually lasts for about 15–20 minutes but can be much quicker and the baby may start to crown before the mother and her doctor recognize that she has entered the second stage.

The third stage of labor—the delivery of the placenta—usually lasts about 10 to 20 minutes, but can be shorter or longer, depending on whether the delivery is actively managed or the placenta is left to expel spontaneously (see p.333).

Someone to lean on

Your birth partner has an important role during the second and third stages of labor. He or she will be able to support you physically as you work hard to push your baby out, as well as being able to tell you what can be seen as the baby's head emerges. He or she can also give you invaluable verbal encouragement and reassurance as you both experience the birth of your baby.

66 *...during the second stage you will probably feel as if you are contracting continuously and that labor has taken on a momentum of its own, which it has.* 99

The second stage of labor

For most mothers, the first indication that they have completed the transition phase and entered the second stage of labor is that they develop a strong urge to push or bear down. However, it is important to refrain from pushing until the doctor or nurse confirms that you are fully dilated.

> *When it is time to start pushing, it is vital that you work as a team with your doctor or nurse.*

Once your doctor or nurse confirms that you are fully dilated and that it is time to start pushing, it is vital that you work as a team with her. Remember that the uterine contractions are going to continue involuntarily and what you now need to do is add the force of your voluntary pushing efforts to help expel the baby. Listen carefully to what the doctor or nurse tells you to do.

The idea is to push as much as you can as each contraction peaks and rest in between them. As each uterine contraction starts, you need to take a deep breath, hold it, brace your feet and push downward to force the baby lower into the pelvis. Try to contract your diaphragm and abdominal muscles and push down into your pelvis—not into your belly, which won't help the baby's head descend. Try to visualize what is actually happening and where you are pushing. The effort needs to be directed specifically into the vagina and rectum, rather than vaguely "somewhere down there." Ideally, try not to hold one long breath, which can end up depriving you of oxygen and making you feel light-headed. During a good contraction you should be able be able to take three separate breaths and achieve three good pushes.

Knowing when to push

If you have had an epidural, the dose increases will hopefully be timed so that the effects are beginning to wear off by the time you begin the second stage. This means you will be aware of the contractions but will not be in pain. If you are totally numb from the epidural (unlikely with the newer continuous slow rate of infusion), your doctor will alert you to the start of each contraction, although you may learn quickly to recognize them because you will feel your uterus tighten when you place your hand on your abdomen. Furthermore, you and your partner will also be able to see from the fetal monitor when the contractions are beginning and ending. However, I think it is best for you to concentrate on breathing and pushing and let your partner and doctor or nurse prompt you when to start and stop. At the end of each contraction try not to relax too quickly—the baby will not continue to move forward unless you relax slowly.

DEALING WITH THE SECOND STAGE

Find a comfortable position that is reasonably upright and avoid lying flat. By this stage most women find that they want to be supported on a bed, but some prefer to squat or use a birthing stool. The more upright you are, the more gravity helps and the quicker your baby will be born.

Many women worry about sounding or looking ridiculous during the second stage of labor. This is not the time to worry about appearances; it is the time to do what feels natural and comfortable for you. If that means grunting and making a lot of noise while you are pushing, that's fine. Similarly, it is pointless feeling worried or embarrassed that you may move your bowels or urinate toward the end of the second stage of labor. This is very common and you need to remember that your nurse and doctor will have seen it all before many times.

Interestingly, once they have reached the second stage of labor, most women lose many of their inhibitions and cease being self-conscious. The combination of concentration and instinct take over and there is no time to think about anything other than doing what is necessary to push the baby out.

The other thing that surprises many women is that, compared to some of the late first-stage contractions, the ones in the second stage often seem more bearable. I think this is because you are now actively participating in the labor and this helps dispel some of the tension caused by painful contractions. Although the pushing is hard work, you will feel a sense of satisfaction as each push brings the birth of your baby closer. Knowing that the finishing line is in sight boosts your energy reserves.

POSITIONS FOR SECOND STAGE

SITTING UPRIGHT Supported by pillows, try to relax your back between contractions.

KNEELING A helper on each side supports you, or kneeling on all fours may be more comfortable.

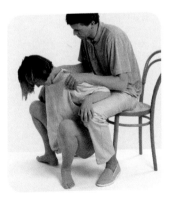

SQUATTING This opens the pelvis wide and uses gravity to help push the baby out.

Descent and delivery

As your baby's head is pushed deeper into the pelvis you will become increasingly aware of pressure in your rectum and may experience pains that radiate down your legs caused by pressure on the nerves in your sacral area. This stage can be extremely painful as the anus starts to bulge and the vagina and perineum are stretched by the emerging head. At the height of a contraction the tip of the baby's head will become visible, although, at first, it will slip back up the birth canal when you are not pushing. Gradually, the head stays in place and starts to crown. It is common to experience the sensation of intense stinging or burning because the vagina is stretched to its limit as the head is on the verge of being delivered. It is now very important to follow your doctor's instructions, particularly when she tells you to stop pushing and to pant instead. This helps avoid the head delivering suddenly, which is dangerous for the baby and may tear your vagina and perineal tissues.

Once the head starts to crown, it usually takes only a couple of contractions for it to be delivered. Your doctor will assess whether you need an episiotomy or whether it is possible to deliver the head safely without one. During the contractions, some doctors will support the perineum, whereas others do not touch it—either method is safe and is the personal choice of the doctor. The doctor will also support the baby's head to help prevent a rapid delivery. Once the head is delivered, the doctor will support it and feel around the neck to make sure the umbilical cord is not wrapped around it. If it is, the cord will be

DELIVERY OF YOUR BABY

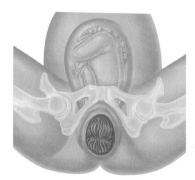 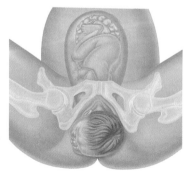 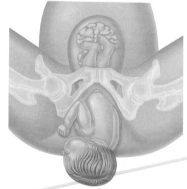

ONCE THE HEAD STARTS TO CROWN it usually takes only a couple of contractions before delivery. Most babies are in an anterior position so the head will emerge with the nose pointing downward.

AS SOON AS THE BABY'S HEAD IS FREE of the perineum, the neck will extend and the baby will automatically rotate to face left or right, so that the shoulders are in the best position to deliver smoothly.

THE FIRST (ANTERIOR) SHOULDER will slip under the pubic bone followed swiftly by the posterior shoulder. The rest of the body slips out and the baby is usually delivered onto the mother's belly.

gently looped over the baby's head and at the same time the baby's nose and mouth will be wiped free of blood and mucus.

During the next contraction, the first, or upper, shoulder will be delivered, usually aided by the doctor applying gentle downward traction on the sides of the baby's head. As the first shoulder slips free from under the pubic bone, the doctor will gently sweep the baby's head and shoulder upward, allowing the second or posterior shoulder more room to emerge during the next few contractions. Once both shoulders are delivered, the rest of the baby literally slides out, followed by a stream of amniotic fluid, which was behind the shoulders. The doctor or labor nurse will be waiting to catch this slippery bundle, covered with blood, fluid, and vernix. If you request, they will place the baby on your belly or chest, covered or wrapped in towels to stay warm. If you intend to breastfeed, you may want to put your baby to the breast immediately.

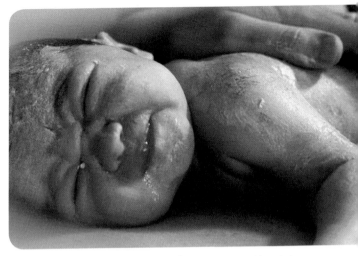

AT BIRTH Your baby will be a slippery bundle, covered in blood, fluid and vernix.

Managing the second stage

Presuming there are no delays or complications, the second stage will usually be completed within two to three hours in a first labor and well within two hours for a second or subsequent birth. Throughout the second stage, the doctor will keep a careful watch on the baby's heart rate after each contraction and maternal push, with an external or internal electronic monitor, depending on what has been happening during the first stage of labor. She will also keep note of the strength and regularity of your uterine contractions. Sometimes they start to fade away during the second stage, in which case you may be advised that a low-dose oxytocin infusion is needed to restore the contractions and enable you to push your baby out promptly.

Every hospital has a policy about how long they will advise a woman actively to push, but because of the risks of fetal distress and maternal fatigue most maternity units will not encourage you to continue pushing for more than two hours in a first labor or one hour in a subsequent labor before suggesting that they assist the delivery with either vacuum extraction or forceps. There are no hard and fast rules if you are both coping with the labor, and careful monitoring is important.

EPISIOTOMIES AND TEARS

Ideally you will give birth without needing an episiotomy or ending up with a perineal tear. However, should the need arise at the time of delivery, careful consideration will be given as to which is likely the better option for you. each has its advantages and disadvantages so find out as much as you can about your hospital's policy on episiotomies and perineal tears during the prenatal period.

EPISIOTOMIES

An episiotomy is a deliberate incision made in the stretched perineum and vagina to prevent uncontrolled tearing of the mother's tissues as the baby's head is delivered. It used to be thought that an episiotomy not only prevented extensive tears of the perineum but also the development of vaginal prolapse in later life. Since the evidence for the latter is now doubtful, episiotomies are no longer performed routinely.

There are, however, several situations when an episiotomy may be advisable:
▶ tight perineum in a first or subsequent labor
▶ large baby
▶ fetal distress that requires immediate delivery
▶ forceps or vacuum delivery
▶ to protect the head of a premature baby
▶ to protect the baby's head in a vaginal breech delivery (however, most breech babies are delivered by cesarean section).

If you have strong views about episiotomies—either that you want to avoid one at all costs or that you positively want to have one—you should make this clear to your doctor early in labor.

THE PROCEDURE

If your doctor judges that you need an episiotomy, they will ask your permission before they perform it. They will then swab the area with antiseptic solution and give you a local anesthetic in your perineum unless you have an active epidural in place.

The most usual incision is a J-shaped mediolateral cut angled away from the vagina and rectum. It is also possible to make a midline cut straight down from the bottom of the vagina toward the rectum. Both types of incision are made with scissors and because the perineum is tightly stretched, almost paper thin, at the time the cut is made, the bleeding is usually minimal. The advantage of a

mediolateral episiotomy is that it keeps the incision well away from the rectal area. This is particularly relevant if a forceps delivery is anticipated because this may extend the cut. The midline episiotomy avoids several blood vessels and is usually easier to repair, but if it extends during the delivery, it is more likely to tear into the rectum.

Once the baby and placenta have been delivered, your doctor will repair the episiotomy with stitches. The doctor and nurse will probably suggest that your legs be placed in stirrups to make the repair easier. You will be given another dose of anesthetic to make sure you feel no pain during the procedure.

The episiotomy will be sutured in layers, ensuring that all of the tissues are brought together properly to repair the vagina and perineum. The sutures may be interrupted or continuous and either external or buried below the skin. Whatever type is used, they all dissolve and do not need to be taken out.

> **❝** *…make your views on episiotomies clear to your doctor early in your labor* **❞**

COMMON CONCERNS

The most common concerns about episiotomies are how much they will hurt afterward and how long they will take to heal. The reality is that they are painful, particularly on the second or third day after the birth, when the stitches often feel tight and uncomfortable. This is the result of the body's natural healing response, which inevitably causes swelling of the traumatized tissues.

Much relief can be obtained from placing maternity cool packs against the episiotomy area and using an inflated rubber ring to sit on. However, the vagina has an excellent blood supply and most cuts will heal in one to two weeks as long as the area is kept as clean and dry as possible. Regular bathing with warm water will help relieve discomfort. It is not necessary to add disinfectant solutions and it is important to avoid highly perfumed soaps and oils, which can irritate the healing wound.

In the long run, most women do not have problems with the episiotomy incision site but some experience continuing perineal pain, which can be a source of distress, particularly when it causes problems with sexual intercourse. Massaging the scar tissue with emollient or estrogen creams can help make the tissues more supple, and Kegel exercises will usually help improve the situation. If the discomfort continues, advice should be sought from your doctor regarding the possiblility of the need for another surgery.

PERINEAL TEARS

There are four degrees of perineal tear.

▶ **First degree:** these are minor tears to the vaginal skin around the entrance to the vagina. Most will heal well without any sutures.

▶ **Second degree:** the posterior vaginal wall and perineal muscles are torn but the anal sphincter muscles remain intact. Most will require several sutures to restore the anatomy of the muscles and more superficial tissues.

▶ **Third degree:** the anal sphincter muscles are torn but the mucous lining of the rectum remains intact. These need careful repair to ensure that the muscle layers are realigned.

▶ **Fourth degree:** the anal sphincter muscles are torn to the extent that the rectal mucosa is opened. Considerable skill is required to repair a fourth degree tear because the apex of the tear must be secured to prevent the development of a rectovaginal fistula (a persistent opening between the vagina and the rectum). These tears are uncommon, occurring in only one percent of births. They usually follow forceps or vacuum extraction delivery for a first baby, persistent occipitoposterior positions, or the delivery of a baby weighing more than 9lb (4kg).

EPISIOTOMY CUTS

The mediolateral cut for an episiotomy—in which the cut is angled down and away from the vagina and the perineum—is the preferred choice in the U.K. and most of the rest of the world. The exception is North America where doctors used the mid-line cut in which the cut was perfomed straight down into the perineum between the vagina and anus. Research suggests that tearing is more likely following a midline episiotomy cut.

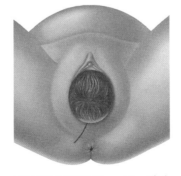

MEDIOLATERAL The cut is angled down and away from the vagina and the perineum into the muscle.

The third stage of labor

The third stage of labor involves the delivery of the placenta and membranes. The minutes after birth are invariably an emotional time and you and your partner will probably be aware only of the fact that you are finally holding your new baby.

Cutting the cord

Many women request that the baby be placed on their belly immediately after delivery to start the bonding process. The cord will still be attached at this point and will continue to pulsate for one to three minutes. Unless the baby has been distressed during the delivery and is in need of prompt attention from the neonatologists, there is no hurry to clamp and cut the cord. Indeed, it is important to wait 2 to 3 minutes to allow blood to redistribute from the placental part of the circulation into the baby's. This provides the baby with a normal volume of blood and extra oxygen until the lungs start working properly. Without this blood transfer, the baby is more likely to have anemia and iron deficiency.

The doctor will place two clamps in the middle of the long umbilical cord, 1–2in (3–5cm) apart, to prevent bleeding from the baby at one end and from the placenta at the other end. She will then cut the cord between the two clamps, or your partner can cut it. Later on, the cord will be trimmed more and a plastic clip will be attached to the stump near your baby's umbilicus. Over the next few days this bit of remaining cord will shrink and the plastic clamp will either fall off or be removed, leaving the baby with a little knot of tissue at the umbilicus, or belly button, that will soon disappear.

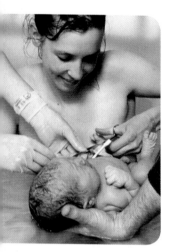

CUTTING THE CORD
This is a straightforward procedure. If your partner would like to do it, let your nurse or doctor know in advance.

Delivery of the placenta

Having cut the cord, the next thing your doctor will do (in addition to checking your baby) is make sure that the placenta is delivered correctly and promptly. After the baby is born, continuing uterine contractions and the shrinking of the uterus considerably reduces the size of the placental bed. The placenta buckles inward, tearing the blood vessels and attachments to the uterine wall. This results in a small hemorrhage behind the placenta which further helps it separate from the uterus. The process starts as soon as the baby is born and is usually complete in five minutes. However, the placenta is usually retained within the uterus because the membranes take more time to strip away from the uterine wall. Once this placental separation has occurred, the uterine muscular wall clamps down on the blood vessels in the placental bed and encourages the formation of clots in the torn ends, reducing further blood loss.

Physiological management

When the placenta and membranes are allowed to separate from the uterus on their own and no attempt is made to deliver them until clear signs of separation are seen, this is described as physiological management of the third stage of labor. These signs include a gush of blood (the retroplacental hemorrhage), which is followed by contractions that make the fundus of the uterus rise in the abdomen. There is also a lengthening of the umbilical cord that is visible outside the vagina and you will experience an urge to bear down, which is a clear sign that the placenta has separated and that the uterus is trying to expel it into the vagina. When these events have occurred (which generally take about 20 minutes), the doctor will place her hand above your pubic bone in order to hold the uterus in place, and will ask you to give a short push downward while she gently pulls the umbilical cord and encourages the separated placenta to leave the vagina, followed by the membranes and the retroplacental blood clot. The nurse or doctor will then massage the uterus firmly to help contract it more and prevent any more bleeding. More massaging of the uterus may be needed at regular intervals during the first hour after the birth to keep it firmly contracted.

If you would prefer not to have medical intervention in the third stage, you can speed up the natural process and try to avoid heavy bleeding by placing your baby at your breast and encouraging breast feeding as soon as possible. Suckling stimulates the release of oxytocin, which causes your uterus to contract and the placenta to separate from the uterine walls. Making sure that your bladder is empty will also help deliver the placenta promptly.

Active management

The reason why many maternity units advise active management of the third stage is because bleeding after the delivery of the baby and placenta can be torrential. Indeed, postpartum hemorrhage (see p.335) remains the most

66 ...you can speed up the natural process by placing your baby at your breast... 99

important cause of maternal death worldwide. During the prenatal period, your doctors will have discussed the third stage with you and explained that active management involves giving you an injection of oxytocin in your thigh muscle as soon as the baby's head and first shoulder have delivered. Oxytocin helps firmly contract the uterus, separate and start to expel the placenta and membranes, and then maintain the uterine contraction without any relaxation for about 45 minutes. Having waited for the uterus to firmly contract, your doctor will place a protective hand above your pubic bone in order to prevent the uterus from being pulled downward when the umbilical cord is gently pulled. This process is called controlled cord traction (CCT) and it usually results in prompt delivery of the placenta and membranes. If undue force is used at this stage it can lead to uterine inversion—the uterus is literally turned inside out.

Your physical response

Immediately before or after the delivery of the placenta, it is common for mothers to experience strange physical reactions in response to the enormous effort of giving birth. You may find yourself shivering and shaking uncontrollably and your teeth chattering violently, as if you have suddenly been exposed to extreme cold. This response is frequently accompanied by intense nausea. Since you are likely to have an empty stomach at this point, this will probably amount only to retching of bile and watery fluids. Rest assured that these reactions are completely normal and common. Your doctor will not be surprised, having seen them many times before, and will help you and your partner deal with them.

CHECKING THE PLACENTA AND MEMBRANES

FETAL SURFACE This smooth side of the placenta has blood vessels radiating out from the umbilical cord.

As soon as the placenta and membranes are delivered, they will be checked to make sure that they are complete. A healthy term placenta weighs about 1lb 2oz (500g), measures about 8–10in (20–25cm) in diameter, and looks like a spongy disk. Any unusual appearances will prompt your doctor to send it to the pathology laboratory for analysis. In the vast majority of cases, the placenta will be unremarkable and, after weighing it and recording all the observations in your notes, it will be disposed of by the hospital or can be stored in special banks. You may want to look at this extraordinary lifeline before it is discarded or stored.

When problems arise

Occasionally, complications can arise during the third stage of labor.

A retained placenta is one that remains within the uterus over 30 minutes after delivery of the baby. About one percent of deliveries are complicated by this and it is more likely to occur after very premature births because the umbilical cord is thinner and may snap more easily during cord traction. Since a retained placenta is often associated with postpartum hemorrhage, prompt action is needed to remove it. This is usually done in the operating room.

Early (primary) postpartum hemorrhage (PPH) is defined as the loss of about one pint (500ml) of blood from the uterus or vagina within 24 hours following a vaginal delivery and 1 liter (about 2 pints) following a caesarean. About 6 percent of deliveries in the US are complicated by a PPH, but it is more likely to follow a prolonged labor if the woman develops an infection during labor or if the placenta is sticking to the uterus. The incidence of PPH has fallen over the last 50 years because of improved awareness of the situations that are likely to cause it and either preventive management or prompt treatment when it occurs. Active management of the third stage of labor is an important preventive measure that has contributed to the the reduction and prenatal diagnosis of placenta previa (see p.428), improved anesthetic techniques, and the realization that prolonged or difficult labors are more likely to lead to a PPH have also contributed to the reduction.

When a serious PPH does occur, strict labor ward protocols, involvement of senior obstetricians and anesthesiologists, improvements in intensive care, readily available blood transfusions, better antibiotics, and a major reduction in the number of women who become severely anemic during pregnancy has significantly reduced the number of maternal deaths.

Late (secondary) postpartum hemorrhage is defined as any sudden loss of blood from the uterus or vagina, regardless of the volume of blood, from 24 hours to six weeks after the delivery. Late PPH occurs in 1 in 50–200 births, and is usually caused by retained pieces of placenta or membranes in the uterus. These frequently become infected when they are left in the uterine cavity and the inflammation that accompanies the infection further contributes to the bleeding. Usually the mother complains of feeling ill with pain and tenderness in the lower abdomen and develops a temperature and a smelly vaginal discharge. The problem needs to be identified quickly and treated with antibiotics. The removal of the retained tissues under general anesthesia is also usually required to resolve most late PPHs.

...active management of the third stage has contributed to the reduction in PPH.

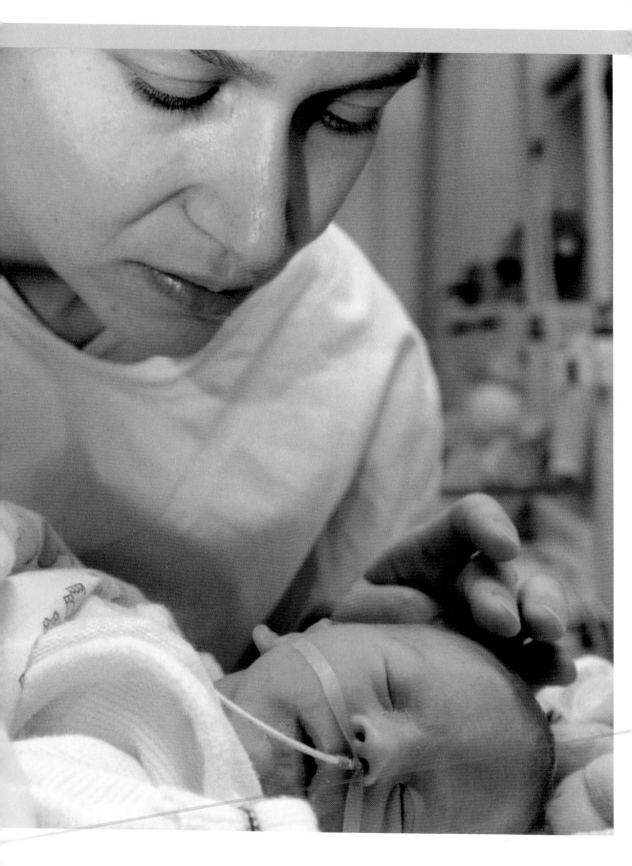

When help is needed

Although all women and their caregivers hope for a full-term pregnancy and an uncomplicated delivery, when events make this difficult or impossible, extra medical help will be needed to provide safety to both the mother and baby. If this should happen to you, being aware of what can occur during labor and birth and the possible outcomes will help you make informed decisions along the way.

CONTENTS

Premature labor

AROUND 11 PERCENT OF BIRTHS IN THE US are classified as premature because they occur before 37 weeks. However, thanks to the enormous advances that have been made in neonatal care over the past 10 years, babies born after 30 weeks who have not had serious complications in utero are unlikely to experience major long-term developmental problems.

The most important thing to remember here is that the longer a healthy normal baby stays in utero and the greater its birth weight, the lower the chances of the baby experiencing problems after delivery and the shorter the time that he or she will have to spend in an neonatal intensive care baby unit. The chances of survival without handicap for a baby born at 23 weeks is only one percent, but with the passage of each week, the chance of survival improves significantly, so that at 26 weeks nearly one quarter will survive unharmed and by 30 weeks the risk of handicap is very small. This is the reason every effort will be made to keep your baby in utero for as long as possible, provided there are no problems that suggest it would be better cared for in the outside world. It is also important to understand that only about 11.8 percent of all births occur prematurely before 37 weeks, and below 28 weeks the figure is less than 1 percent.

Doctors spend a lot of time to identify those pregnant women who are at greater risk of going into premature labor. In the Journey section of this book I have included many mentions of the possible symptoms that may help you, your doctor to recognize that you are at risk of delivering your baby prematurely.

...every effort will be made to keep your baby in utero for as long as possible...

Causes of premature labor

There are many reasons for premature delivery, but despite the research to try to predict why a pregnant woman might go into premature labor or why her membranes may rupture weeks before the due date, we are still unable to prevent the vast majority of preterm births. Indeed, we do not know exactly what triggers labor itself, let alone what the exact mechanisms are that set this trigger off too early. One theory focuses on the role of hormones secreted by the baby, the mother, or the placenta, while another indicates that the levels of a specific protein called fibronectin that is present in the vagina and cervix will increase markedly when a woman is about to go into labor. It appears that infection plays a part in

20–40 percent of all premature births, and if you have already had a premature baby but there was not a clear medical reason for the early birth, then statistically you are more likely to have another premature birth.

However, babies sometimes have to be delivered early for medical reasons. These include preeclampsia, high blood pressure, diabetes, placental insufficiency, placental abruption, and bleeding from placenta previa (see pp.426–28). If you have already had a preterm baby as a result of any of the above, doctors will keep a close eye on you to try to avoid the same thing happening again.

Signs of premature labor

If your water breaks before 37 weeks or you experience abdominal pain or vaginal bleeding or you start to develop uterine contractions, you should contact your doctor and make sure that you are promptly examined by a her at the hospital. She will check whether your uterus is contracting and the baby's position and then perform an internal examination to assess the cervix, determine the presenting part, and check that the umbilical cord has not prolapsed (see p.430). They will also be looking for signs of infection, now that the protective membranes have gone. If there is a significant risk of infection, they will want to induce or augment uterine contractions with an oxytocin IV (see p.297) to deliver your baby as quickly as possible. Sometimes you will need to deliver by cesarean section, especially if you or your baby are showing signs of distress, the presentation and lie of the baby are not optimal, or the cervix is very unripe.

You must also call the doctor immediately if your water has not broken but you think you are getting contractions, or if you have had a bloody show. You will almost certainly be advised to go to the hospital, after which you will be monitored closely and advised to have complete bed rest. So long as there are no contraindications you may also be given a tocolytic drug in an attempt to stop the uterine activity. Nifedipine (a calcium channel blocking drug) is cheapest and can be given by mouth but has some maternal side effects. Terbutaline is a Beta adrenergic receptor agonist which is also used. These are usually reserved for woman with uterine contractions and a positive fetal fibronectin test (see box, left) who require steroids, in utero transfer or have another reason to delay delivery. If your membranes are intact and your contractions are only mild, then bed rest, with or without drugs, may halt the beginning of labor, even if your cervix has begun to dilate. You will be able to go home as soon as your contractions have stopped,

FETAL FIBRONECTIN TEST (fFN)

▶ **Indications**
Threatened Preterm Labor (PTL)
Baby 24 to 34 weeks
Membranes intact
Cervix < 3cm dilated
Healthy fetus

▶ **Method**
Take vaginal swab from mother
Place swab in test fluid
Machine reads Fibronectin + or −

▶ **Interpretation of results**
Positive (+): 1 in 6 women will deliver within next 14 days.
Action: start treatment with tocolytics, steroids, arrange in utero transfer.
Negative(−): 1 in 25 women will deliver in next 14 days.
Action: reassure mother, discharge home if contractions settle.

although you will need to take it easy for the remainder of the pregnancy and avoid having sex.

If contractions are well established, however, it is difficult, even with the use of drugs, to delay delivery by much more than 48 hours. Even this short time can be of critical importance for your baby. It allows you to be transferred to a hospital with a NICU and gives time for a dose of prenatal steroids to be given to you to help mature your baby's lungs (see p.342). Before 34 weeks, the advantages of the baby remaining in utero are usually greater than those of being delivered, so you may be advised to take one of the drugs mentioned earlier, in an attempt to stop or reduce the uterine contractions. However, none of the tocolytic drugs (those used to help stop contractions) currently available can be thought of as reliable once it has been triggered to start early in pregnancy.

Giving birth to a premature baby

If doctors cannot stop your labor, then, for all intents and purposes, you will go through a normal delivery as long as your baby does not show signs of distress, in which case a cesarean will be done. The good news is that your labor is likely to be a little shorter than that for a full-term baby. The head will be just that little bit smaller and this can make all the difference between needing an episiotomy and having just a little minor tearing or nothing at all. Despite this, doctors may decide to use forceps (and therefore do an episiotomy) to protect the baby's head as it descends the birth canal, because a premature baby's skull will be softer than that of a full-term baby. You will probably be advised to avoid morphine as a means of pain relief, because this can slow down an already fragile baby in its attempt to move down the birth canal and it causes problems after delivery because the respiratory system is depressed.

It is unlikely that you will be allowed to attempt a vaginal birth if your baby is breech, even if you are already in labor, because this is considered too risky (see p.356), so a cesarean will be done instead. The exception is a gestational age of less than 26 weeks when a vaginal breech delivery is preferable to a cesarean. Similarly, if you have developed a condition, such as placental abruption, bleeding from the placenta, or preeclampsia, vaginal delivery will be too dangerous for your baby.

If your premature baby is delivered vaginally, a pediatrician will be at the birth as well. Immediately after the birth, the baby will be assessed and given help breathing if necessary. Depending on the degree of prematurity, you may have the opportunity to hold your baby briefly before he is taken to the NICU. If your hospital does not have a dedicated NICU, you will either have been transferred to the closest maternity unit before the delivery that does, or your baby will be transferred there in a special ambulance immediately after the birth.

...a preterm labor is likely to be shorter than that for a full-term baby.

Problems your baby may encounter

Most healthy babies born at or before 35 weeks will need help from a neonatal intensive care unit (NICU) because they are likely to experience breathing difficulties and require help with feeding. The breathing problems result from the fact that their lungs are not yet sufficiently developed and elastic enough to breathe unaided.

As discussed in the Journey section (see p.231) the fetal lungs continue to develop more tiny airways and alveoli well into the third trimester of pregnancy and only start to produce surfactant from about 26 weeks. This substance coats the developing alveoli, allowing them to remain open and available for oxygen exchange after birth. When there is little surfactant present, the lungs are rigid and collapse down easily, which makes every breath the newly born premature baby takes more difficult. This is why a mechanical ventilator is often needed to push the air into and out of the immature lungs, and the neonatal doctors may suggest spraying some artificial surfactant into the baby's lungs. After 35 weeks there is usually enough surfactant available to make full ventilation unnecessary, although your baby may require some little tubes placed in his nose for a short time to ensure that sufficient oxygen is readily available.

If you have to be delivered at or before 35 weeks, you will be given an injection of steroids (either betamethasone or dexamethasone) to speed up the production of surfactant in the baby's lungs. The steroids need 24 to 48

IN SAFE HANDS
Don't be alarmed by the equipment you encounter in the intensive care unit. The machinery and tubes are there to monitor your preterm baby and to help her breathe and feed until she is able to do it herself.

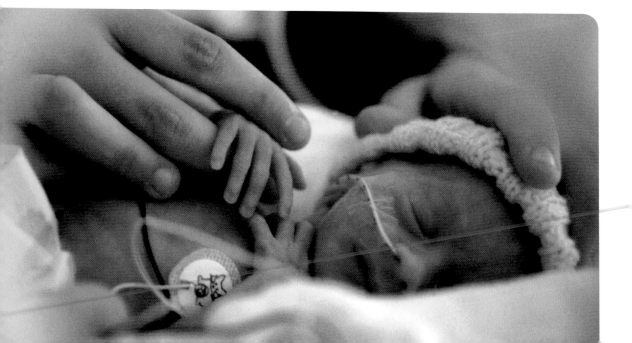

hours to reach their full effect, so in cases of threatened preterm labor, the doctors will always try to delay the delivery for enough time to allow the steroids to work. Similarly, your baby's sucking reflex is poorly developed before 35 weeks (see p.231) and the digestive system is often too immature to cope with large liquid feedings. This is why many premature babies need to be tube fed with small quantities of milk (preferably expressed breast milk) at frequent intervals. They soon get used to larger meals and I often find myself telling anxious mothers who have a baby in the NICU that when their baby can deal with 2fl oz (60ml) of tube feeding every three to four hours, she is likely to be discharged from the NICU very soon. This is a clear indication that she will now be capable of suckling on her mother's breast or from a formula milk bottle.

After the birth

Rest assured that, although the birth of a premature baby is invariably very high-tech, it is vital to assemble all the machines, equipment, and staff to make sure that your baby is given the best care possible at this delicate stage in her life. The medical staff attending the birth will be well aware of how upsetting the whole situation usually is for parents—particularly first-time parents with no previous experience in childbirth—and they will do their best to make things easier for you. They will take a photograph of your baby within minutes of birth, so that you can put it beside your bed and get to know your baby's face. They will also give you all the time you need to ask questions, both during and after the birth, and will do everything in their power to reassure you and to explain the situation clearly to you.

On pages 404 and 405, I provide more detailed information on caring for premature babies, particularly during their stay in the NICU. Remember, though, that however distressing the early days and weeks can be, the vast majority of premature babies go on to become strapping toddlers who are every bit as healthy as their full-term peers.

My daughters were delivered by emergency cesarean section after I went into premature labor at 33 weeks. They were both a reasonable size for twin babies of their gestational age and had not had any problems prenatally. Nonetheless, they both needed assistance with breathing and feeding and were in the NICU for four weeks. The pediatricians were very careful to perform numerous checkups on them before they allowed me to take them home. It took only a month or two for them to catch up in their growth and by the time they started nursery school they were taller and heavier than their classmates, which I found very reassuring.

EXPRESSING MILK
A positive way of getting involved with helping care for your premature baby is to express milk. You will be providing her with the best nutrition available.

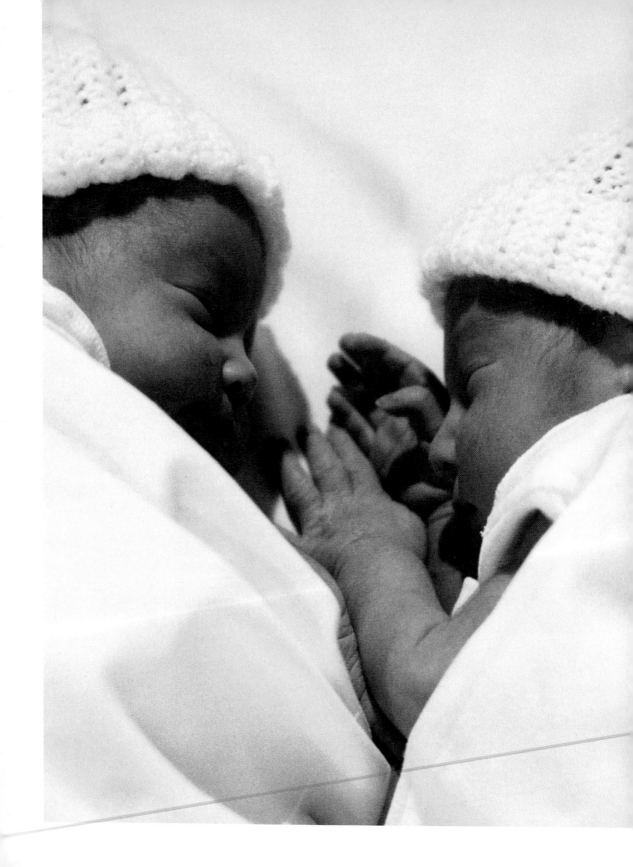

Multiple births

THE NUMBER OF TWIN AND TRIPLET PREGNANCIES has increased over the
last 10 to 20 years and currently some four percent of all deliveries
in the US are multiple births. This rise is mainly due to the wider
availability of assisted fertility treatments, which increase the chances
of more than one egg being fertilized at the time of conception. Another
important factor is that more women are taking on pregnancies later in
life, and rising maternal age increases the incidence of nonidentical
(dizygotic) twin pregnancies.

Multiple pregnancies are at greater risk of a variety of complications, in
particular premature delivery, fetal growth restriction, preeclampsia, anemia,
placenta previa (see pp.424–29), and twin-to-twin transfusion syndrome (see
p.346). Furthermore, the incidence of cerebral palsy is dramatically higher for
multiple births compared to singletons. As a result, prenatal care for mothers
with a multiple pregnancy is more closely monitored and the delivery is usually
planned in a hospital unit where emergency help is always available.

Even if the pregnancy has been uncomplicated, 50 percent of twins will be
born prematurely, before 37 weeks, and these babies are more likely to need
help from a neonatal intensive care unit (NICU). This is because they tend to
be smaller than singleton babies and, irrespective of their actual birthweight,
also behave less maturely than singletons and often require some assistance
with breathing and feeding during the first few days or weeks of their lives.

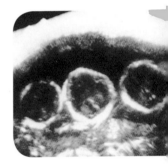

TRIPLETS The three
babies visible in this scan will
be delivered in a hospital unit
where emergency help is
readily available.

Delivering twins

The main concern in a vaginal twin birth is the delivery of the second twin.
Even if the first baby is head down and the labor starts spontaneously and
progresses smoothly, there is no way of knowing how the second twin will cope
with the mechanics of a hasty descent through the birth canal until the first
twin has been delivered. No woman wants to go through labor and the vaginal
delivery of her first twin, only to be told that she needs to undergo an emergency
cesarean section to deliver her second baby. As a result, an increasing number
of twin and all triplet (and higher number) pregnancies are now delivered by
cesarean section. This may be an emergency procedure if the labor is very
premature or if complications develop, or an elective procedure because the

risks of a vaginal delivery are generally considered to be too great. If there are no maternal or fetal reasons to perform the elective cesarean section sooner, delivery should be delayed until 37–38 weeks to avoid neonatal respiratory problems. A cesarean delivery of twins is likely to be planned when:

• the mother requests this rather than vaginal delivery
• the first twin is not presenting head first
• placenta previa is diagnosed on an ultrasound scan
• intrauterine growth restriction (see p.429) is identified
• the birthweight of the second twin is estimated to be 1lb (500g) greater than that of the first
• one or both of the twins has a physical abnormality
• twin-to-twin transfusion syndrome is present. This disorder in the blood supply to identical twins only affects monochorionic pregnancies and has serious consequences because the blood vessels in the shared placenta favor one twin over the other—early delivery is usually required to save the life of the smaller twin.
• the babies are conjoined twins. An attempt to separate them surgically may be attempted after delivery, depending on which organs they share.

Labor may be induced for twin pregnancies in which the first twin is cephalic at 37–38 weeks, as more complications arise after this gestation, including an increased risk of stillbirth. Recent studies have shown there is no significant increase in the rate of emergency cesarean section or reduction in the likelihood of delivering a healthy baby if labor is induced at 37 weeks.

Vaginal delivery of twins

If you are planning a vaginal delivery for your twins, you will be carefully reviewed in the early stages of labor, to ensure that nothing has happened since the last office visit to suggest the plan should be revised. The labor ward staff will monitor both babies and will also perform an ultrasound to check their sizes and positions. If you have a scar in your uterus from a previous cesarean section or surgery, a vaginal delivery may be possible if the first twin is a cephalic presentation.

Since there will need to be a lot of people on hand during a twin delivery (one or more obstetricians, an anesthesiologist, two labor nurses, and two pediatricians), you will be cared for in a larger than normal delivery room equipped to perform emergency procedures. You will most probably have an epidural so an emergency cesarean can be performed without delay at any stage. A good epidural block is particularly important during the second stage of labor when external or internal manipulation to turn the second twin into a longitudinal lie may be required.

During labor, continuous electronic fetal monitoring (see p.292) is advisable and attaching a scalp clip to the first twin ensures that the readings from the abdominal monitor on the second twin can be read without confusion. The first stage of labor is often a little shorter than for a singleton. If progress is slow, this is usually seen as an ominous sign prompting cesarean delivery. The use of oxytocin to augment labor is rarely considered to be the best option. The second stage of labor, before the delivery of the first twin, is essentially the same as for a single baby, but there will always be an anesthesiologist and obstetrician present in addition to the labor nurses and pediatricians. Immediately after the delivery of the first twin, the umbilical cord is clamped in two places (near the baby and at the end of the cord leading to the placenta) to prevent the transfusion of placental blood away from the second twin, who may remain in utero for some time.

The first stage of a twin labor is often a little shorter than for a singleton.

Delivery of the second twin

The obstetrician will now palpate your abdomen to establish the lie of the second twin. If it is transverse, gentle external pressure will be applied to achieve a longitudinal lie (the baby lying parallel to your spine), and the nurse will be asked to maintain this position by gentle manual pressure.

If there is any uncertainty as to whether the presentation is cephalic or breech, a quick ultrasound scan can be performed. External cephalic version (turning a baby from presenting breech to head first, see p.271) for a breech presentation of a second twin is uncommon practice because it often results in further complications, which may then necessitate an emergency cesarean section. A smooth assisted breech delivery (see pp.356–59) is preferable.

There are no strict rules as to how long the second stage for twin two should be allowed to continue, but if delivery cannot be achieved within 30 minutes it is highly likely that an emergency cesarean section will be needed. Because the uterine contractions frequently diminish after the delivery of the first baby, most obstetricians will already have an oxytocin infusion in place and this will be started as soon as a longitudinal lie has been confirmed, in order to help drive the presenting part of twin two into your pelvis.

Ideally, the membranes around your second baby are left intact until the baby has descended farther through your cervix and into the vagina, which helps prevent the cervix from closing. If the membranes rupture and there is a delay in the delivery, the obstetrician will insert a hand into your vagina and up into your uterus to guide the head downward (toward the helping hands of a pair of forceps or a vacuum extractor) or grasp hold of the breech or legs and assist the vaginal delivery. Occasionally, it may be better to perform an internal version or manipulation in the uterine cavity, turning a breech presentation through 180 degrees, but more usually the delivery continues as an assisted breech (see pp.357–59). Because of the limited experience of modern-day obstetricians, the incidence of cesarean section for twin two has increased.

Third stage of a twin birth

Active management of the third stage of labor (see p.333) is especially important in twin births since the risk of postpartum hemorrhage is higher because the uterus is more distended. As soon as your second twin is delivered, the doctors will increase the oxytocin infusion and you will be given an intramuscular injection of methylergonovine maleate or ergonovine. The infusion may need to be continued for some time after the delivery to make sure that your uterus remains contracted.

TWIN BIRTHS These can be more complicated than singleton births, but they are also very special with the end result invariably being the safe delivery of two perfect babies.

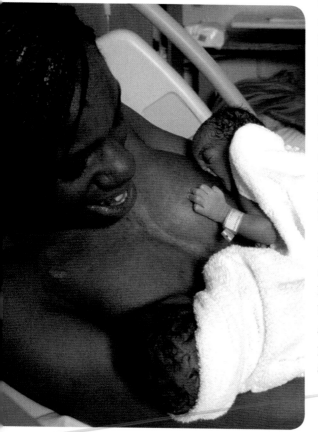

Twin births are special. I speak here as a mother and an obstetrician, having been blessed with twin daughters. Since the birth is not always straightforward and often involves babies that are premature and small for dates, pediatricians will monitor them carefully and will not hesitate to admit them to the NICU if they have any concerns. This is distressing and alarming for parents, but do remember that in the vast majority of cases this is a short, almost routine visit and the outcome is a happy one. Rest assured that the pediatric staff will keep you fully informed about your babies' progress and will be anxious to reunite them with you as soon as possible.

Support for the parents of multiple pregnancies is important, and detailed information and practical support to help them prepare for caring for their babies is best started in childbirth classes. After the birth, joining a local twin group will offer the great advantage of being put in touch with experienced parents who are undoubtedly one of the best sources of advice for others in the same situation.

BIRTH STORY

Deborah, 32, first pregnancy
Nicholas and Patrick, born 37 weeks + 5 days,
Nicholas weighed 5lb (2.25kg), Patrick weighed 7lb (3.15kg)

I discovered I was carrying identical twins at my 12-week scan and, after I recovered from the shock, I found that my pregnancy progressed well. I was seen regularly by my consultant from 28 weeks onward and was told there was no reason why I could not attempt a vaginal delivery.

My 36-week scan showed that the babies were growing well. But the following scan, at 37 weeks + 4 days, showed that one twin had stopped growing. This was because, like all identical twins, the babies were sharing a placenta but had separate amniotic sacs. In other words, they were monochorionic but diamniotic twins. The scan took place at 4pm and I was immediately scheduled for a cesarean section the following morning. There was no question of inducting a vaginal birth, especially since the doctor could see from the scan that one baby was lying transverse. Nobody could know what

would happen during labor. My biggest fear all along had been that I might have one baby vaginally, only to end up with a cesarean section for the second twin. My top priority was to have both babies safely and it didn't matter to me how they came out. I had been present at my cousin's vaginal birth and I couldn't see what was so appealing about all that pain. As a result, I was very happy at the prospect of the cesarean section.

The birth itself was very straightforward and calm. My husband was in the operating room with me, and both babies came out without any problems. Little Nicholas was born first and it turns out he had been cephalic and very much ready to come out. But he was the one who had stopped growing. Patrick was lying transverse, with one hand nonchalantly behind his ear, and he was born five minutes later. I don't think he felt like leaving his

warm home. I asked to see the placenta: Nicholas' side was all crusty and dried up, while Patrick's was red and healthy. It was clear what had happened and I'm so grateful that the doctors spotted the problem and were able to act on it immediately.

The staff was fantastic and helped me through the first days after the delivery, although I was more or less sent home from the hospital after four days of recovery because the maternity unit was short on beds. After two months, I have learned to be very organized. I actually breast-fed for about one month but it was very difficult, so I now bottle-feed and the twins almost sleep through the night. Nicholas' weight is catching up with his younger, heavier brother and I feel blessed to have them both in such good health. Looking back, I feel this was the best possibly delivery for me.

> 66 *My top priority was to have both babies safely and it didn't matter how they came out.* 99

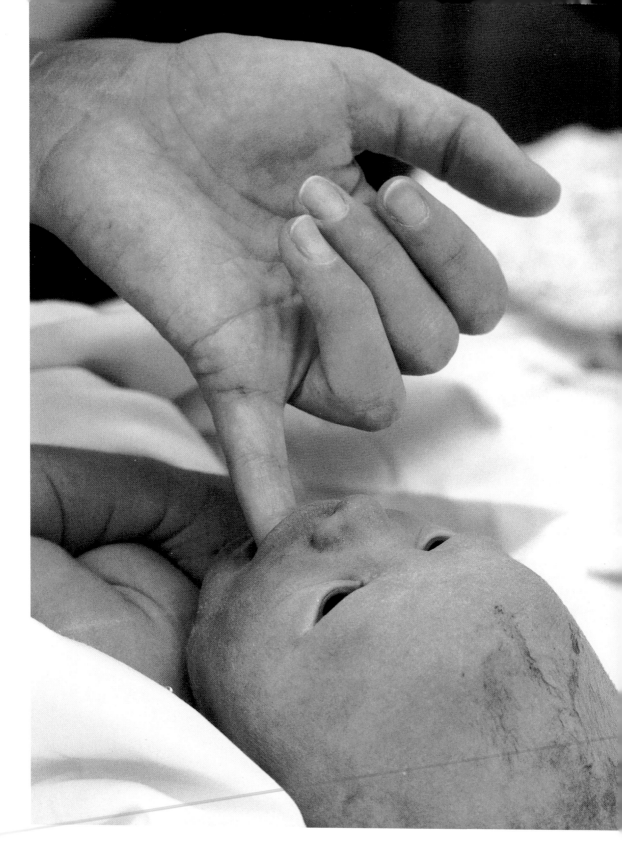

Assisted deliveries

THE TERM ASSISTED OR INSTRUMENTAL DELIVERY can sound somewhat frightening and I want to remind you that the vast majority of babies are still delivered vaginally without any need for assistance or medical intervention. However, if labor has been prolonged or the second stage is not progressing smoothly, it may be necessary to assist the vaginal delivery with the use of instruments such as forceps or a vacuum extractor.

It is important to understand that the aim of an assisted delivery is to guide the baby out of the birth canal with the help of your uterine contractions pushing the baby from above. The forceps or vacuum equipment are not designed to pull the baby out on their own. Assisted deliveries are performed by an experienced obstetrician who have received special training and are able to perform vacuum extraction deliveries and lift-out forceps. For either a forceps or a vacuum extraction, your doctor will request that you are in the lithotomy position with your legs supported by stirrups so that they have maximum visibility and access to your baby during the delivery. Whether or not you need an episiotomy will depend on individual circumstances and considerations (see pp.330–31). Most forceps deliveries will require an episiotomy but if you are having a vacuum extraction you may not need one.

Effects of instrumental deliveries

Babies born by assisted vaginal delivery often bear the marks of the instruments after the birth, but don't worry because these usually disappear within a few days. After a vacuum extraction, there will always be a swelling on the scalp where the suction cup was placed and this can sometimes lead to extensive bruising and may even cause the baby to become jaundiced. With forceps, the skull or face may be bruised and appear slightly misshapen at the sites where the blades were placed. However, it is important to remember that a baby's skull is designed to deal with being put under pressure or compressed in some way during the birth, so these instruments are highly unlikely to pose any serious threat to your baby's long-term well-being.

> *...a baby's skull is designed to deal with pressure during birth...*

Vacuum extraction

The use of the vacuum extractor has become increasingly popular over the past few years and, in many obstetric units, this is now the instrument of choice for assisted vaginal deliveries, virtually replacing the use of forceps.

How the vacuum extractor works

Vacuum extraction equipment essentially consists of a cup made of metal or soft plastic that has a chain or a handle. The cup is attached to a tube, which connects with suction apparatus. The cup is positioned over the crown (occiput) of the baby's head and is held firmly against the scalp while a vacuum is gently built up, using a hand or electrical pump. This sucks some of the scalp tissue into the cup and the swelling that develops on the baby's head effectively produces a firm attachment.

When suitable vacuum pressure has been achieved, the edges of the suction cup are checked to ensure that no maternal tissues have been trapped. Traction is applied by pulling on the chain or plastic handle attached to the cup, while the mother bears down during a contraction. Once the head has crowned, the vacuum is released and the cup is removed. After allowing time for the baby's head to rotate externally, delivery of the shoulders and body proceeds as normal. The principle of vacuum extraction is that the line of traction should follow the mother's pelvic curve, since this is the pathway of least resistance for the baby's head as it descends through her birth canal.

Advantages and disadvantages

The greatest advantage of the vacuum is that, if the baby's head is not directly in the occipito-anterior position, it can still rotate automatically as it descends through the maternal pelvis. Hence it is easier for the diameters of the head to negotiate the various diameters of the maternal pelvis (see p.328). Another important advantage of the vacuum extractor is that it takes up less space in the vagina than forceps would. As a result, there is less risk of damage to the vagina and perineum compared to a forceps delivery and some women may not require an episiotomy. In addition, the requirements for pain relief are usually less, although, in an ideal world, an effective regional block should always be in place before an instrumental vaginal delivery is attempted. The disadvantage of the vacuum is that the delivery tends to be slower, because

...the vacuum extractor is often the instrument of choice for assisted vaginal deliveries...

time is required to set up the equipment and achieve good pressue and also because the cup can become dislodged. However, in experienced hands, a good vacuum application pressure can usually be developed in about two minutes, about the same as the gap between contractions during the second stage of labor. If careful attention is paid to the positioning of the cup, the number that become dislodged is reduced and the number of smooth, swift vacuum deliveries increases.

If the baby has not been delivered after three or four good contractions and there has been no descent of the fetal head, an alternative means of delivery needs to be considered. A smaller vacuum cap made of very soft rubber is now available.

Possible complications

Although maternal complications after vacuum delivery are uncommon, fetal complications, including superficial scalp injuries, bruising, and bleeding into the head, can occur even after an apparently simple procedure but are more likely when the cap has become dislodged or the delivery is prolonged. A scalp swelling is always present after a vacuum extraction and is usually more extensive when a metal cup has been used. It invariably subsides in a few days with no harmful effects.

Superficial scalp injuries occur in about 12 percent of vacuum deliveries and again it is rare for them to lead to any long-term complications. However, cephalohematomas (a collection of blood beneath the top layer of the skull bones) develop in about six percent of vacuum extractions. The hematomas usually resolve by themselves within two weeks but, if they are extensive, may cause jaundice in the baby and presumably result in a nasty headache!

Bleeding into the head (intracranial hemorrhage) is uncommon (about 1 in 300–400 cases) but is potentially extremely serious. However, recent studies have shown that this figure is no greater than the figure for babies delivered by forceps or emergency cesarean section during labor, suggesting that the abnormal labor is the cause of the problem rather than the vacuum application itself.

VACUUM EXTRACTION
The cup is attached to the baby's head and a vacuum is built up within it. Then, synchronized with the mother's contractions, the baby is helped down the birth canal.

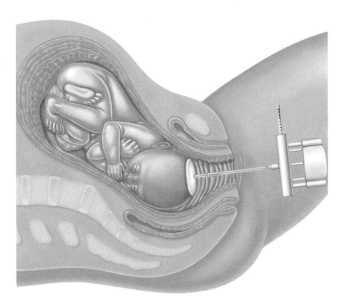

Forceps delivery

Forceps have been used by obstetricians for nearly 400 years. Until the latter half of the 20th century, cesarean section was a dangerous surgery and forceps prevented many mothers and babies from succumbing to life-threatening events during delivery.

Now, thanks to the enormous medical advances that have been made in anesthesia and the routine availability of antibiotics, blood transfusion, and adult and neonatal intensive care units, cesarean section is a relatively safe procedure and, as a result, the potential complications of a difficult forceps delivery are considered to be a greater threat to the mother and her baby.

There are three types of forceps: lift-out, straight traction, and rotational. Forceps deliveries can be described as high, mid, low and outlet, depending on where the baby's head is positioned in the pelvic cavity when the forceps are applied. High-cavity forceps deliveries are no longer performed because of the significant risk of causing maternal and fetal damage.

Lift-out (outlet) forceps forceps are used to lift the baby's head out when the scalp is visible at the mother's vulva, which means that the entire baby's head has already reached the pelvic floor and is distending the vagina but the perineal muscles are holding it back. They are only appropriate if the baby's head is positioned directly occipito anterior (see p.270) or slightly rotated to the left or right. An injection of local anesthesia into the perineal tissues or the pudendal nerve (see p.316) may be all that is needed for a lift-out forceps delivery. An episiotomy is not always necessary.

Straight traction forceps are longer and are used for low- and mid-cavity deliveries when the baby's head has engaged and descended to more than 2cm below the ischial spines (low) or just above this station (mid-cavity). (See p.302 for diagrams of stations and head descent). The forceps lock together easily to form a protective cradle around the baby's head. At the height of a

FORCEPS DELIVERY
The curved blades of the forceps are inserted one at a time and cradled around the baby's head. They are then used, in time with your contractions, to help draw the baby down the birth canal and deliver it.

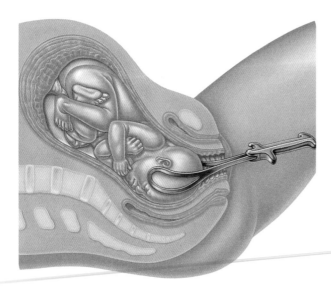

contraction, the doctor applies gentle downward traction. The head descends with each pull even if it slips backward again between contractions. Delivery is usually achieved with three good contractions and moderate traction, but if there is no obvious head descent by this time it suggests that there may be a degree of disproportion present and the decision as to whether to proceed with a vaginal delivery should be reviewed. Because the forcep blades take up additional room in the vagina, an episiotomy is often required to prevent uncontrolled perineal tearing as the baby's head stretches the perineum. An epidural or spinal block should be used for this type of forceps delivery to make sure the mother does not experience pain or discomfort when the blades are inserted and during the traction, delivery, and subsequent repair of the perineum.

Rotational forceps are rarely used nowadays to turn the baby's head in the mid-cavity from a transverse or occipito-posterior position to direct occipito-anterior. Following this downward traction on the long handles of the forceps allows the smooth descent of the baby's head and the delivery is then completed similar as with straight forceps. The application of rotational forceps requires considerable skill and experience and this type of delivery is likely to be performed in an operating room, if any difficulties are encountered, a cesarean section can be performed immediately. It is essential for the mother to have an effective epidural or spinal block in place, not just for the delivery, but also afterward so that her vagina and cervix can be carefully examined for any potential tears or damage that the rotational forceps may have caused.

...forceps have prevented many mothers and babies from succumbing to life-threatening events...

Forceps or vacuum extraction?

Despite the current trend favoring use of vacuum over forceps, there is still debate as to which is the best method of instrumental delivery. It is generally thought that the use of the vacuum results in less damage to the mother's vagina and perineum. On the other hand, the vacuum may be more traumatic for many babies compared to a smooth forceps delivery, because it leaves a swelling on the baby's head where the cup was placed.

There are pros and cons to each type of instrument, and I think it is more useful to consider the two methods as complementary, designed for different circumstances, rather than rivals to each other. The final choice should be determined by the circumstances that have led to the need for an attempted instrumental vaginal delivery and the experience and skills of the individual who performs the delivery.

Breech births

The number of babies presenting as breech (bottom first) is closely related to their gestational age. At 28 weeks some 20 percent of all babies are breech, but most of them turn of their own accord during the third trimester to become cephalic (head-down) presentations, leaving less than 4 percent of babies in a breech position at term.

If your baby is breech it will be in one of three positions. These are: a complete breech, in which the bottom is still lowest but the thighs are extended upward and the knees are flexed so that the legs are folded against the baby's body; a frank breech, in which the bottom presents to the cervix and the legs are extended upward along the baby's body; or a footling breech, in which the legs are presenting below the baby's bottom.

Vaginal breech deliveries are more prone to risks and complications, the most important being that the largest diameter of the baby (the head) is the last to deliver. Furthermore, the breech does not fit into the mother's pelvis as well as a head, so there is always the risk that the umbilical cord will prolapse or fall through the cervix along the bottom or the legs (see p.430). Invariably, a prolapsed cord results in acute fetal distress because when the umbilical cord is exposed to air, it constricts or closes down, cutting off the oxygen supply to the baby.

BREECH PRESENTATIONS ···

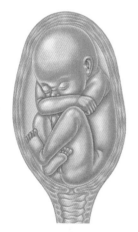

COMPLETE FLEXED BREECH   **FRANK EXTENDED BREECH**   **FOOTLING BREECH**

COMPLETE FLEXED BREECH
The buttocks are presenting but the thighs are extended upward and the knees are flexed.

FRANK EXTENDED BREECH
The baby's buttocks are presenting and legs are extended upward against the baby's body.

FOOTLING BREECH The legs present below the baby's bottom. They drop down once the membranes have ruptured.

Trial of labor

A vaginal breech birth is most likely to be considered if your baby is in a frank breech position with the back of his bottom (the sacrum) positioned anteriorly in the birth canal. It is always considered to be a trial of labor and will be allowed to continue only if no problems develop. The doctor will want you to be monitored continuously, often using an external monitor. They may also advise you to have an epidural anesthesia in place early in your trial of labor, in order that any necessary interventions can be performed quickly. The other advantage of having an epidural is that it will protect you from the urge to push before the cervix is fully dilated.

Breech labors are often slower than cephalic labors because the presenting part, the baby's bottom, is softer and does not exert the same downward pressure on the cervix. The first stage of labor can be longer and more tiring and since most obstetricians will be reluctant to use oxytocin augmentation when progress is poor, it is quite possible that you will be advised to have a cesarean section during the first stage.

> *...the first stage of a breech labor can be longer and more tiring for the mother...*

Second stage

Assuming that all has gone well and that you have now reached full dilation, the mechanics of the second stage of a breech labor are best thought of as the reverse of a cephalic delivery. The bottom and legs travel through the pelvis first, followed by the trunk and shoulders. An obstetrician will always be present during the second stage of a breech labor. They will request that you have your legs in stirrups so that they have easy access to your emerging baby. They will also want to ensure that your epidural is well dosed so that they can rotate your baby, apply forceps to the baby's "after-coming head," or resort to a cesarean section if the delivery proves to be difficult.

The buttocks will be delivered first with the help of your contractions and pushing efforts. The obstetrician will then gently guide the delivery of the two legs. This frequently involves rotating the baby's buttocks to the left or right to allow the doctor to insert a finger into the vagina and hook it around the first and then the second leg, to encourage their smooth entry into the world.

When the buttocks and legs have been delivered, the baby's back and trunk are then allowed to emerge up to the shoulders, in their own time. The shoulders usually need to be rotated to one side and then the other, so that once again the obstetrician can insert a finger into the vagina and hook a finger around the upper limbs to aid delivery of the arms. The key to a successful vaginal breech delivery is that it should not be hurried and there should never

BIRTH STORY

Nathalie, 34, has one daughter, four years old
Second baby Enzo, born 40 weeks + 4 days, weight 8lb 6oz (3.8kg)
Length of labor from start of contractions: 21 hours

My 34-week scan had shown the baby to be transverse but at subsequent examinations I was told by my doctor that he had turned and was cephalic, although his head never engaged, even at term. The day before my labor started, I remember a doctor getting me to feel the baby's bottom high up in my abdomen.

My labor started at around 1am with regular, periodlike cramps that were generally perfectly bearable, so much so that I was able to get some sleep. By midday, the cramping stopped completely and only started again at around 3pm. I decided to go shopping with my husband for some last-minute food items, even though the contractions were now coming every 10 minutes or so. At 5pm I had a bloody show, so I called the hospital but was told to wait until the contractions were longer and closer together. By 5:30, my water broke; the contractions were coming every five minutes and were more painful so we left soon after for the hospital.

A labor nurse examined me on arrival and, palpating my abdomen, remarked on how hard the place was where the baby's bottom should be. I was taken to a delivery suite and waited while things progressed. At around 7pm, a doctor came and did a scan and revealed that in fact the baby was breech. What we had assumed was his firm little bottom had obviously been his head. We were shocked. He explained to us what a breech vaginal birth entailed and informed us of the risks, and left us to digest the information. By now the contractions were very strong and I still didn't have pain meds.

At 8pm, a new doctor came in, did an internal examination, and told us we needed to hurry up and make a decision since I was already dilated 3cm. He strongly advised me to have a cesarean because the baby was big. Not only were we really disappointed but we were still coming to terms with the turn of events. Soon after, we agreed to a cesarean delivery, simply because I did not want to risk my baby's health for the sake of misguided hesitation and stubbornness.

Enzo was finally born at 10pm and, although the birth itself went fine, I had to be given morphine regularly for the first couple of days. I was a bit frustrated and depressed not to be able to take care of my baby properly because of the pain from the incision. I was also just exhausted, partly because of the inevitable noise on the hospital ward. After three days, I discharged myself, knowing I would get more rest (and better food!) at home.

66 *...the doctor explained in detail to us what a breech vaginal birth entailed and informed us of the risks...* **99**

be any pulling or tugging of the baby, just guidance and gentle rotation as the baby emerges.

Delivering the head

If all is going well, the weight of the baby's body will help with the remainder of the delivery, encouraging the neck to be well flexed, which results in the head being best positioned to be smoothly and safely delivered. If the baby's neck remains extended, with the face looking upward into the uterine cavity (called star gazing) it is highly likely that problems will develop in delivering the "after-coming head." The baby's head is the largest part of its body and when the neck is extended an even wider diameter is presenting to the cervix, which may not yet have been fully dilated by the delivery of the buttocks, trunk, and shoulders. Performing a cesarean section in this situation is traumatic for everyone concerned. This is why everyone involved in your delivery will have been careful to heed earlier signs that a breech labor may not be progressing well and will advise changing to a cesarean delivery if it is thought the "after-coming head" is in danger of getting stuck at the last moment.

On a positive note, if all continues to go well, your obstetrician will now gently sweep the baby's body upward, over your pubic bone, and may insert a finger into the baby's mouth and gently pull downward, further flexing the head and helping it emerge smoothly. At this point, it is often best to apply a pair of lift-out forceps to the breech baby's head to guide it out in a controlled manner as it escapes the constrictions of the lower part of the birth canal. As you will now realize, several pairs of hands are needed to deliver a breech baby vaginally and an episiotomy is usually performed.

Vaginal or cesarean breech delivery

In the last decade, several studies have been published that have changed most obstetricians' views about the best way to deliver a breech baby. Cesarean section is now recommended as the optimal way to deliver a full-term breech baby in a first-time mother, if attempts to turn the baby (see external cephalic version, p.271) have failed. The risk of your baby dying or having a serious problem are reduced by planned cesarean section when compared to planned vaginal birth. However, some 10 percent of women with a breech presentation who are scheduled for a cesarean section will deliver vaginally because labor starts earlier than expected and is already well advanced when they arrive at the hospital. In addition, a small number of women will find themselves in an advanced stage of labor before it is realized they have a breech baby—the undiagnosed breech.

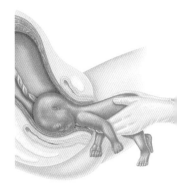

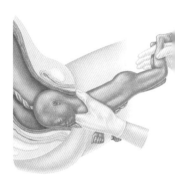

DURING A BREECH DELIVERY The baby's buttocks are delivered first, followed by the legs. The baby then turns so that the shoulders can be delivered. The baby's weight draws the head down and the legs are lifted to allow safe delivery of the head.

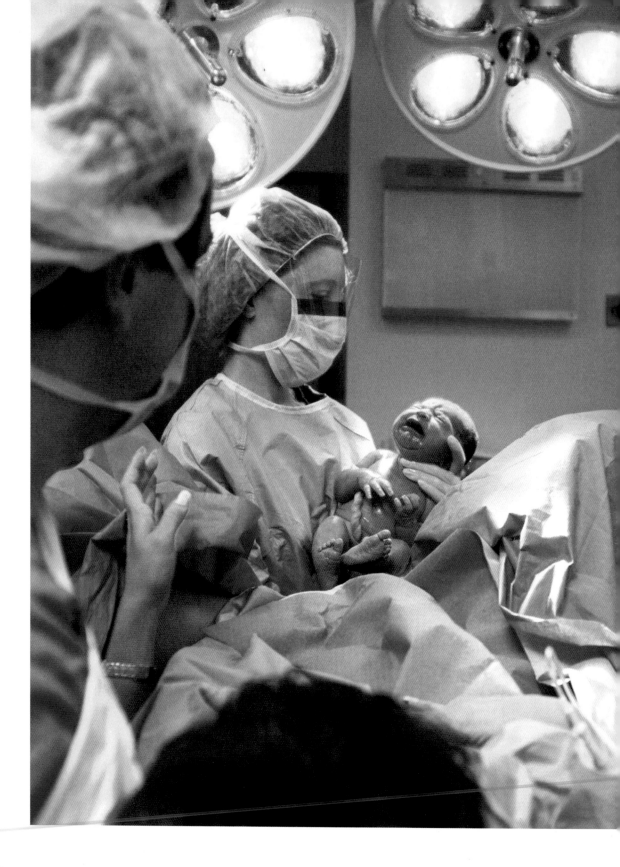

Cesarean birth

I AM VERY CONSCIOUS OF THE FACT **that most childbirth classes, pregnancy books, and publications focus most of their attention on what labor will be like. Yet, in reality, a sizable percentage of women (at least one in five births in the US) now deliver their babies by cesarean section and this is why I have included a detailed account of what you can expect to happen if you turn out to be one of them.**

Despite the marked increase in the numbers of babies that are born by cesarean section, there are many individuals, both men and women, who continue to view this mode of delivery as the poor relation in childbirth. Indeed, some people feel that it is the method of last resort. Of course, these people are entitled to have their own viewpoint, but my concern is that all too often this leaves some women who have needed to have a cesarean delivery feeling that they have in some way failed or not done their part in natural childbirth.

I feel strongly that pregnant women should not be subjected to pressure or disapproval about the way their baby is delivered. There is no way that even the most experienced doctor can accurately predict what will happen to a woman in labor, so to suggest that, "if all else fails you will have to be subjected to a cesarean section" is, in my opinion, not only insensitive but also unkind. Every labor is different, and it is impossible to consider yourself a failure when you have successfully nurtured a baby in your uterus for nine months and then delivered a healthy baby into the outside world. If cesarean sections allow more women a guarantee of going home with healthy babies, then this mode of delivery has to be a good thing and quite the opposite of "failed." I really do believe that the route by which you deliver your baby is of secondary importance, just as long as both of you are safe.

A cesarean section may be performed electively—meaning that this method of delivery was decided before labor started—or it may be an emergency procedure carried out after labor has already began. Of course, the surgery itself is exactly the same, whether it is performed electively or as an emergency, but the underlying reasons for the cesarean section may be different.

…women should not be subjected to disapproval about the way their baby is delivered.

Elective or emergency sections

There are a wide variety of reasons why a cesarean section may be advised or chosen. It is important to remember that only a few of these are absolute indications; most are relative, depending on individual labor and circumstances.

...elective cesarean sections are performed when a vaginal birth is considered potentially risky...

Most "elective" cesarean sections are performed when there are existing medical reasons to indicate that a vaginal birth is potentially risky, for the mother and/or her baby. That is not to say that a vaginal birth cannot be attempted, but in the opinion of the mother and her prenatal caregivers, cesarean section is the safer option. This may be the case when the baby is presenting breech or in another position which may make vaginal delivery difficult, if you have a placenta previa, if you are expecting twins or more, if you are suffering from an illness such as kidney, lung, or heart disease or diabetes, or if you have developed preeclampsia or severe high blood pressure in pregnancy.

The term "emergency" cesarean section may sound as if the baby needs to be delivered in a matter of minutes, if not seconds, in order to avoid a catastrophe. Very occasionally this is the case, but more often it means that it has become obvious to everyone involved that a cesarean section is required in the next hour or so.

The indications for an emergency cesarean section are dependent on many complex factors, which include unpredictable events that occur in labor, such as prolapse of the umbilical cord or signs of fetal distress, maternal and medical perceptions of the progress of that labor, and practical considerations such as the staffing and expertise available to achieve the safe delivery of a healthy baby.

Safety of surgical deliveries

Cesarean section is now considered to be a relatively safe procedure. When complications develop they are almost always due to the fact that the cesarean is performed under emergency circumstances or because of an underlying maternal or fetal problem. Thanks to the medical advances in anesthesia, antibiotics, blood transfusion, and adult and neonatal intensive care units, the risks for the mother have become a secondary consideration in the vast majority of cases, and the risk of physical damage to the baby during the delivery is small.

However, any surgical procedure is associated with a degree of risk and there are factors that significantly increase the risk associated with cesarean section. When the pregnant woman is seriously overweight, a smoker, has a personal or

THE RISE IN THE RATE OF CESAREAN SECTIONS

The number of cesarean deliveries has increased dramatically over the last 10 to 15 years and I think the most useful way to understand this change is to consider the contributing factors.

ADVANCES IN GENERAL MEDICAL CARE

Advances have allowed some women with medical problems that were previously thought to preclude them from having children to now get pregnant and, with specialized help, remain healthy during pregnancy. However, an underlying medical disorder may make a planned cesarean section the safest way to deliver the mother and her baby. Similarly, mothers-to-be who develop severe diabetes or preeclampsia may require early delivery by cesarean section to protect their health.

ADVANCES IN OBSTETRIC CARE

Prenatal and intrapartum (labor) care have become so much more sophisticated. The routine availability of ultrasound scanning has resulted in our being able to identify many mothers and babies who are likely to experience problems in labor, before they are exposed to serious complications. The widespread introduction of regional anesthesia (see pp.311–16) for cesarean deliveries needs a special mention here, since it allows the woman to remain conscious throughout the surgery, supported by her partner, and avoids the significant risks of a general anesthesia. There are three other obstetric reasons that have contributed to the increase in the cesarean section rate. First, while rates of babies born prematurely have stabilized, premature birth rates still remain high. Second, the increase in the numbers of older women giving birth makes an additional contribution to the cesarean section rate, since these women experience more complications during labor. Third, the use of high-cavity obstetric forceps in the second stage of labor has virtually disappeared over the last five to 10 years, and an emergency cesarean section is usually performed instead.

SOCIAL CHANGES, VIEWPOINTS, AND PERCEPTIONS

The views that pregnant women now hold about the way in which they would prefer to deliver their babies have also made a significant contribution to the increased rate of cesarean sections. Many of the women I meet in my office have already developed clear views about the method of delivery they want. Some are passionate about achieving a vaginal delivery. Others are equally clear that they want to be delivered by cesarean section. Most recently, the trend for celebrity mothers to opt for a cesarean section (invariably in the private sector), because they want to dovetail their baby's birth into a busy work schedule, has resulted in the procedure being perceived as a fashionable option or the must-have accessory in childbirth. It must be emphasized that all cesarean sections carry an element of risk.

MEDICOLEGAL CONSIDERATIONS

On some occasions when complex vaginal deliveries have resulted in babies suffering from brain damage or other physical injuries, expensive and lengthy legal cases have followed. Inevitably, this leaves doctors more likely to err on the side of caution when there is choice to be made between a cesarean section or a complicated vaginal delivery.

family history of thrombosis, has a pregnancy-related problem such as preeclampsia, or is unable (for whatever reasons) to have an epidural, her risks of developing complications because she has undergone pelvic surgery are markedly higher.

Vaginal delivery versus cesarean section

It is difficult to make direct comparisons of the likely post-delivery complications experienced by women who have delivered vaginally and abdominally, but recent figures show that, for the most part, the increased risks associated with elective cesarean deliveries are marginal. The risk of postpartum hemorrhage is only slightly increased, as is the risk of endometrial or urinary infection. Breast-feeding is often quicker to establish after a vaginal birth, but there is no difference in the mother's risk of postpartum depression or pain during intercourse after three months. After a cesarean birth you will need to stay in the hospital longer, and recovery from major surgery will take longer than recovery from a vaginal delivery. There is a slightly increased risk of being admitted to an adult intensive care unit and of needing more major surgery, such as a hysterectomy, after a cesarean section. On the other hand, the incidence of urinary incontinence is greater after a vaginal birth, as is the incidence of uterovaginal prolapse in later life. So, overall, there are advantages and disadvantages to both types of delivery.

Vaginal birth after caesarean delivery (VBAC)

In the past most cesarean sections were performed using a vertical incision in the uterus, leaving the muscle weakened along its whole length and as a result doctors were reluctant to advise an attempt at subsequent vaginal delivery. The majority of cesarean sections are performed using a transverse or horizontal incision in the lower segment of the uterus, which is much thinner and usually heals more effectively and is less likely to rupture during a subsequent labor. Nevertheless, the risk of uterine rupture is higher in a vaginal birth after cesarean, and significantly increased when labor is induced or augmented with oxytocin. Generally speaking, if a woman chooses to have a trial of labor after a previous cesarean section performed for a nonrecurring cause, she has a more than 70 percent chance of achieving a successful vaginal birth after cesarean delivery (VBAC). The single best predictor of vaginal delivery is a previous vaginal birth and is associated with a 87–90 percent success rate.

Similarly, the old adage that a woman could undergo a maximum of two cesareans before her uterus became too fragile to risk further pregnancies has been overturned. Although the uterus is inevitably weakened by scar tissue, theoretically, there is no limit to how many pregnancies a woman can have delivered by cesarean, as long as every case is individually assessed.

> *...the majority of cesarean sections are performed using a horizontal incision...*

What to expect in the operating room

Once the decision has been made to deliver your baby by cesarean section, your labor nurse will help you prepare for the surgery. You will be asked to change into a loose hospital gown.

You will also be asked to remove your jewelry, apart from any rings that cannot be taken off easily. These will be covered by sticky tape to ensure that they do not act as a conductor of heat, since the surgeon will probably use diathermy during the surgery—an electrical instrument that cauterizes bleeding blood vessels—and uncovered pieces of metal next to your skin could result in a superficial burn or skin blister. You will probably be asked to remove your makeup and nail polish as well, so that in the unlikely event that you become unwell during the surgery, the anesthesiologist will be able to assess your true skin color immediately.

WHO WILL BE PRESENT DURING THE BIRTH?

I know that many women are surprised and may be a bit shocked by how many people are present in the operating room for a cesarean-section birth. However, every person is there for a specific purpose: to be sure that you and your baby are delivered safely. This (below), i s the usual list of people, which may increase if you are delivering twins, triplets, or more:

▶ anesthesiologist
▶ assistant to the anesthesiologist
▶ obstetrician—who performs the surgery
▶ assistant surgeon
▶ sterile operating room nurse—to pass instruments, sutures, and other items to the surgeons
▶ nonsterile nurse—to get all of the above
▶ nurse—to collect the baby when delivered
▶ pediatrician—to give the baby a checkup after the birth
▶ porter—to transfer you to and from a ward elsewhere in the hospital
▶ medical and/or nursing students —you may request that they leave at any time, but remember that the only way these people can be trained is by practical experience.

Your partner can stay beside you and hold your hand to provide comfort and reassurance throughout the entire procedure. The only time your partner will be asked to leave the operating room is if you have to have general anesthesia. In this situation you will be unconscious and your partner cannot communicate with you, which means that he becomes an extra person taking up space in a crowded room, not to mention the fact that he may start to feel distressed by being unable to contribute to the procedure. It is important for you to realize that the request that he leave the room is not because anything is being hidden and that you both fully understand and agree with the proposal.

...the anesthesiologist will do everything possible to make you feel at ease and to explain what is going on, so you are sufficiently relaxed.

If this is an elective cesarean you will probably walk into the operating room and lie down or sit on the operating table, ready for the epidural/spinal anesthesia to be inserted (see p.313). If you are already in labor, you will be wheeled into the operating room on your delivery room bed and, once in the room, you will be transferred onto the operating table. Looking around the room, you will see lots of pieces of equipment, much of it on mobile stainless-steel carts. There will be an anesthesia machine at the head of the operating table you are lying on, covered in instruments, monitors, dials, cylinders of different types of gases, and drawers full of useful equipment.

There will be a baby resuscitation cart in the room, equipped with an overhead heater to keep him/her warm, a piped oxygen supply, and lots of drawers containing equipment that the pediatrician may need. As you lie down, the nurse will be opening up sterile packs of instruments onto several carts that will be moved into position beside the operating table when they are needed. The walls of the operating room will be covered by open shelves containing sterile packs of instruments, gloves, gowns, syringes, needles, swabs, and sutures.

Cesarean anesthesia

Once in the operating room, the anesthesiologist will insert an intravenous line into your arm so that you can receive fluids during the surgery, and start to perform the epidural or spinal block. It is common at this stage for women to get anxious about the whole situation. Some start to hyperventilate and feel light-headed and nauseous. Others start to shake with nerves, not necessarily at the prospect of the surgery, but more at the thought of the needle going into their spine. They can also worry about the anesthesia not working properly and being in pain during the surgery. Let me reassure you on all these fronts. The anesthesiologist will be used to the physical signs of anxiety and, if you let him or her know you are feeling faint, you will be given oxygen through a mask. Any obstetric anesthesiologist will be skilled at inserting the needle, with or without maternal shakes. Indeed, they will do everything possible to make you feel at ease and to explain what is going on, so you are sufficiently relaxed throughout.

If you are having an elective cesarean, the anesthesia will take effect in a few minutes but the positioning of an additional epidural catheter (to provide you with pain relief after the surgery) will take more time, usually around 20 minutes. If you are having an emergency cesarean section, you may already have an epidural in place, in which case you will need only an increased dose of anesthesia, which usually requires only a few minutes to take effect. As for the idea of the epidural or spinal block not working properly, the anesthesiologist will make sure that it is 100 percent successful by making a few checks: he or she will assess your sensation above and below the skin line where the anesthesia has been introduced and will only be satisfied that you are properly anesthetized when you confirm that you cannot feel below the line.

Final preparations

When the anesthesiologist is confident that you are completely pain-free, a urinary catheter will be placed into your bladder. This has a dual purpose. First, it makes sure that your bladder remains empty during the surgery and does not get in the way of delivering your baby. Second, because it will be left in for 24 hours or so after the surgery, it will eliminate the need for you to get out of bed and struggle to the bathroom during those first few uncomfortable hours after the surgery.

The next thing to happen is that your pubic hair will be shaved away from the site of the planned skin incision. Your abdomen will be thoroughly cleaned with an antiseptic solution and sterile sheets will then be draped over your upper abdomen and legs, leaving just the space where the incision will be made uncovered. The top end of the sterile sheets near your head will be hooked up to the anesthesiologist's IV stands to form a screen, so that you will not have to see what is going on during the surgery, unless you tell the doctor that you want to be able to watch.

Cesarean delivery

Once everything is in place, the doctor will make an incision through the skin on your lower abdomen, at the top of where your pubic hair used to be, so that the scar will be mostly hidden once the hair has grown back. Incisions vary slightly in shape and length but, generally speaking, they are about 8in (20cm) long and are either straight or slightly curved, like a smile. The surgeon will then cut through several layers of fat, fibrous, and muscular tissue before making an incision into the lower part of the uterus.

Once the uterus is opened, the membranes will be ruptured (if they have not already done so) and amniotic fluid will come pouring out. For practical reasons, most of the fluid will be suctioned away before delivering the baby, so that all of the sterile sheets, not to mention the surgeon's clothes and feet, are not completely soaked with fluid. The surgeon will then check the exact position of the baby's head and insert a hand into the uterus, around the top of the head, and gently disengage it from the pelvic brim, in order to deliver the

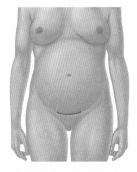

CESAREAN INCISION
The cut will be made just above the line of your pubic hair (bikini line). When the incision heals, the scar is very discreet.

head through the uterine incision. This is often a tight fit and the assistant surgeon may be asked to apply some pressure at the top of the uterus to help the delivery. Sometimes small forceps will also be required, particularly if the baby's head is in an awkward position.

If this is an emergency performed during the second stage of labor, it may even be necessary for another assistant to examine you vaginally and help push the baby back up the birth canal, to deliver the head smoothly through the incision. This may seem a bit alarming to the parents, but I promise you that it is not dangerous for the baby. As the head gently emerges from the uterus, the baby's mouth and nose are immediately suctioned to clear the mucus and then the shoulders, followed by the trunk, are delivered very quickly.

After the birth

Most babies have started crying and protesting before their legs are free of the uterine cavity. Indeed, the baby positively bursts into the world during a cesarean section. The cord is clamped and cut and the baby is then free—to be shown to the expectant parents for a first kiss and then quickly wrapped in towels to dry the fluid and prevent her from becoming cold. It is very likely that the pediatrician will choose to move the baby to the warmed rescusitation machine for a short time, in order to check her breathing and heart rate, clean all the vernix off her face and body, and perform the Apgar scores (see pp.375–76). They will then bring the bundled baby over to you and your partner to hold, unless of course there is a problem that necessitates moving the baby to the neonatal intensive care unit (NICU). If this is necessary, you will still be shown your baby beforehand.

Many partners choose to move in front of the screen so they can watch the delivery and take photographs of the birth. Others will want to remain

FIRST EMBRACE
Your newborn baby will be handed to you to hold while the surgeon stitches together the layers of tissue that were cut to reach the uterus.

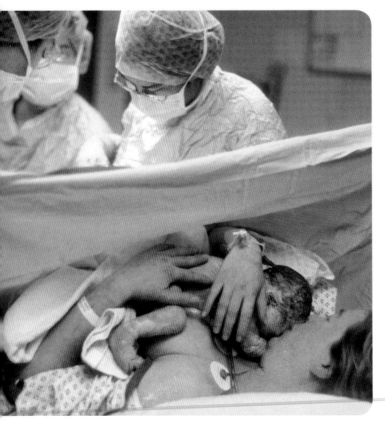

shielded from the view of the surgery and the delivery. For the mother, the cesarean delivery will be a strange and sometimes rather amusing experience. There won't be many times in your life when you have no sensation of pain but are conscious of the fact that someone is rummaging around your belly.

As soon as the baby is delivered, the anesthesiologist will give the mother an injection of methylergonovine maleate or ergonovine to contract the uterus and help deliver the placenta. Just like in a vaginal delivery, active management of the third stage of labor helps reduce bleeding from the uterus or placental bed (see p.333). The placenta will be carefully checked to make sure it is complete. Meanwhile, the surgeons will clean the uterine cavity before starting to repair the uterine incision with one or two layers of stitches. They will then start to stitch up all the layers of tissue that were cut through to reach the uterus, using soluble stitches, and finishing with the skin stitches or staples. The skin stitches or staples will be removed during the next three to five days.

From the time the anesthesia is given to the time when the stitching is complete, a cesarean will take around one hour, of which only five minutes involves the actual delivery of your baby. The rest of the time is taken up by anesthetizing you and stitching you up afterward.

…the procedure takes about one hour, of which only five minutes involves the delivery of your baby.

Classical cesarean sections

A classical section, in which the uterine incision is made vertically into the muscles in the upper segment of the uterus, is now performed only rarely. This is because horizontal lower segment incisions are preferred for most cesarean deliveries, since the uterine muscle heals better and the skin incision also heals more quickly and is less unsightly. However, there are occasionally reasons to perform a classical incision, most usually for a premature baby of less than 30 weeks, when the lower segment may be so narrow and poorly developed that trying to deliver a fragile and already compromised baby through such a small aperture will undoubtedly lead to physical trauma for that baby.

When a baby is in a transverse lie and the membranes have ruptured it may be impossible for the surgeon to manipulate the baby to deliver through a lower segment incision without risking serious trauma to the uterus or baby. Similarly, when the lower segment cannot be approached easily because of large uterine fibroids or dense adhesions from previous surgery, it may prove necessary to perform a classical uterine incision. Because the risks of uterine rupture in a subsequent labor are high, individuals who have undergone a classical section will be advised not to have a trial of vaginal delivery for future pregnancies.

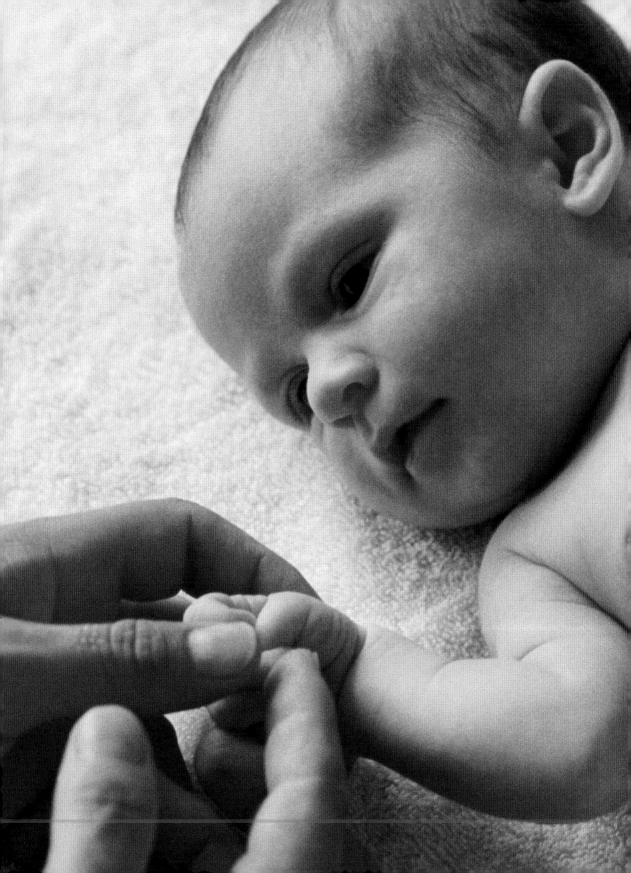

Life After Birth

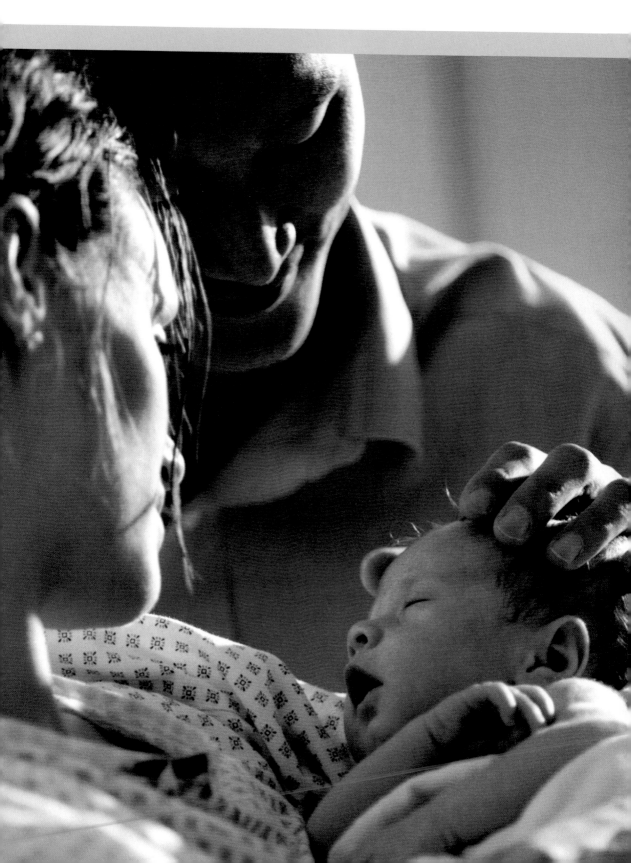

Your newborn baby

The busy hours you spent in labor and giving birth are often followed by a brief period of reflective calm. The staff who has been involved in your delivery seem to melt away and you and your partner are left to marvel at and savor the arrival of your beautiful new baby. It is an emotional time—after nine months of anticipation, you now meet the tiny new person you have created together.

CONTENTS

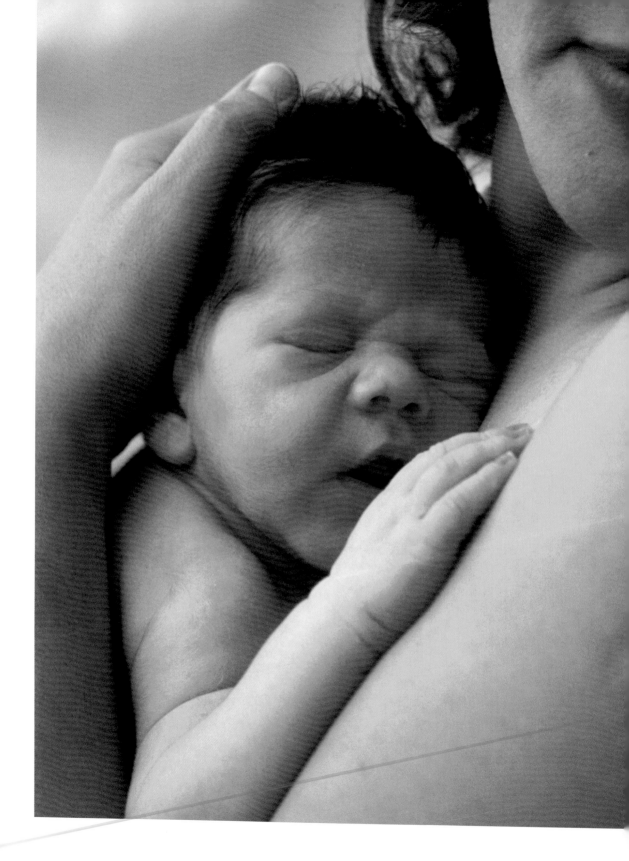

The hours after birth

MOST TERM BABIES WILL TAKE THEIR FIRST BREATH or gasp within 30–60 seconds of the head emerging from the birth canal and before the umbilical cord is cut. This gasp is stimulated by the light and colder temperature of the delivery room compared to that of the environment inside the uterus. It is an extraordinary event: the baby's chest is often still trapped in the pelvic cavity. Nevertheless, this first gasp is usually strong enough to inflate the lungs.

As your baby takes his first breath, your doctor may need to make sure that his upper airway is clear by suctioning mucus and amniotic fluid from the mouth and nose. Once the baby's cord has been cut, the lack of oxygen supply from the mother acts as further stimulus to establish breathing on dry land.

Successful inflation of the baby's lungs is achieved by the presence of surfactant in the alveoli or air sacs. Surfactant determines the stability of the lungs after birth by lowering the surface tension in the alveoli, thereby allowing efficient gas exchange. Without enough surfactant, the alveoli become completely airless at the end of each breath and the baby then has to struggle against the continuing high surface tension to draw in the next breath. Just after birth, the baby's breathing rate increases, the nostrils flare, expiratory grunting is noted, and the tissue between the ribs is pulled in with each breath. This is respiratory distress syndrome (or surfactant deficient respiratory disease, SDRD) and occurs in 1 in 100–200 of all deliveries, but is usually mild. Premature babies frequently require help with breathing since they often have insufficient surfactant in the alveoli of their lungs and may need surfactant to reduce the surface tension (see p.342).

Apgar scores

The doctor will assess your baby's overall condition one minute after birth and again after a five-minute period of observation, using the Apgar scoring system. This is a simple and highly effective tool developed by American doctor, Virginia Apgar, after whom it is named. The maximum score is 10, two points for each of the signs that are assessed: skin color, breathing, heart rate, muscle tone, and reflexes (see table, p.376). In black or Asian

> *…your baby's overall condition will be assessed using the Apgar scoring system.*

APGAR SCORING SYSTEM

Apgar score	2	1	0
Skin color	Pink all over	Body pink; extremities blue	Pale/blue all over
Breathing	Regular, strong cry	Irregular, weak cry	Absent
Pulse/heart rate	Greater than 100 bpm	Less than 100 bpm	Absent
Movements/muscle tone	Active	Moderate activity	Limp
Reflexes after given certain stimuli	Crying or grimacing strongly	Moderate reaction or grimace	No response

babies the color of the mouth, palms of the hands, and soles of the feet is checked. A score of seven or more at one minute indicates a baby in good condition, a score between four and six usually means that the baby will need help to breathe, and a first score of less than four means that resuscitation and life-saving procedures are required. At the five-minute assessment, a score of seven or more indicates a good prognosis and a lower score will require careful monitoring of the baby.

Apgar scores provide an excellent short-term diagnosis of the baby's well-being immediately after birth. However, they are of little help in assessing a baby's long-term development, so don't worry if your baby's first score is low because it invariably picks up by the second assessment. Even if this is not the case, it is unlikely that your baby will have any serious long-term problems.

Measurements and identification

While the Apgar scores are being performed, your nurse will be busy cleaning off the blood and fluids from your baby's skin. The body temperature of newborns falls by 1.8–2.7°F (1–1.5°C) immediately after birth because they rapidly lose heat from their wet skin and also because they have a relatively large surface area to body weight. This is why it is so important to dry babies as soon as possible after delivery and make sure that they are warmly wrapped.

Next, your delivery-room nurse will weigh your baby, measure the baby's head circumference and body length, and attach plastic identification bracelets to the wrists or ankles with your name and the baby's hospital number and date of birth. It is essential that your baby can be clearly identified before leaving the delivery room so that there is never any confusion or baby mix-up fears at a later date. The baby's crib will also be labeled clearly. Some maternity units now attach an electronic tag to the baby which sets off an alarm if they are removed from the ward.

Physical checks

In addition to assessing your baby's condition with the Apgar scoring system, your doctor will do a preliminary checkup to make sure that your baby has no obvious physical abnormality. She will look at the baby's face and tummy, listen to the heart and lungs with a stethoscope (a newborn baby's heart rate is normally about 120 beats per minute), turn him over and look at his back, run their fingers down his spine, check that his anus is open, note whether he has or has not urinated, and count the number of fingers and toes present. At a later date, a pediatrician will check your baby again and perform a more thorough physical examination before you go home (see p.387). It is quite common for newborn babies to have mild inflammation or conjunctivitis after the lengthy trip through the birth canal. If your baby has a sticky eye, you will be shown how to bathe the eyes with sterile water. Antibiotic eye drops will also be given to all babies. After all of the checks have been completed, your warmly wrapped baby will be handed over so that you can start to get to know him.

…your doctor will do a preliminary check to make sure that your baby has no obvious physical abnormality.

Vitamin K

This vitamin is present in food, especially in liver and in some vegetables. Vitamin K is essential because it helps blood clot and prevents internal bleeding. However, newborn babies receive very little vitamin K because they are fed entirely on milk. In addition to this, their livers, which are responsible for producing other essential blood-clotting substances, are relatively immature and, as a result, they run a small risk of developing vitamin K deficiency bleeding (VKDB) or hemorrhagic disease of the newborn. Consequently, the American Academy of Pediatrics (AAP) recommends that all babies receive vitamin K supplements soon after birth. This is done by injection—one dose of intramuscular vitamin K prevents VKDB in virtually all babies. A single dose will be given to your baby in the delivery room This is the most effective method of administration.

Formula is fortified with vitamin K, and as a result of this, bottle-fed babies are at an even lower risk of VKDB. However, the advantages of feeding with breast milk over formula are considered to far outweigh the marginal increase in the risk of VKDB.

Adaptations at birth

Throughout pregnancy, your baby receives a continuous supply of oxygen and nutrients from the placenta, which also removes all the baby's waste products. Within a few minutes of delivery, your newborn has to change from being completely dependent on the placenta to taking independent control of the entire metabolism.

The first thing is that your baby's lungs need to receive blood to oxygenate and pass to the left side of the heart to pump to all of the body organs. In utero, 90 percent of the blood supply by-passed your baby's lungs because there was no need to oxygenate it and hence the right side of the heart and the pulmonary (lung) vessels were at a higher pressure than the left side. Thus, the blood returning to the heart was either shunted directly from the right to the left side through a hole (the foramen ovale) between the two upper chambers (atria), or passed via the lower right chamber of the heart (ventricle) into the pulmonary artery. Because of the high pressure in the lungs, most of this blood was forced to flow into a duct, the ductus arteriosus, which diverts blood into your baby's aorta to reach the rest of his body.

As your baby takes a first breath and fills his lungs with air, the pressure in the pulmonary blood vessels falls and the ductus arteriosus closes, which results in all of the blood from the right ventricle entering the lungs to receive

CIRCULATION BEFORE AND AFTER BIRTH ·····································

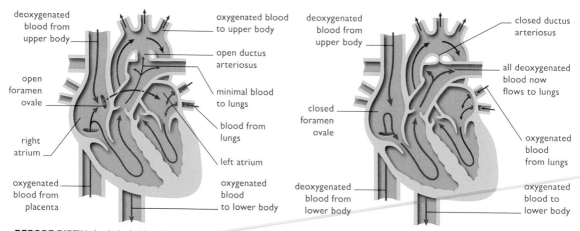

BEFORE BIRTH the baby's blood supply is shunted from the right to the left side of the heart through the foramen ovale.

AFTER BIRTH the blood supply passes through the lungs where it receives oxygen and begins to circulate around the body.

oxygen. From the lungs, this massive increase in blood flow passes to the left side of the heart, ready for pumping around his body. At the same time the flow of blood to the right side of the heart is reduced as the vessels in the umbilical cord start to constrict. As the pressure on the left side of the heart increases and the pressure on the right side decreases, it is no longer possible to shunt blood through the foramen ovale, which closes over like a flap. The baby now has an "adult" blood circulation (see diagram).

After all these cardiovascular changes are in place, your infant's liver receives much larger quantities of blood. As a result, the liver can start to metabolize the food or glycogen reserves that it has built up over the last eight weeks of intrauterine life in order to meet the energy needs of the first few days of life until feeding becomes established.

The body temperature of most newborn infants falls by 1.8–2.7°F (1–1.5°C) after birth (see p.376). Term babies have laid down brown fat which they can now use for heat production without even needing to shiver.

How your newborn looks

Many couples are surprised by the appearance of their newborn baby. Their newly delivered infant can be a far cry from the cherublike individual depicted in magazines. But it will only take a few days for the blemishes and visible signs of the traumas of delivery to disappear.

The first glimpse of your baby will arouse all sorts of different emotions, not all of which are necessarily positive.

Many babies are born with blue eyes and the final color of their eyes may not be evident until the baby is six months old. The eyelids will probably look puffy—another consequence of the pressure effects of labor. Your baby may squint or appear to be cross-eyed for several months after birth, but this is rarely anything to worry about. Your baby's focus is poorly adjusted at birth but when you hold him at about 8in (20cm) from your face, he will be able to see you and start to understand the details of your face.

The head often appears pointed or cone-shaped after a vaginal delivery, particularly if the labor was prolonged. This is because the bones of the skull overlap to allow the head to be subjected to pressure and progress smoothly during the descent through the bony birth canal. Within a week, your baby's skull will return to its normal shape. Sometimes the pressure also causes swelling on the sides of the face and if forceps or a vacuum extractor were

Many babies are born with blue eyes. The final color may not be evident until the baby is six months old…

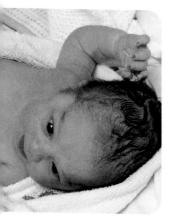

YOUR BABY'S HEAD
This may appear cone-shaped or pointed for a few days after a vaginal delivery.

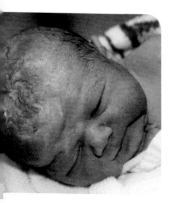

AT BIRTH Your baby may be covered by a thick greasy vernix which has protected his skin from the watery environment of the uterus.

needed, there may be some bruising on the face or scalp. You will notice a soft diamond-shaped area on the top of the head (the anterior fontanelle) where the baby's skull bones have not met together. This will not close completely until about 18 months of age.

Some babies are born with a thick coating of white greasy vernix caseosa, which has protected their skin from the watery environment in the uterus. Others will have none or just a few residual patches. Some doctors will clean it off the baby's skin soon after the delivery while others will leave it on to fall off or rub off in the next few days. Most newborn babies have very blotchy skin, not just because of the rigors of labor but also because it takes some time for the circulation to the arms and legs to become well established. Dry, flaky areas of skin on the arms and legs are common. In the uterus, your baby was covered in fine, downy lanugo hair. At birth, some babies have lots of it left on the scalp and shoulders, whereas others have none. Any that remains rubs off during the next week or two. Small white spots on the face, called milia or milk spots, are very common. They are caused by blocked sebaceous glands that lubricate the skin. They will disappear quickly after delivery. The color of your baby's scalp hair at birth may change over the next few months.

Some babies are born with long fingernails, and the problem with this is that they tend to lead to scratches on the face and elsewhere as the baby starts to explore his own body. Avoid cutting your baby's young nails with scissors, which can damage the nail beds. Instead, nibble the nails off gently and painlessly in your mouth. Placing protective mittens on your baby's hands will help prevent more scratches from developing.

Birthmarks are skin blemishes caused by clusters of small blood vessels under the surface of the baby's skin. They do not usually require any treatment. Caucasians are commonly born with pink skin patches (called stork bites) on the nose, eyelids, forehead, base of the skull, and under the hairline on the neck. Most of them disappear within a year. Strawberry birthmarks (nevi) start off as small red dots on the skin and may continue to increase in size for a year after birth. The majority of them have disappeared by the time they are five. Most babies with dark skin tones have birthmarks called Mongolian spots. These are blue-gray patches of skin on the back or the buttocks. They are completely harmless and usually fade away in the first couple of years. Port-wine stains are large reddish purple skin marks usually found on the baby's face and neck. Since these are permanent birthmarks you may want to seek expert advice from a skin specialist.

Babies of both sexes frequently have swollen breasts at delivery and may even leak a little milk. This is entirely normal and is a result of the mother's pregnancy hormones that take time to clear from the baby's body. The swelling and milk secretions will subside in a few days. Both girls and boys frequently have swollen genitals at birth. Again, this is the result of maternal hormones engorging the tissues and will settle down quickly. In girls, high estrogen levels produced by the placenta can also cause the uterus lining to become thickened while they are still in utero. If this is the case you may notice that your baby daughter experiences some vaginal bleeding after delivery (like a light period) as the thickened uterus lining breaks down. It will only last for a day or two and is nothing to worry about. The testes of a baby boy may still be in the groin at birth but usually descend without complication at a later date.

Putting your baby to your breast

There are enormous benefits of skin to skin contact with your baby in the first hour after birth as a method of promoting breast-feeding. While you are holding your new baby it is a good idea to try putting him to your breast because the hormones oxytocin and prolactin are produced when the nipple is touched or stimulated. Oxytocin helps the uterus contract, so even if you plan to bottle-feed, it is useful to put your baby to the breast just after the delivery. Prolactin makes the milk come in and, although you will produce only colostrum for the first few days, the sooner you get the milk or "let-down" reflex working, the better. You are simply getting the baby used to being on the breast, so don't worry if he does not seem very interested in feeding. Most term babies already have a sucking or rooting reflex (see p.387), so when you touch the corners of their mouths with a finger or the nipple, they will turn to the stimulus and attempt to nurse.

Term babies already have food reserves built up when they are born. Many women feel anxious if their baby does not start to feed immediately but babies are often more interested in sleeping after delivery. Premature babies, however, need to be given small bottle or tube feedings of expressed milk or formula during the first 24–48 hours of life because they have less reserves and the sucking reflex is rarely developed before 35 weeks of gestation.

FIRST FEEDINGS
These help your baby get used to the idea of breast-feeding as well as stimulating the release of hormones, causing the uterus to contract.

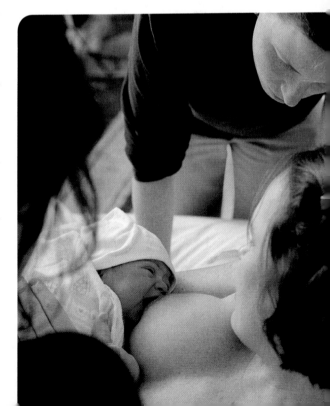

The first six weeks

AFTER NINE MONTHS OF ANTICIPATION, excitement, and probably some apprehension, you and your partner can now start the next stage of this journey as you discover and explore your roles as parents. This section of the book will guide you through the first few weeks of your life after your baby's birth.

During this period of enormous change, you will inevitably experience a huge range of emotions. You will feel wonder and awe at the tiny person you have created and will be fascinated as you get to know the little quirks of her personality. You may also feel overwhelmed by her vulnerability and complete dependency on you.

In addition to all of this, you will be recovering from the physical effects of giving birth, as well as adjusting your relationship with your partner as you learn to accommodate your new little family member. Coming to terms with the practical aspects of taking care of your baby can also be challenging, especially when you still have your usual household work to do as well.

Adjusting to change

At times it may seem like there are a lot of balls in the air to juggle and the pressure on women not to drop any of those balls is probably greater than it has ever been. It seems that this is made worse by the media coverage of celebrity mothers who, within 10 minutes of having their babies, are zipped back into their size 2 jeans, starring in their next blockbuster movie, and at the same time, they seem able to be the perfect mother. Faced with these kinds of images, many "normal" women dare not admit to having a difficult time during the first few weeks of parenting.

Becoming a mother is a time of great joy for both you and your partner but it is also a time when the parenting learning curve is steep. I hope that by being honest with you about what life after birth can be like, you will realize that all the physical and emotional changes that you are experiencing are very common, and regardless of whatever you are doing, or not doing, with your baby, you are still going to be a great mother.

> *...many women dare not admit to having a difficult time in the first few weeks of parenting.*

Your physical recovery

The postpartum period is the six weeks following the birth of your baby. How you recover physically will depend on a variety of factors, including the type of labor and delivery you had, your general state of health, and your domestic support and social circumstances.

Here, I will discuss the most common physical aftereffects of labor and delivery and some of the problems that can arise.

As your uterus starts to shrink back to its prepregnant state in the first few days after the birth, you will have a very heavy bloody vaginal discharge, called lochia. Lochia is made up of blood, mucus, and tissue debris, all of which needs to be expelled from the uterus. You will need extra-thick sanitary pads for the first few days and disposable panties because the blood loss is likely to be heavy. The flow usually calms down after the first week and the color of the blood gradually changes from bright red to brown.

After pains are the periodlike pains that many women experience after delivery, particularly if they are breast-feeding. They are uterine contractions caused by the hormone oxytocin, which encourages the uterus to shrink back down into the pelvis more quickly. Since oxytocin is released when the baby nurses at the breast, it is common for women to experience after pains or pass some small blood clots while they are breast-feeding. After pains should last only a few days after the delivery, but if they are making you feel uncomfortable ask your doctor for an appropriate form of pain relief. Depending on the type of delivery you have had and the extent of your discomfort, you can choose from injections, pills, or rectal suppositories.

Some degree of breast engorgement may occur as the milk starts to come into your breasts. The breasts become swollen, hard, and sore, and this normal inflammation commonly raises your temperature slightly. Fortunately, the problem usually resolves spontaneously in a day or two as breast-feeding becomes established (see p.398) but in the meantime it can be helped by improved positioning and attachment (see p.399).

If you have stitches they will become tighter as the skin surrounding them swells and the wound starts to heal. This can make sitting down uncomfortable. Sitting on a rubber ring helps during the first few days since this avoids any direct pressure being placed on the perineum. Similarly, maternity cool packs or local anesthetic creams and sprays on the perineum can help. You may also find that urinating produces a burning or stinging

After pains are uterine contractions caused by the hormone oxytocin...

sensation as the urine flows directly over the incision. If possible, try to stand up or crouch over the toilet with your legs as wide apart as possible to help direct the flow of urine straight into the toilet. Gently wash the area with a cool sponge or washcloth afterward and pat the incision dry.

The bladder undergoes a particularly stressful time during labor and delivery, which can result in difficulties urinating. If this happens, you may need to have a catheter inserted into your bladder to rest the muscles and allow them to regain their normal tone. The physical trauma of delivery can also encourage the development of a bladder infection. Prompt treatment with antibiotics and drinking plenty of water usually resolves the problem.

Many women fear that opening their bowels for the first time after the delivery will be a painful experience. Rest assured that your stitches are unlikely to burst open even if you do find that you are straining. To help prevent constipation, start drinking plenty of fluids (ideally water) as soon as possible and eat plenty of high-fiber foods such as cereals, fresh fruits and vegetables, and dried fruits. Gentle exercise will also help.

Toning your abdomen as quickly as possible after the birth is probably a key concern. If you had a vaginal delivery, you can try gentle abdominal exercises during the first few weeks after delivery. If you had a cesarean

POSTPARTUM EXERCISES

It is important to perform Kegel exercises (see p.165) after a vaginal delivery, particularly if you had a long, drawn-out labor, that will have stretched the muscles considerably. Make sure you do these exercises for a short time on a regular basis. Try setting yourself the target of doing them by lunchtime every day, for example, which is much better than doing lots of them once a week. You can start by doing a few pelvic squeezes the day that you give birth, then gradually build up your regular exercise program.

After giving birth, you can also use deep breathing to tone your lower back and abdominal muscles.

KEGEL EXERCISE PULL in and tense your pelvic floor muscles as if you are holding back urine, hold for a few seconds then relax gradually. Repeat 10 times.

ABDOMINAL STRETCH Lie on your back with your hands clasping your bent knees. Breathe deeply, pulling your abdominal muscles inward and upward as you exhale.

delivery you may be advised to wait until after your six-week checkup. My view is that you can try some gentle exercises after a cesarean delivery as long as they do not make you feel uncomfortable.

After a cesarean section

The lochia is frequently less heavy after a cesarean section because the surgeon usually cleans out the uterine cavity before stitching up the walls of the uterus, thereby removing blood clots, pieces of membrane, placenta, and other debris. Nonetheless, you will have lochia for several weeks and may pass small blood clots and experience some after pains when breast-feeding.

Most women will need strong analgesia for the first 48 hours after a cesarean section.

Most women will need strong, effective analgesia for the first 48 hours after the cesarean section because this is major abdominal surgery. Some hospitals will offer patient-controlled analgesia (PCA)—these are handheld pumps that allow you to give yourself small doses of intravenous morphine through an IV line. Pain relief can come in the form of more intramuscular injections of morphine (which may leave you feeling somewhat hazy) or pills. Pills are the slowest-acting form of pain relief and are usually more appropriate after the first couple of days.

Any person who has an abdominal operation and subsequently requires bed rest is at risk of developing thrombosis (see p.424). Pregnant women have additional risk factors because of their hormonal status and the fact that they have been carrying a heavy weight pressing on their pelvic and lower leg veins for many months. Your doctor will be encouraging you to get up and move around as early as possible after your cesarean section. If your first attempt to stand up and walk leaves you feeling dizzy, just remember that in a few hours you will feel much stronger. The more active you are in the first few days, the shorter your overall recovery time will be.

Your abdominal incision will be covered by a sterile dressing that may remain in place for about 24 hours. You may not realize what type of stitches or clips have been used to close the skinlayer until one of the nurses or doctors takes the dressing off to inspect the incision. Staples or clips are usually removed on about day three, whereas individual or continuous sutures are usually left in place until about day five. It is rare for the removal of the skin sutures to cause much more than a little discomfort, and rarely is pain relief given. Your doctor will give you pain relief beforehand if you do need it. Your scar will appear red and raised at this stage and will also be tender to the

EARLY POSTPARTUM CHECKUP

Before being discharged from the hospital, both you and
your baby will undergo a postpartum checkup. Many
hospitals prefer to have a pediatrician check your baby,
but in some units specially trained senior nurses
perform this task.

Your doctor will want to know how much lochia you are passing, whether you are having any problems urinating or defecating, and how you feel emotionally. They will measure your temperature, pulse, and blood pressure, examine your breasts, check that your uterus is well contracted, inspect your perineum, and make sure your calves are not tender or swollen. Your hemoglobin level will be measured and you will be given iron tablets if it is low. You will be offered a rubella vaccination if you are not already immune. Your doctor will make sure that you have enough analgesics to take home with you. She will also discuss your plans for contraception since the majority of women ovulate since six to eight weeks after giving birth, even if they are still breast-feeding.

Your baby's physical condition will be reviewed by examining her head, eyes, skin, limbs, breasts, and genitals. The heart and lungs will be carefully listened to with a stethoscope. Her hips will be checked for any signs of dislocation by gently bending the legs upward and then rotating the hips outward. The abdomen will be palpated to exclude enlargement of any organs such as the liver or spleen and the baby's spine will be carefully checked to confirm that all of the vertebrae are complete. Your doctor will also look for more generalized problems, such as signs of infection, jaundice (see p.388), or low blood sugar. They will check your baby's temperature, skin color, and muscle tone, and look for evidence of lethargy or irritability. All babies now undergo a routine hearing checkup before leaving the hospital.

YOUR BABY'S REFLEXES

Newborn babies have several important reflexes which will be tested during the discharge examination:

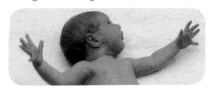

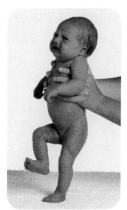

STARTLE REFLEX Your baby's arms and legs will stretch out when her head is allowed to flop backward.

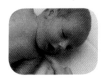

ROOTING REFLEX
Her head will turn
toward a finger
stroking her cheek,
her mouth open
ready to suck.

GRASP REFLEX
Your baby's fingers
and toes will be
able to grasp
your finger
very strongly.

STEPPING REFLEX
When supported
under the armpits
your baby
makes stepping
movements.

touch. You may find it helpful to place a soft pad over it when you are dressed. However, the incision does not need to be covered all the time, and indeed being exposed to the air will help it to heal more quickly. Unfortunately, taking a bath is recommended against for those who had a cesarean or vaginal delivery. Always dry the incision gently with a clean towel.

You may find that the skin around the incision becomes dry and itchy after about a week—gently rubbing in emollient cream helps relieve this. The area of skin around the incision may be numb because the nerves that innervate the skin have been cut. This superficial numbness is normal and tends to continue for several months while the nerves grow back. Another cause of concern is that the upper edge of the scar tends to be somewhat bumpy and sometimes hangs over the lower edge when you are standing upright. Again, this is normal and just reflects the fact that the surgeon cut through several muscle layers and these take time to knit together and provide a flat muscular wall.

NEONATAL JAUNDICE

Jaundice is common in newborns because the excess red blood cells that the baby required in utero have to be broken down and eliminated. A yellow pigment called bilirubin is produced and needs to be processed by the liver before it can be excreted. When the bilirubin levels are high, the pigment is deposited in the skin and the whites of the eyes, which then appear yellow.

Physiological jaundice is very common, occurring in as many as 60 percent of all newborns, particularly premature babies because their livers are immature. The yellow discoloration affects the skin and is visible within 24 hours of birth. The jaundice usually peaks at day four and

disappears without treatment within 10 days. However, if the levels of bilirubin become very high there is a risk that the pigment will be deposited in the brain causing permanent damage (called kernicterus). To prevent this, heel prick samples will be taken to monitor bilirubin levels in your baby's blood. If they reach a certain threshold ultraviolet light or phototherapy will be given for a few hours every day. The UV light breaks down the bilirubin in the skin, which can then be excreted in the urine without needing to be processed by the liver. You will be encouraged to feed your baby regularly, because the calories and fluid intake will also help resolve the problem. As soon as the

bilirubin levels fall below the threshold levels, the phototherapy will be stopped.

Breast-milk jaundice affects about five percent of breast-fed babies who remain mildly jaundiced for up to 10 weeks, probably because hormones in the breast milk interfere with the liver's ability to break down bilirubin. The jaundice is not harmful and disappears if bottle-feeding is started, but there is no need to stop breast-feeding if your baby is otherwise well. After two to three weeks your doctor will order blood tests to confirm that your baby's liver and thyroid function are normal. (For pathological jaundice see p.435.)

Leaving the hospital

The length of your stay in the maternity unit will depend on the type of birth you have had. It can range from a few hours (24 hours after giving birth is usually the earliest a woman will be allowed home) to more than a week if you have experienced complications. The average stay in the hospital is one to two days after a normal vaginal delivery, and five days after a cesarean section.

How long you stay in the hospital is an individual decision. It is important to remember that the purpose of you being in the maternity unit is to get advice and help you look after your newborn baby and also make sure that you make a fast recovery from labor and delivery. If you have opted to go home as soon as possible, make sure that you feel confident about how to change a diaper and bathe your baby and have talked to the nurses about obtaining practical help with feeding after you have left the hospital.

Your doctor will discharge you from the maternity unit with your newborn baby if all is well. Details of your baby's postpartum progress and any concerns or problems that have occurred or that may need special attention during the next few weeks will be given to your doctor. Your baby will have a heel prick blood test, before you two leave the hospital, to exclude phenylketonuria (a rare metabolic disorder), cystic fibrosis, and thyroid deficiency. The tiny sample of blood is obtained by pricking your baby's heel and applied to a card and sent to a laboratory for analysis.

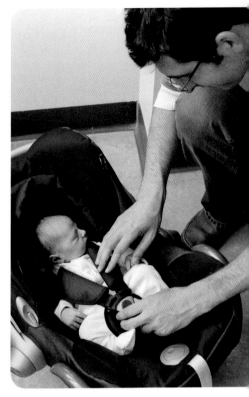

REAR-FACING CAR SEAT
A correctly installed car seat is a must if you are driving your baby home from the hospital.

On the road

If you are returning home by car, law requires that your baby travels in a special infant car seat that needs to face the rear of the car. If you have a passenger-side airbag in the front, your baby seat must be installed in the back seat of the car. Wrap up your baby warmly, because newborn babies are not good at maintaining or regulating their body temperature. As a rough rule of thumb, they should wear one more layer than you are wearing, plus a hat and mittens in winter, or a sun hat in summer.

If you have had a cesarean section, you technically should not drive for several weeks because, due to your abdominal incision, you are less capable of making a sudden stop. For this reason, you are more likely to cause an accident or injure a third party.

Your emotional recovery

The dramatic changes in hormone levels that occur immediately after the birth of a baby frequently result in emotional peaks and valleys. So don't be surprised if, during the first few days and weeks, you find yourself bursting into tears for no obvious reason.

TIME FOR ADJUSTMENT The days and weeks following the birth of your baby are ones of major adjustment. Allow yourself time to get to know your baby and to adapt to your new responsibilities.

Giving birth is an enormous achievement and most women will feel physically and emotionally exhausted. Instead of being able to catch up on much needed sleep and recover from the events in peace and quiet, you now find yourself on call day and night for your new baby. The realization that you are completely responsible for this helpless new human being hits you. These are powerful and difficult emotions to deal with, particularly for first-time mothers, and it is not at all surprising that they leave you feeling vulnerable and weepy. Let me reassure you that these feelings and responses are entirely normal and also temporary. They will start to subside in the next few days or weeks as your hormone levels become more stable and you adjust to the new demands and responsibilities of being a parent.

Bonding

Many new mothers I talk to are worried about whether they are bonding well with their newborn baby. I firmly believe that there is no right or wrong way to get to know and learn to love your baby. Some women fall immediately and unconditionally in love with their newborn, whereas others will be so shell-shocked by the delivery that it may take some time to adjust to the fact that they have just become a parent. Just because the so-called bonding process starts a little more slowly, this does not mean that they are going to be bad mothers or that their baby will suffer in the future. Please do not fall into the trap of feeling guilty or inadequate if this is the case for you. You will bond with your baby in your own time and unnecessary anxiety and distress can be avoided if you keep reminding yourself of this fact.

Another common problem in the early postpartum period is that many women strive for "imaginary perfection" as soon as they reach home and subsequently feel distressed and frustrated when they realize that this is virtually impossible to achieve. The reality of your new lifestyle is that it will be unpredictable. Young babies rarely understand how to fit in with your perception of an ideal daily routine and it will take time and patience to reach some sort of acceptable compromise.

Baby blues

The demands of an infant are endless and often tedious and many women, especially when they have their first child (less so with subsequent children), suddenly find that they are left, literally, holding the baby. In the past, women were surrounded by an extended network of female relatives who helped out when a new baby arrived (and continued to do so throughout the subsequent years). Today, women are much more isolated and, until they have made a few friends locally, they often have to struggle by themselves.

The vast majority of women experience some degree of the "baby blues" in the week following the birth of their baby. The baby blues usually start around day four or five, just when your breast milk is beginning to come in, and you are feeling physically uncomfortable. However well prepared you may think you are for the postpartum period and no matter how many people have warned you about what is likely to happen, the baby blues will invariably hit you by surprise. You are expecting to continue to feel euphoric and exhilarated that you have successfully delivered a healthy baby, but suddenly and inexplicably you find yourself weeping uncontrollably. I think the most disturbing aspect for most women is that they are powerless to do anything about these extraordinary surges of emotion. Make sure to tell your doctor how you are feeling.

The period of baby blues usually disappears naturally within a couple of weeks. You start to recover physically, your hormone levels settle down, and you learn to take care of your baby and find ways to make sure that you are not left coping with this by yourself. Yet, for some mothers, these feelings of mild depression do not improve and postpartum depression can ensue.

Postpartum depression

Exactly how many women suffer from postpartum depression has always been difficult to establish. Depending on who you ask, the answer could be anything from 5 to 20 percent of all women during the first year of their baby's life. I am sure that this uncertainty is because many women feel ashamed by their feelings of distress and are reluctant to reveal the problem and ask for help. It is also because family, friends, and doctors are not very good at recognizing that

SIGNS OF DEPRESSION

If you are experiencing some of the following feelings it may be that you are suffering from postpartum depression:
▶ overwhelming fatigue, disturbed sleep and early-morning wakening
▶ persistent anxiety and low self-esteem
▶ lack of concentration
▶ weepiness
▶ your mouth is dry, you lose your appetite or suffer from constipation
▶ loss of libido
▶ rejection of your partner

...symptoms of postpartum depression can develop at any time during the first year after birth.

early baby blues has now developed into a more serious problem. Postpartum depression (PPD) is an illness, and when you are not well it is difficult to be objective about the problems you are experiencing. As a result, you may not even know that you have it.

The symptoms of postpartum depression (see p.391) may not become apparent until after the six-week postpartum checkup and, in fact, can develop at any time during the first year after the birth. Postpartum depression may be short-lived, lasting for just a few weeks, but if unrecognized or untreated, can persist for a long time and be seriously debilitating. Mothers who have had complicated deliveries or a multiple birth are more likely to develop postpartum depression. For mothers of twins or triplets, the diagnosis is often delayed because it is assumed that the woman's symptoms reflect the fact that she has even more reason to be tired and distressed and to experience difficulties in coping with life after labor.

For milder cases of PPD, treatment may simply involve making sure the woman is supported, both emotionally and practically, by those who are around her. However, severe cases may need to be treated with antidepressant medication (which is not contraindicated during breast-feeding). Counseling and therapy, with or without drug therapy, play an important role in treatment.

No one knows exactly what causes postpartum depression. It may be the sudden change in hormonal balance after the birth, but the fact that this affects some women more than others suggests that there are other triggers such as genetic and environmental factors. Women who have suffered from depression in the past are more likely to develop postpartum depression. Furthermore, of those women who have had postpartum depression following a previous birth, 1 in 4 are likely to have a recurrence after each pregnancy. Although not directly linked, it is important to remember that thyroid disorders are very common postpartum and that this may lead to symptoms that are very similar to those of postpartum depression. It is useful to perform postpartum thyroid function tests for women who become severely lethargic or hyperactive.

Postpartum psychosis

This acute psychotic illness differs from severe postpartum depression because it usually appears within two weeks of delivery and involves schizophrenic or manic depressive symptoms. Psychosis is thought to affect 1 in 500 women, but if there has been a previous episode, the recurrence risk may be as high as 25–50 percent. Occasionally, the mother is at risk of suicide or of harming her baby and will need to be taken care of in a secure mother and baby unit.

COPING STRATEGIES

First and foremost, every mother needs to remind herself that she is doing her best and that there is no such thing—fortunately—as the perfect mother, regardless of what alleged child-care experts and those around you might say.

If you do become depressed there are several things you can do to help yourself and make the illness as short-lived as possible. Start by reminding yourself that there is no such thing as the perfect mother. As long as you try to do your best, that is all that anyone can reasonably expect of you at this difficult time.

Expectations placed on new mothers are often unrealistically high, but when the woman is seen to fall short of being the ideal mother, either in her emotional response or her practical everyday care of the new baby, she often feels guilty, inadequate, and confused. It is easy to see this lead to women experiencing some or all of the symptoms for postpartum depression.

You should remember that you need to reserve some time for yourself during the postpartum period. Everyone is so focused on the new baby that the mother's emotional or physical health are often overlooked.

Here are some tips to help you cope, practically and emotionally, with being a new mother.

▶ **Avoid becoming isolated** and try to get out of the house at least once every day.

▶ **Actively seek out other new mothers.** Many of them will be experiencing the same emotions and can provide a useful support network.

▶ **Make sure you receive as much domestic help as possible.** Pay for it, if necessary.

▶ **Don't suffer in silence.** Talk to your partner, friends, and family and make sure they understand your feelings and help support you practically and emotionally.

▶ **Seek medical help early.** Don't hesitate to talk to your doctor if you are feeling down. You may benefit from a short course of antidepressants (which do not interfere with breast-feeding) and/or seeing a therapist.

▶ **Regular gentle exercise** and plenty of fresh air will do wonders for your sense of well-being.

▶ **Try to eat regularly and sensibly.** This is particularly important in helping with breast-feeding.

▶ **Women, and mothers in particular, are very good at feeling guilty.** Don't. You are allowed to complain and to feel unhappy about your situation.

▶ **Schedule regular treats** or things to look forward to. Accept offers from family or friends to take care of the baby, giving you some free time to yourself.

▶ **Contact organizations and support groups** that deal with postpartum depression, such as Depression After Delivery or Postpartum Support International (see Contacts, pp.437–38).

66 *...remember to reserve some time for yourself during the postpartum period.* 99

First days and weeks at home

Now that you are at home with your tiny baby, all kinds of anxieties may start to surface. Remember that babies are a lot tougher than they look. Short of accidentally dropping him, there is not much you can do that will actually harm your baby.

GRANDPARENTAL ROLE Accept any offers of help that come your way. Grandparents in particular are often enthusiastic supporters as you adjust to your new life.

During the first few weeks, try to find someone to do the household chores, particularly those that involve lifting, carrying, or bending, since you will need your energy to take care of your baby. The more help you can get, the quicker your recovery will be. You also must be sure that your return home does not result in your house becoming a coffee shop. Family and friends want to see the new baby, but make sure that they pull their weight when they visit.

If a home-care nurse is offered under your medical insurance coverage, take advantage of it. Your nurse will visit you the first few days after delivery to check how you and your baby are progressing. She gives you and your baby a thorough examination to make sure you are both doing well and will assess your recovery, checking for signs of infection, including newborn jaundice. The nurse will also make sure your milk is coming in for breast-feeding and that your baby is latching on successfully.

Pace yourself

Many women find that once they have returned home they don't feel like venturing out again for some days. For many women, caring for a newborn baby takes up all of their time and waking thoughts and the last thing they want to do is expose themselves to the fast pace of life beyond the front door. This is an entirely normal reaction and my advice is to aim to do only what you feel like doing so that you can recover from the birth and get to know your baby in your own time.

During the next few months you will experience interrupted sleep and are likely to become tired as you provide round-the-clock care for your baby. Pace yourself from the beginning. Instead of trying to do the housework while your baby is sleeping (he will sleep an average of 16 hours a day) use the time to get some rest and much-needed sleep for yourself.

COMMON CONCERNS ABOUT YOUNG BABIES

UMBILICAL CORD

The stump of the umbilical cord usually stays in place for about 10 days after birth, during which time it should be washed and carefully dried on a daily basis to avoid infection. If the cord becomes sticky or smells offensive, contact your pediatrician.

VOMITING

Young babies frequently bring up some of their feeding, particularly when they are trying to get rid of gas. This is nothing to worry about unless the vomiting is forceful and occurs after every feeding (see pyloric stenosis p.435) in which case you should talk to your pediatrician.

TRAPPED GAS

Trapped gas gives rise to abdominal cramps and pain and not surprisingly the baby cries and is difficult to settle down after a feeding. If simple measures such as placing your baby against your chest and massaging his back in order to encourage the gas to escape do not work, then ask your pediatrician to recommend some medication that will ease your baby's discomfort.

LOOSE STOOLS

Your baby should pass meconium (a sticky green-black mixture of bile and mucus) for the first 24 hours, after which the stools will turn yellow-brown. Breast-fed babies commonly pass looser stools than bottle-fed babies, but if your baby starts to pass watery green stools it is likely that he is suffering from diarrhea. Young babies can quickly become dehydrated and this needs to be treated immediately. Give your baby some cool, boiled water. If he has constant diarrhea, develops a dry mouth, and a sunken anterior fontanelle, contact your doctor immediately.

DIAPER RASH

The ammonia in urine irritates the baby's sensitive skin, so it is not surprising that most babies experience a degree of diaper rash, even if their diapers are changed regularly. However, diaper rash can be exacerbated by perfumed toiletries, creams, and wipes that are not hypoallergenic. It is best to wash your baby's bottom with water and nonperfumed baby soap and gently pat it dry. Zinc- and sulfur-based barrier creams applied to the reddened area encourage the skin to heal and help protect against further irritation.

STICKY EYES

This is usually due to a mild eye infection called conjunctivitis. It is common immediately after the delivery when blood or other fluids may have come into contact with the baby's eye. Gently wiping each eye with a fresh piece of a cotton pad soaked in cool, boiled water usually resolves the sticky eyes, but if the

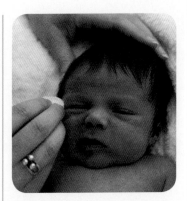

CLEANING A STICKY EYE Always wipe from the inside outward.

problem continues your pediatrician will prescribe an antibiotic eye cream to use on the affected area.

FACIAL SPOTS

The small white spots (milia) that many babies are born with usually disappear in a few weeks without any treatment. If they become infected and red, bathe them in cool, boiled water before applying an antiseptic cream.

Contact your doctor immediately if your baby:
▶ is vomiting continually
▶ is passing watery green stools
▶ is very lethargic
▶ is irritable and feeding poorly
▶ starts to wheeze or develops a cough
▶ is breathing very quickly, very slowly, or irregularly
▶ develops a high temperature
▶ shows signs of an infection or skin rash

Feeding your baby

Most women have decided before the birth whether they want to breastfeed or not. The decision is a personal matter and I believe that women should not feel inadequate if, for whatever reason, they choose to bottle-feed from the beginning.

However, it is important to be aware that you can offer your baby significant long-term health benefits by breastfeeding, even for a few weeks. Breast-fed babies are likely to experience fewer respiratory, gastrointestinal, urinary, or ear infections and less likely to develop allergies and childhood obesity than bottle-fed babies. Furthermore, breastfeeding for a minimum of two months is thought to reduce a woman's risk of developing breast and ovarian cancer later in life. On a practical note, breastfeeding can be done at any time and in any place. Vitamin D supplementation (10mcg per day) is now recommended to breastfeeding mothers.

Breastfeeding

When a baby sucks on a nipple and areola (the brown skin around the nipple) two things occur. First, the mother's pituitary gland (at the base of the brain) is stimulated into releasing the hormone prolactin that produces milk. Secondly, the pituitary gland releases oxytocin, which stimulates the alveoli to contract and forces the milk into the milk ducts and toward the nipple. This process is called the let-down reflex.

MILK PRODUCTION ···

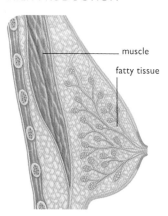

muscle
fatty tissue

BEFORE PREGNANCY

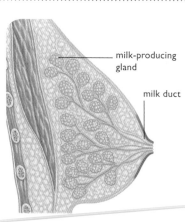

milk-producing gland

milk duct

DURING BREASTFEEDING

STRUCTURE OF THE BREAST
Your breast is made up of a mixture of adipose (fatty) and secretory tissue. Each breast contains about 15–25 lobes and each lobe is drained by a milk duct that leads to the nipple. The lobes are made up of individual alveoli (sacs) that swell up and contain the milk.

For the first few days after the birth, your breasts will produce only small quantities of colostrum (about 3–4 teaspoonfuls daily). This is a concentrated clear yellow secretion which provides your baby with all the water, protein, and minerals required until you are producing proper milk. Colostrum also contains high levels of maternal antibodies and a substance with natural antibiotic activity called lactoferrin, which helps combat infection. If, for any reason, your baby is not with you in the days immediately after the birth, try to express some colostrum and ask that it be given to your baby.

Your breasts start to produce white breast milk in increasing quantities from about day three onward. Breast milk contains fat, carbohydrate, protein, and other nutrients in exactly the right proportions to ensure the healthy growth of your baby. After the milk comes in, you may find you are feeding up to a dozen times over 24 hours. Soon, you will build up to around 20 minutes per feed every two to four hours.

How to breastfeed

When breastfeeding, the whole areola should be in your baby's mouth in order for him to "latch on" and feed correctly. If he is latched on correctly, he will have his mouth wide open and you should feel a suction effect over the whole area. His top lip is turned upward and creased over and you will see his ears and jaw moving rhythmically as he sucks. If he is not latched on correctly, start again. Do not leave him to suck on just your nipple because it will become sore and cracked. Your baby should empty one breast at each feeding to ensure he gets both the dilute and thirst-quenching foremilk and the thicker, more nutritious hindmilk.

It is essential to be comfortable when you are feeding. Support your back properly and put a pillow under your baby so you are not bending down. Your baby should lie facing you, rather than just with his head turned toward your breast. You can also try lying down with your baby alongside you.

Expressing breast milk

You can express your milk by hand or with a breast pump, although expressing by hand can take longer. Breast pumps have a funnel that you place over your areola, forming an airtight seal. Pumps can be either hand- or battery-operated. Before starting to express you will need to use the bottle sterilizer that comes with the breast pump. You can keep the milk in the refrigerator at 39°F for 3–5 days or freeze it at 4°F in a sterile freezer bag, where it will keep for up to three months.

LATCHING ON Your baby should have all of your nipple and as much of the areola as possible in his mouth, which then forms a tight seal. As he squeezes the nipple against the roof of his mouth milk will be drawn out.

BREASTFEEDING

▶ My breasts are very engorged and painful. What should I do?

When the milk starts to come in, between the third and fifth day after delivery, it is common for the breasts to swell and become engorged due to overproduction. Finding yourself with a fever and swollen, rock-hard breasts can be a painful experience, but it is entirely normal and usually settles down within 24 hours or so. As your baby starts to nurse regularly and empty your breasts more effectively, your body will adapt to producing the right quantities of milk needed to nourish him.

If you develop breast engorgement, it is essential that you drain the milk from your breasts regularly to prevent the milk from seeping into the surrounding breast tissue, which can lead to mastitis (see breast infection pp.433–34). There are several things you can do to drain swollen breasts and prevent severe engorgement.

• Feed your baby a little and often so that your breasts are drained regularly.

• Express a little milk before you start to feed—this will soften the nipple and will help your baby latch on.

• Even if you have a cracked nipple, you must try to drain that breast—try using a nipple shield or express milk regularly from that breast.

• If your baby is not feeding very well, express the milk and either store it or discard it. Milk production works on a supply-and-demand basis, so if you do not empty a full breast regularly, future production will be poorer.

▶ How can I cure blocked ducts?

If a red, tender patch develops on the breast, you have a blocked milk duct. This is extremely common, but to prevent it from developing into mastitis:

• start each feeding from that breast, because your baby's suction is strongest at the beginning of a feeding.

• place a warm washcloth or a cold compress inside your bra over the red patch.

• feed your baby on all fours, so that your breasts hang down directly over him— this allows the breast to drain much more quickly.

• express milk from the breast with the blocked duct to help clear up the blockage.

▶ My nipple is cracked and really sore. What should I do?

Try to keep feeding on that breast to avoid engorgement. If necessary, express milk from that side while your nipple recovers and feed your baby from the other side. After a feeding, smear a little milk or saliva on the nipple and let it dry naturally. Expose your breasts as much as possible to the air and change breast pads after each feeding to help your nipples heal.

▶ I don't think my baby is putting on as much weight as he should. What can I do?

Breastfed babies often put on weight at a slower rate than bottle-fed babies and the growth chart you will be shown at the health center gives a wide spectrum for so-called "normal" weight. If your doctor is worried about your baby's weight gain ask yourself the following questions:

• Are you eating enough? Successful breastfeeding requires an extra 500 calories per day, 1,000 for twins, for your body to produce enough milk.

• Are you drinking enough? Plenty of fluids are needed to aid milk production. Add an extra two pints a day of water to your normal intake.

• Are you resting enough? If you are tired your milk supply will be lower.

If you have breastfeeding problems seek advice from your midwife, lactation consultant, or doctor or contact an organization such as the La Leche League International (see pp.437–38).

Bottle-feeding

Bottle-feeding does offer some advantages, the least being that your partner can help feed your baby. Infant formulas are made with cow's milk (soy alternatives should not be fed to low birth weight preterm infants.). They are fortified with essential vitamins and minerals and are a close imitation of human milk.

If, like many women, you breastfeed first and then switch to bottle-feeding, the switch should be gradual, starting with one bottle a day, so that your baby gets accustomed to the nipple and to the taste of formula. This way, you will also prevent your breasts from becoming engorged. If you have been able to express breast milk into a bottle while breastfeeding, you should find the switch to formula relatively effortless. If your baby complains, it can sometimes help if someone other than you offers the bottle.

If you bottle-feed your baby from the start, you will probably find that your milk does not come through very strongly and gradually dries up. Formula-fed babies invariably need fewer feedings a day and wake less during the night. This is because cow's milk forms a much more solid curd, which takes longer to digest, hence the tendency for formula-fed babies to go longer between feedings.

It is important to be hygienic and organized when bottle-feeding. Wash the bottles thoroughly and use boiled water cooled to 158°F to mix the formula. To reduce the risk of gastrointestinal problems, it is best to make up bottles as and when you need them rather than storing them for many hours. Cool the feed further before giving it to your baby. Bottle-fed babies need to drink additional water since formula is not as thirst-quenching as breast milk. Once again, cool, boiled water is best.

The temperature of formula milk is a matter of habit, with some babies happy to drink cold milk straight from the refrigerator. If you do warm it up, always test it on the inside of your wrist and, if you heat it up in a microwave, shake the bottle to mix the formula before testing the temperature.

Burping your baby

Breast-fed babies take in very little air during a feeding, particularly once they have learned to latch on correctly. Bottle-fed babies, on the other hand, tend to take in more air—their mouths form a less airtight seal around a nipple—so may need more burping. The two main positions for burping a baby are over the shoulder and sitting on your lap. When your baby is sitting on your lap, make sure that the head does not slump down. The esophagus needs to stay relatively straight in order for the air to escape easily. Rub rather than pat your baby's back and place a cloth under your baby's chin to catch any milk that is brought up with the burp. You will eventually figure out the most effective way to deal with gas.

BURPING YOUR BABY
Sit him on your lap using one hand to support his neck and prevent his head from slumping down while you use your other hand to firmly rub his back.

Family adjustments

However much you vowed—as many first-time parents do—that your baby would not change your life, reality will be very different. Having a first baby causes an enormous emotional, practical, and financial change in your lives.

Partners

Fathers are often overlooked during the adjustment period after the birth because the baby and mother are usually receiving most of the attention. Yet your partner is probably feeling physically tired, too, and is expected to be supportive and understanding of you and appear delighted by the demands that a new baby has suddenly placed on his lifestyle. The key thing here is to make sure that you both clearly communicate your individual needs.

Try to get your partner involved in the care of the new baby, since this will help him understand some of the difficulties that you may be experiencing and prevent him from feeling excluded. This will probably require you to step back a bit and allow him to do things in his own way, even if this is slightly different from how you do them. Try to resist the temptation of continually criticizing his diaper changing or dressing technique. Babies are very adaptable and the last thing you want to do is undermine your partner's ability to look after his baby.

Your partner may find that he ends up feeling physically distanced from you during the first few weeks, particularly if you are breastfeeding. It is not difficult to understand how this can become another source of resentment. You are locked into a close relationship with your new baby and seem to have an endless supply of cuddling and kisses for him, but not enough energy left for anyone else. It is easy for your partner to start feeling neglected. This situation will change in time, but recognizing that your partner may need some reassurance that he is not going to be permanently left out in the cold is an important consideration and one that you need to make sure you convey.

BEING A DAD Allow your partner to be as involved as possible in the care of his new baby —even if his way of doing things is slightly different than yours.

RESUMING YOUR SEX LIFE

Sex is an issue that is rarely discussed openly by new parents or their doctors. However, it is estimated that more than 50 percent of couples have not returned to their prepregnancy levels of sexual activity one year after the birth of their first child.

This fact suggests that it is common for couples to experience a significant change in desire, frequency, and the quality of their sexual relationship at this time. There are many factors that could account for this change. An appreciation of the most likely reasons may help you and your partner discuss and hopefully improve the situation.

▶ **Many mothers are so exhausted** by birth and by the continuous demands of their baby that the only thing they are interested in doing in bed is sleeping until the next time their baby wakes them.

▶ **An episiotomy scar** (or perineal tear) can make penetrative sex painful for weeks after the birth.

▶ **Vaginal dryness** is a consequence of breastfeeding (the result of high levels of prolactin and low levels of estrogen) and can make sex painful.

▶ **Some women feel unattractive** to their partner because they have gained weight during pregnancy, because their lactating breasts leak milk as soon as they are touched, or because they now have a large red abdominal scar after having a cesarean delivery.

▶ **Women often feel unsupported,** lonely, worried, or taken for granted after having a baby, all of which can result in a lack of libido. Postpartum depression (PPD) is much more common than most people realize (see p.391) and mothers may experience symptoms for up to a year following the birth, which will invariably have an adverse effect on their sex lives.

▶ **Men, too, can suffer a temporary loss of desire.** This may simply be caused by fatigue, or the fact that they are adjusting to their new role as a father, but it may have a more deep-rooted cause. For example, some men now view their partner more as a mother than a lover, or they feel distressed having witnessed their partner's difficult vaginal delivery and the pain she suffered.

It is important to understand that you are not unusual or alone if you find your desire for sex has deserted you in the weeks or months after birth. It is also important that you discuss the physical or emotional problems that you are experiencing with your partner at the earliest possible opportunity. You will both need to be open with each other in order to prevent resentment and anger from further complicating a sensitive and potentially explosive situation.

Most couples will find that they are better able to deal with the situation if they receive regular reassurances that they are still loved and valued. Staying in regular physical contact, albeit only with cuddling and kisses, during the first months after the birth can help reinforce the fact that this is not a permanent state of affairs.

Eventually, you will resume your sex life but it may take time and it may not be with the same frequency. However, many couples discover that the change in their relationship and lifestyle results in a greater level of sexual intimacy in the long term.

Siblings

The arrival of a new baby in the house can be an unwelcome shock for other young children, who may find it difficult to accept the permanence of this new situation. Preparing a child psychologically and trying to involve him in the practical preparations for the arrival of the baby can be helpful. Friends and family can also ease the situation by bringing a little gift for siblings as well as for the new baby, playing with them, or taking them out for a day trip.

In the weeks following the birth, try to provide as much continuity as possible in the lives of other siblings so that they don't feel that family life now revolves completely around the new baby. Keep up with playgroups and after-school activities, invite their friends to play and, in particular, try to continue the normal bedtime routine—including reading a story—whenever possible.

NEW ARRIVAL In the weeks following the arrival of a new addition to the family, try to involve older siblings as much as you can.

Dealing with jealousy

Many children will exhibit attention-seeking behavior by being more clingy, whiny, or naughty in the early weeks. There is no doubt that the addition of a new baby to the family frequently causes feelings of jealousy in other siblings and this needs to be addressed rather than swept under the carpet. With older children, encouraging them to talk about their feelings and telling them that you love them just as much as before will help reassure them. However, with younger children it is likely that plenty of extra cuddling and some time spent alone with you each day will also be needed. Your partner and close family friends can really help by taking care of the new baby for you while you offer your other children some individual and much-needed attention.

Don't be surprised to hear your young children talk about putting the baby in the garbage or returning him to the hospital or back into mom's belly. You also need to be aware of the fact that it is common for young children to "accidentally" attempt to harm the new baby, by pinching or hitting him, when they think your back is turned. These are all entirely normal and predictable responses and I can assure you that your older children will come around to loving their new sibling in time. Nevertheless, you do need to take the precaution of never leaving your baby unattended with a young child in the room.

The six-week postpartum checkup

You and your baby will go to your pediatrician about six weeks after the birth. This provides an opportunity for you to talk about any concerns you have about yourself or your baby.

Your baby's checkup will involve a full physical examination together with an assessment of his developmental progress since birth, including:

• growth—size, length, and weight, which will be plotted on a growth chart
• head circumference and assessment of the anterior and posterior fontanelles
• eyes, ears, and mouth (sight and hearing will be formally assessed at later visits)
• heart, chest, and breathing
• abdominal organs and genitals
• hip alignment and stability
• reflexes—degree of head control, grasp reflex, and muscle tone.

In addition to asking you questions about your baby's general well-being, feeding, and toilet habits, the staff will explain the recommended times for your baby's vaccinations. If you have concerns about these, talk to your doctor or pediatrician or obtain up-to-date information from the American Academy of Pediatrics (see p.437). Although many scary stories relating to vaccinations have now been refuted and new vaccination programs promise even greater safety, there are still babies at risk as a result of mothers deciding against some vaccinations.

Your examination will make sure that you have fully recovered from delivery.
• Your blood pressure will be measured.
• Your urine will be tested to check if there is protein or blood present.
• You will be weighed and advised about diet if necessary.
• Your breasts and nipples will be checked.
• Your abdomen will be examined to check that the uterus is well contracted and if you had a cesarean section, the scar will be examined.
• A pelvic examination, if you had an episiotomy or perineal tear, to ensure the vagina has healed and you have no pain or discomfort. An internal examination, after a cesarean or complicated vaginal delivery, ensures that the uterus is well contracted, not tender, and that you have no vaginal bleeding or discharge.

Now is the time to discuss any problems you may have with pain, bleeding, bowels, or bladder. You will also be reminded to think about contraception.

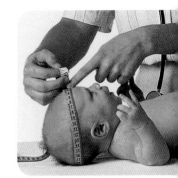

THE HEAD CIRCUMFERENCE of your baby will be noted.

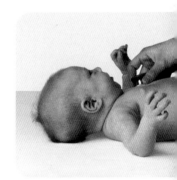

YOUR BABY'S HEART RATE and breathing will be monitored.

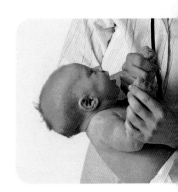

HIS DEGREE OF HEAD CONTROL will be checked.

CARING FOR A PREMATURE BABY

Around 11 percent of babies are born before the 37th week and are classified as being premature. Many of these babies, although smaller than a full-term baby, can be cared for like a normal newborn and do not need any special care. Another two–three percent of babies are born at term but with a low birthweight relative to their gestational age and need special care.

Generally speaking, any baby whose birthweight is less than 3lb (1.5kg), either because they are premature or small-for-dates, will be taken to the Neonatal Intensive Care Unit (NICU). Other premature babies may be above this weight but may have other problems. Usually, they simply need more time to grow or for their lungs to mature so that they can breathe unaided.

NEONATAL INTENSIVE CARE UNIT (NICU)

This special unit cares for vulnerable babies in the best possible environment. The unit protects your baby from infection by limiting access to all but essential people—medical staff, parents, and the immediate family of the baby. There is also a high ratio of caregivers to babies. While in the unit the babies are constantly monitored to make sure any problems are dealt with immediately. The staff always encourages parents to participate in the care of their baby and they take care to explain what

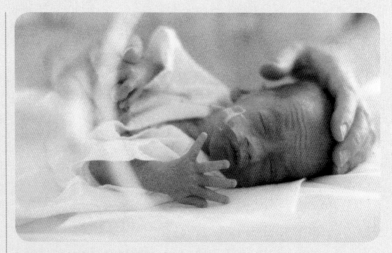

TOUCH IS VERY IMPORTANT Studies have shown that cuddling and touching premature babies can help them to gain weight and grow.

is going on so that the parents understand and feel involved. Additionally, all NICU's have counselors for parents to talk to and many have baby feeding advisers to help with breastfeeding.

If your baby is taken to an NICU, she may be placed in an incubator and attached to various monitors and wires. She may be ventilated to enable her breathing. The first visit

to an NICU is invariably a shock for parents. It is difficult to see your tiny baby lying helplessly in an incubator, surrounded by high-tech machinery and a tangle of wires. The medical staff will always be on hand at this time to reassure you and provide answers to any questions you might have. They will show you how to express milk and feed your baby, how you can stroke her inside the incubator, and will encourage you to talk to her. In addition, even the smallest baby can usually be cuddled, so they will help you take your baby out of the incubator.

Gradually, you will learn how to care for her, including changing her diaper, giving her a bath, and helping with her feedings. Once she is stronger and no longer ventilated, you will be able to spend as long as you want holding her. There are no visiting hours in an NICU—parents can spend as much time as they want with their babies.

COPING EMOTIONALLY

One of the major hurdles to overcome when your baby spends more than a few days in the neonatal intensive care unit is the realization that you will be going home without her. This can be a severe blow, particularly if you are leaving behind a very premature or sick baby. Yet, it is vital not to feel guilty—although

I realize that this is easier said than done. It is not likely that anything you have done has actively contributed to the premature birth and you can be sure that your baby will now receive the best care available to enable her to thrive and go home as soon as possible.

When you go home make the most of this opportunity to regain your strength and make the emotional and practical adjustments to having a baby that you and your family were perhaps not able to do because of the unexpected early birth.

Furthermore, you do not have to stay at the hospital every waking hour, particularly if you have other children. This could lead to feelings of resentment on their part once the baby is back home. The time that your new baby spends in the hospital can be a useful adjustment period for other children, and although you may

not be spending more than a few hours a day with your baby in the NICU there is not a shred of evidence to suggest that this will negatively impact your ability to love your child.

GOING HOME

Generally speaking, a baby will go home when she is able to feed fully, either with breast milk or formula; she weighs at least 4lb 7oz (2kg) and gaining weight; she is more than 34 weeks gestation; and she maintains her own body temperature.

When the time to go home is imminent, many units ask the mother to come in and stay for at least one night in the bedrooms available for this purpose. This gives mothers the confidence to take care of their baby when they are back at home.

Premature babies are usually home within two to three weeks of the time they were due. By then, unless there is an underlying health problem, the baby should be treated just like any full-term newborn. She should be given plenty of cuddling to make her feel secure and it is unlikely she will display any differences in her behavior from a full-term infant.

The stages of development and the weight gain of a premature baby are calculated from the date that the baby was due. By the time your baby is two years old, she will have caught up developmentally and physically with full-term children.

TAKING CARE OF A SPECIAL-CARE BABY

BABY CARE You will be encouraged to care for your premature baby, including changing her diapers.

MAKING CONTACT There are many ways to interact with your baby while she is in the incubator.

Concerns and Complications

Existing medical conditions

This section includes most of the disorders that I encounter on a day-to-day basis and those that I am asked about most frequently. If you know that you have a medical condition or are diagnosed with one in pregnancy, it is important to make sure that you get specialized medical help.

Epilepsy

If you suffer from epilepsy, it is particularly important that you are under close medical supervision when trying to become pregnant and during pregnancy. Some epileptic medicine (particularly sodium valproate) can cause abnormalities in the baby, such as heart and limb defects, developmental disability, CLEFT LIP AND PALATE, and are best changed to a different type of drug. Your doctor will advise you of the best drug to be taking and, early in pregnancy, will arrange for you to have a specialized ultrasound scan to detect any fetal abnormalities. Pregnancy can change the way epileptic drugs are metabolized in the body and you may need a higher dose to ensure that you do not have a seizure. Some epileptic drugs may reduce your folic acid levels so you need to take high-dose folic acid supplements before becoming pregnant and during the first 12 weeks of pregnancy to minimize the risk of having a baby with a neural tube defect such as SPINA BIFIDA. Since the drugs induce your liver enzymes, you will be advised to take daily vitamin K supplements from 36 weeks.

Any seizure in a pregnant or postpartum woman needs to be carefully assessed to establish whether it is due to epilepsy or eclampsia. Breastfeeding is safe, although some drugs (primidone and phenobarbital) may have a sedative effect on the infant.

Diabetes

There are two types of diabetes in pregnancy: preexisting diabetes mellitus and GESTATIONAL DIABETES (pregnancy induced). Diabetes mellitus affects 3 percent of the population and since pregnancy usually exacerbates the disorder, specialized prenatal care is needed to minimize the maternal and fetal complications associated with diabetic pregnancy. Women with preexisting diabetes mellitus should aim for optimal control of their blood sugar levels before they become pregnant, since hyperglycemia at conception and during the period of embryogenesis increases the risk of miscarriage and may cause major fetal abnormalities including cardiac, skeletal, and neural tube defects. High-dose folic acid (5mg) supplements should be prescribed for all diabetic women planning to become pregnant and continued until they reach the second trimester. A careful search for fetal abnormalities will be performed at the 20-week ultrasound scan. Careful control of the mother's sugar levels should be maintained throughout pregnancy because maternal glucose, but not insulin, crosses the placenta. If your diabetes is usually managed with oral hypoglycemic drugs you may need to change or add insulin injections. The additional demands of pregnancy make it more difficult to accurately control your blood sugar levels on oral drugs alone, since they are long-acting, less predictable than insulin, and cross the placenta. Maternal hyperglycemia prompts the secretion of additional insulin from the fetal pancreas, resulting in macrosomia (big babies), poor lung maturation, polycythemia (too many red blood cells), and problems for the newborn including hypoglycemia, respiratory distress syndrome, jaundice, and poor temperature control. Babies exposed to high and low blood sugar levels in utero are at risk of intrauterine death and stillbirth. Ultrasound scans for fetal growth and well-being will be performed regularly. Remember that babies of some diabetic mothers are growth restricted.

Women with diabetes are more prone to developing pregnancy-induced hypertension, preeclampsia, polyhydramnios, and urinary tract and vaginal candida infections. You will be advised to have an eye checkup since retinopathy worsens in pregnancy. A combination of careful dietary measures and regular adjustments of the mother's insulin dosage is usually needed to maintain stable blood sugar levels in later pregnancy. You will be shown how to check your own blood sugar levels. The timing of delivery, which should be in a hospital with neonatal facilities, will be dictated by the presence or absence of fetal and maternal complications. Most women will require IV insulin infusions during labor and delivery.

Asthma

Some 3 percent of pregnant women will have symptoms of asthma, which may be missed because most pregnant women develop a degree of breathlessness. Asthma is often triggered by allergies to foods, chemicals, dust, pollen, and smoke, or it follows a viral chest infection. If possible, pregnant women should avoid exposure to these triggers. One-third of women find that their symptoms improve in pregnancy, one-third are about the same, and in one-third their asthma gets worse. Inhaled steroids and bronchodilator drugs have no effect on the fetus, but women who require oral steroids during pregnancy have a greater risk of developing PREECLAMPSIA and having a baby with INTRAUTERINE GROWTH RESTRICTION (IUGR). During labor and delivery, which need to be covered with IV

steroids, epidural is the best option for pain relief. Breastfeeding reduces the risk of the baby developing future allergies (atopy).

Inflammatory bowel disease

Inflammation of the small intestine (Crohn's disease) and large intestine (ulcerative colitis) usually causes profuse diarrhea, with blood and mucus in the stool, and is accompanied by severe abdominal pain. Pregnancy does not usually alter the course of IBD, but women with IBD have higher rates of preterm delivery and low birthweight infants. A flare during pregnancy is most likely if your disease is active at the time of conception, so you will be advised to avoid pregnancy until your symptoms are well controlled so the amount of steroid medication needed is minimal during pregnancy. Vaginal delivery is preferable since women with this disease are at greater risk of postoperative complications.

Celiac disease

This is a common disorder in which gluten-containing foods (wheat, barley, and rye, but not oats) trigger inflammation of the small intestine, which improves when gluten is excluded from the diet. The symptoms are variable and include fatigue, malaise, anemia, diarrhea, abdominal pain or bloating, weight loss, mouth ulcers, and vitamin deficiencies. Celiac disease affects 1 in 70 pregnant women. The diagnosis should always be considered in pregnant women with iron deficiency anemia or unexplained

folate, B_6 or B_{12} deficiency, and is confirmed by a finding of antibodies to gliadin or endomysin. Untreated celiac disease is associated with high rates of miscarriage, IUGR, premature birth, and maternal anemia. Gluten-free dietary control along with folate and vitamin B supplements should be started preconceptually.

Gallstones

Gallstones are a common finding in pregnant women and acute inflammation of the gall bladder (cholecystitis) complicates 1 in 1,000 pregnancies. High estrogen levels increase the secretion of cholesterol and raised progesterone makes the drainage of bile into the small intestine sluggish. Conservative management with bed rest, fluids, and antibiotics are the first option to avoid the risks of miscarriage and preterm labor associated with surgery in the first and third trimesters respectively. However, many pregnant women relapse and may require surgery at a later date.

Heart disease

Maternal heart disease during pregnancy is uncommon but potentially serious and should always be cared for in units with specialized expertise. Rheumatic heart disease used to be the most likely underlying cause of heart disease in pregnant women, but today this is rare. However, there are now a significant number of women of childbearing age who underwent surgery for congenital heart disease as a child. The dramatic improvement in the life expectancy of these women has allowed many of them to now request help to achieve successful pregnancies of their own. Detailed management is beyond the scope of this book, but preventing sudden blood loss, controlling blood pressure, ensuring that the second stage of labor is short, and that delivery is covered by antibiotics are all important issues.

It is increasingly common to see older pregnant women with ischaemic heart disease. It is essential that lifestyle changes are encouraged, such as eating healthily and stopping smoking. Some types of heart disease are associated with very high mortality, such as pulmonary hypertension; these women may be advised not to conceive.

Essential hypertension

Preexisting high blood pressure (essential hypertension) needs to be under good control when you plan to become pregnant. The risk of developing PREECLAMPSIA (toxemia) and other severe problems, such as kidney damage, are greater if your blood pressure is high at the start of your pregnancy. About 3 percent of pregnant women are taking hypertensive drugs but some are not suitable for use in pregnancy, so you should talk to your doctor about your plans to become pregnant well in advance, or as soon as the pregnancy test is positive.

Renal disease

Occasionally renal disease presents for the first time during pregnancy, triggered by the extra filtering load on the kidneys and additional problems of HIGH BLOOD PRESSURE and PREECLAMPSIA. Women with existing renal disease need to understand that pregnancy can result in a marked deterioration in their renal function and an earlier requirement for dialysis. If the disease is progressive it is best to try to get pregnant sooner rather than later. In relapsing disease it is better to aim for a remission before getting pregnant. Prepregnancy counseling should include advice on the risks of various medications to the fetus and the woman's fertility, the need for early medical care, and a possible change in medication, strict blood pressure control, and details of the obstetric problems that may be encountered (premature delivery, preeclampsia, IUGR). Women with good kidney transplant function can usually achieve a successful pregnancy with help. Immunosuppressive drugs do not significantly increase the risk of fetal abnormalities, but early delivery by cesarean is common. Women with transplants should also be warned that they run an increased risk of an episode of graft rejection after pregnancy.

These pregnancies need careful team management by renal physicians and obstetricians.

Connective tissue disorders

Improved medical care for women with CT disorders has led to a dramatic increase in the number of women trying to become pregnant and achieving successful outcomes. However, these pregnancies require prenatal care by a multidisciplinary team since the risks of PREECLAMPSIA, IUGR, ABRUPTION and prematurity are increased and may be exacerbated by the steroid medication that many of these women need to continue taking during pregnancy.

Systemic Lupus Erythematosis (SLE)

This multisystem disease may affect the kidneys, skin, joints, nervous system, blood, heart, and lungs. Pregnancy increases the risk of flare-ups, which are most common in late pregnancy and postpartum, particularly in women with active or new onset disease. If the mother carries anti Ro or La antibodies, the fetus is at risk of congenital heart block and neonatal lupus. In antiphospholipid syndrome, the presence of cardiolipin antibodies or lupus anticoagulant leads to RECURRENT MISCARRIAGE, late pregnancy complications, and an increased risk of maternal THROMBOSIS. Treatment with aspirin and heparin significantly improves pregnancy outcome.

Scleroderma

Women with scleroderma usually deteriorate during pregnancy and are at risk of serious maternal complications if they have heart, lung, or kidney involvement, and poor fetal outcome.

Rheumatoid Arthritis (RA)

Rheumatoid arthritis usually improves during pregnancy but frequently relapses postpartum.

Thyroid disease

If you suffer from an under- or overactive thyroid it is unlikely that you will get pregnant until your thyroid function is controlled. Women with thyroid disease need careful supervision during pregnancy because alterations in their thyroid function can be masked by the symptoms of pregnancy. If you are taking drugs for your thyroid, your doctor may advise you to switch to another type when you become pregnant and the doses may need to be altered as pregnancy progresses. Women with Graves' disease who have had a thyroidectomy will be taking thyroxine replacement, but they may still have thyroid antibodies present that can affect the fetus. Scans will be needed to check for fetal goiter (enlarged thyroid gland). Hypothyroidism in the baby results in cretinism (a severe form of mental retardation), which is why all babies in the US are screened during the first week of life (see p.389).

Skin disorders

ECZEMA

Atopic eczema affects 1–5 percent of the population and is the most common cause of an itchy rash during pregnancy. It is treated with regular topical applications of emollients and use of bath additives. Hand and nipple eczema are a common postpartum irritation.

ACNE

Acne usually improves during pregnancy, but can flare in the third trimester, and acne rosacea often worsens. Treatment with vitamin A (retinoids) and tetracycline antibiotics can cause abnormalities in the unborn baby and should be discontinued before you try to conceive. If you find yourself pregnant unexpectedly, stop the treatment immediately. Oral or topical erythromycin is safe to use.

PSORIASIS

This chronic autoimmune disorder affects 2 percent of young adults and behaves unpredictably during pregnancy. It tends to improve or remain unchanged, but for 15 percent of women the symptoms deteriorate. Many treatments are harmful to the fetus, so careful preconception planning is important. Topical steroids can still be used during pregnancy.

Psychiatric disorders

Mental health problems affecting pregnant women can be divided into psychotic and depressive illnesses. Schizophrenia affects 1 in 1,000 individuals and presents problems for pregnant women since people with schizophrenia are usually single, socially isolated, and likely to be heavy cigarette, alcohol, and drug users. The effect of antipsychotic drugs on the fetus, the capacity of the woman to give informed consent for procedures, and the likelihood that she will relapse postpartum all have serious safety implications for mother and baby. These problems are increasing because modern-day antipsychotic drugs no longer reduce fertility. Bipolar affective disorder or manic depression is usually controlled with a combination of drugs (lithium and sodium valproate) all of which have well-recognized teratogenic effects on the developing baby. If the illness is stable, reducing the dose or replacing these drugs with an antidepressant may reduce the risk of serious relapse. A postpartum relapse will occur in half of all women with bipolar disease and their drugs should be restarted immediately after delivery. At least 1 in 10 women will suffer some form of depression in their lifetime, and women with serious depressive illness often experience a deterioration during pregnancy or the postpartum period, which is worsened if their medication is withdrawn or reduced suddenly. Depression has a negative impact on pregnancy outcomes, affecting diet, attendance of prenatal care appointments, levels of smoking, alcohol consumption, self harm, and domestic violence. It is possible that depression directly affects placental function and predisposes to IUGR and preterm birth. Women are at greatest risk of mental health problems during the postpartum period, which is when most maternal suicides take place. Some 15–30 percent of women will experience a depressive episode, and for 1 in 10 this will be a major illness. Although postpartum depression is no longer a taboo subject (see p.391), preexisting psychiatric disease in pregnant women remains stigmatized and poorly cared for. Most importantly, women who develop psychiatric problems during pregnancy are highly likely to need mental health services in later life.

Infections and illnesses

I am regularly asked by pregnant women about the possible effects of an infectious illness on their own health and that of their baby. Common illnesses such as colds and flu are unlikely to cause harm (see p.33) but others can have damaging effects. Information on these is provided here.

Chickenpox

Chickenpox is caused by varicella zoster virus and is transmitted by droplets spread at the time of face-to-face contact. The incubation period is between 10 and 21 days, during which a fever usually develops along with an itchy rash of watery blisters that burst and crust over in a few days. An individual with chickenpox is infectious for 48 hours before the blisters appear until they crust over. Contact with chickenpox in pregnancy is common, but only 1 in 300 women will become infected because 90 percent of children have had chickenpox before they reach adolescence. If you are not immune, and your exposure to the virus occurred within the last 10 days, you will receive an injection of VZIG (varicella zoster immune globulin).

If you develop your first chicken pox infection in the first eight weeks of pregnancy it is unlikely to cause a miscarriage, but if you catch it between eight and 20 weeks your baby may develop the congenital varicella syndrome with abnormalities affecting the limbs, eyes, skin, bowel, bladder, and brain, together with growth problems in later pregnancy—but the risk is small (1–2 percent). If you are

over 20 weeks pregnant and have chickenpox, you will be given the antiviral drug acyclovir if you are seen within 24 hours of symptoms developing. Between 20 and 36 weeks your baby will not be affected but the virus will remain in the body and may show up as shingles in the first years of life. However, if you develop chickenpox after 36 weeks and up to 21 days after the delivery, your baby may develop chickenpox. This may be a severe infection (neonatal chickenpox) if it starts within five days of delivery or within the next three weeks, because the newborn baby's immune system is not mature enough to deal with the virus. If you are over 36 weeks pregnant and have chickenpox or have been exposed, your baby will be given VZIG at the time of delivery, which reduces the severity of the attack if given before symptoms appear. Acyclovir may reduce the symptoms if it is started within 24 hours of the rash appearing.

Rubella

Ninety percent of pregnant women are immune to rubella because they have been previously infected or vaccinated as children (part of the MMR program—measles, mumps, rubella). Only a few

women will be infected for the first time in pregnancy, but this has serious implications for the fetus, who may develop congenital rubella syndrome. Rubella is spread by breathing infected air particles. Maternal symptoms develop two to three weeks after exposure—a rash of flat pink spots on the face and ears that spread to the torso, with pain, joint swelling, fever, and swollen glands. Infected individuals can infect others for one week before symptoms begin and for a few days after they have passed.

If you develop a rash in pregnancy, your doctor will do a blood test to establish whether you have rubella and repeat it two weeks later to see whether you have developed an antibody response. If rubella is confirmed before 12 weeks your baby has an 80 percent risk of congenital abnormalities ranging from eye cataracts, deafness, heart defects, and learning difficulties. Between weeks 13 and 17, a primary rubella infection may cause deafness in the baby. After 17 weeks there is no risk to your baby. Babies born with congenital rubella may have a low birthweight, skin rash, enlarged liver and spleen, and jaundice and may remain contagious for months. If you are rubella susceptible, you need to be vaccinated after the birth and you should use contraception for three months.

Parvovirus

In adults, the mild flulike symptoms of infection with parvovirus B19 may go unrecognized. In children it often causes a characteristic rash across the cheeks. Parvovirus is spread by droplets (coughing and sneezing) and contact with bedding, clothes, and carpets. This virus does not cause congenital abnormalities and most infections during pregnancy are followed by healthy live births. It is most common in pregnant women who work with children. It may cause late miscarriage and intrauterine death, which is usually associated fetal anemia leading to HYDROPS, and can be seen on ultrasound scan.

Cytomegalovirus (CMV)

CMV is one of the herpes viruses and is so common in young children that 60 percent of women are immune when they become pregnant. The infection may cause a vague flulike illness with sore throat, mild fever, aching limbs, and fatigue. It is usually acquired by close physical contact or from infected blood, urine, saliva, mucus, or breast milk.

Only a few susceptible women will experience a primary CMV infection during pregnancy and among these women the chance of transmitting the virus to the baby is about 40 percent. These babies are at risk of congenital CMV, which may result in being developmentally disabled and hearing and sight problems. There

stop or reverse the effects of congenital CMV, but new antiviral drugs are being investigated. Since CMV is the most common infectious cause of developmental disability in the US, the search for a vaccine is ongoing but routine screening for CMV is not recommended. CMV is spread by droplets and excreted in the urine. At-risk groups (hospital, laboratory, and preschool workers) need to take simple aseptic precautions, such as handwashing, when pregnant.

Toxoplasmosis

Most of the population is immune to toxoplasmosis due to a previous infection, which may have been so mild that the symptoms of a flulike illness with low-grade fever and swollen glands went unnoticed. Infection for the first time during pregnancy is rare (1 in 1,000) but can cause serious problems for the baby. During the first three months of pregnancy the risk of the baby becoming infected is low but the risk of injury is high and includes early or late miscarriage and live-born babies with severe neurological problems (cerebral calcification, HYDROCEPHALY, and damage to the eyes). Near delivery, the baby is more likely to become infected, but congenital toxoplasma infection is less likely to cause neurological damage.

Typically, pregnant women are not routinely tested for toxoplasmosis, but if a blood test suggests that a woman is infected she will be offered antibiotic treatment to reduce transmission risk to the baby. If an abnormal ultrasound scan suggests the baby has been infected, amniocentesis

will confirm the diagnosis and some women will opt for termination of pregnancy.

Tuberculosis

TB infection used to be extremely rare in pregnancy since the prevalence of pulmonary (chest) TB in developed countries is low and in underdeveloped countries more widespread disease involving the pelvis frequently led to infertility. However, increasingly mobile populations have resulted in a significant number of pregnant women with pulmonary TB. Individuals with HIV infection are more susceptible to infection with TB due to alterations in their immune response. Active TB in the first half of pregnancy is usually treated with isoniazid antibiotic, but after 20 weeks it is safe to use rifampicin. If the mother's TB is inactive at the time of delivery, the baby needs to be vaccinated with BCG, does not need to be isolated, and can be breastfed.

Listeria

Listeria is a food-borne bacteria. Infection during pregnancy is uncommon but can have serious consequences for the baby including late miscarriage and intrauterine death. When pregnant, women have lowered resistance to listeria, which multiplies rapidly in the placenta. Typically the mother experiences a brief flulike illness with malaise, nausea, diarrhea, and abdominal pain. Penicillin antibiotics are a rapid cure but the best way to avoid problems during pregnancy is to take careful measures to prevent any exposure to infection (see above).

Streptococcus B

Some 25 percent of women carry this bacterium in their vagina. Most have no symptoms although it can produce a discharge or urinary tract infection. If the infection is present during labor the baby may be affected. Only 1 percent of at-risk babies develop strep B infection by swallowing or inhaling vaginal secretions, but this neonatal infection can be fatal. Classically, signs of septicemia and meningitis develop some two days after birth. Premature babies are at greater risk of infection particularly if the mother's membranes are ruptured. Prenatal screening is done between 35–36 weeks of pregnancy. If you go into premature labor you will be tested then, but will be given IV antibiotics during labor. If you have a positive culture, you will be treated with IV antibiotics in labor to protect the baby from neonatal infection.

Malaria

Malaria is a parasitic infection transmitted by female mosquitos. Pregnant women are more likely to catch malaria and develop severe disease with cerebral complications due to the plasmodium falciparum species. It is far more dangerous for you and your baby to risk catching malaria than it is to take antimalarial drugs to prevent or treat an infection: the high fever puts you at risk of miscarrying and your baby is more likely to be of low birth weight, premature, or even stillborn. The parasites may be found in the placenta in large numbers and pregnant women may develop severe anemia with hemolysis leading to jaundice,

heart, and renal failure. Expert advice should always be sought promptly regarding the best choice of drug for an established infection and the risks of teratogenicity balanced against the poor prognosis for the mother and baby if inadequately treated.

Sexually transmitted diseases (STDs)

HERPES

There are two types of herpes infection. Type 1 (HSV1) causes cold sores of the mouth or lips. Type 2 (HSV2), or genital herpes, causes painful ulcers of the vulva, vagina, or cervix. If a mother develops her first genital herpes infection within six weeks of delivery, there is a 10 percent chance the baby will be infected during delivery. The consequences of this can be severe, including herpes encephalitis or meningitis, which is why cesareans are advised and after delivery the baby will be given antiviral drugs. Following her primary infection the mother produces antibodies, which protect a future fetus but do not prevent her from suffering more episodes. Secondary genital herpes infection during pregnancy may be unpleasant for the mother but only rarely causes the baby any problems. A vaginal birth is the best option as long as you're not having an outbreak.

GONORRHEA

Gonorrhea is a highly contagious bacterial infection of the cervix, urethra, rectum, or throat. It is often accompanied by CHLAMYDIA, TRICHOMONAS, and SYPHILIS. Unprotected sex

with an infected individual leads to transmission of the infection in 90 percent of cases. The infection may be asymptomatic or accompanied by vaginal discharge, pain, and urinary discomfort. It is an important cause of pelvic inflammation, which damages the fallopian tubes, leading to ECTOPIC PREGNANCY and infertility. Infection during pregnancy is associated with premature rupture of membranes and preterm delivery. The risk of postpartum pelvic inflammatory disease and systemic spread (painful joints and skin rash) is increased. The diagnosis is best made from culture of cervical swabs and effectively treated with penicillin antibiotics. Although the baby is not at risk of infection during pregnancy, exposure to the organism during delivery may cause neonatal conjunctivitis and occasionally septicemia requiring intensive treatment.

CHLAMYDIA

Chlamydia trachomatis is the most common STD in the US. Some 40 percent of infected men complain of penile discharge, testicular inflammation, and urinary discomfort, but only 15 percent of infected women have symptoms such as vaginal discharge, pelvic pain, or urinary problems. Even when there are no symptoms, infection may be present in the vagina, cervix, uterus, anus, urethra, or eyes and can have serious consequences. Damage to the fallopian tubes increases the chance of ectopic pregnancy and may lead to infertility. If present at the time of delivery, 40 percent of babies will become infected. Chlamydia is the leading cause of neonatal conjunctivitis (eye infection) which may result in blindness or pneumonia. Early diagnosis is important because the infection can be effectively treated with antibiotics.

SYPHILIS

Syphilis infection is caused by Treponema pallidum and was rare in pregnant women in the US. The incidence has increased in the last 10 years and damage to the baby can be prevented by early treatment with penicillin. This is why all pregnant women have routine screening for syphilis. In the primary stage of infection, an ulcer (chancre) appears, which is similar to herpes but less painful and lasts three to six weeks. Untreated, the infection progresses in a few months to secondary syphilis with fever, rash, swollen glands, weight loss, and fatigue. Without treatment, tertiary syphilis develops some years later with damage to the brain, nerves, and heart. The bacteria are able to penetrate the placenta after 15 weeks and 70 percent of infected women will transmit the infection to the fetus, with 30 percent being stillborn. If the fetus survives, it will be in the second stage of the illness at birth, and a further 30 percent of these babies will be born with congenital syphilis and suffer from seizures, developmental delay, skin and mouth sores, infected bones, jaundice, anemia and microcephaly. Penicillin usually cures the maternal infection and prevents fetal infection. At birth

the baby will be given more antibiotics. Diagnosing syphilis should prompt an additional search for chlamydia, gonorrhea, HIV, and hepatitis B and C infections.

HIV INFECTION

HIV infection is most commonly transmitted through sexual contact, the use of contaminated needles or infected blood and blood products. In Western countries most cases of HIV infection are found in homosexual/bisexual males and heterosexual drug abusers. In some African countries the prenatal prevalence is as high as 40 percent and is an important cause of infant death. Routine prenatal screening and treatment of HIV-positive pregnant women has dramatically reduced the risk of transmission to the baby and development of full-blown AIDS in the mother thereby increasing her life expectancy. HIV transmission to the baby can be reduced from 25 percent to less than 2 percent by treating positive pregnant women with a combination of antiretroviral prenatally, and during delivery of the baby by an elective cesarean (if her viral load remains high), avoidance of breastfeeding, and actively treating the new baby for four to six weeks. Sadly, the majority of pregnant women with HIV live in countries where the high cost of drugs and medical interventions excludes them from being offered life-saving treatment.

TRICHOMONIASIS

This infection is caused by the organism *Trichomonas vaginalis*, which inhabits the

urinary tract and vagina and is frequently accompanied by chlamydia and gonorrhea. There may be no symptoms or a vaginal discharge with a fishy smell, together with inflammation and pain in the vagina and urethra. Infection during pregnancy is occasionally responsible for pneumonia in the newborn baby. The diagnosis can be made on a Pap smear or vaginal swab. Infections should be treated with metronidazole antibiotic, which is safe in late pregnancy and when breastfeeding.

BACTERIAL VAGINOSIS (BV)

This is a common cause of vaginal discharge, but may be symptomless. The discharge is usually thin, gray-colored and nonitchy, with a strong fishy odor. It is diagnosed by the presence of clue cells on a Pap smear. In pregnancy, the altered hormonal environment in the vagina is less acidic, favoring the growth of the many organisms that are present in bacterial vaginosis.

BV infection during pregnancy has been strongly linked with late miscarriage and premature birth. Although treatment with clindamycin or metronidazole antibiotics clears BV in a few days, recurrence in pregnancy is common. Screening and treating all infected pregnant women has not reduced the preterm birth rate. Pregnant women with a history of premature delivery are particularly susceptible and may benefit from regular screening and antibiotic treatment when the infection is present.

Fetal abnormalities

Congenital abnormalities are present at birth and are often due to genetic causes. Others are acquired as a result of environmental factors during pregnancy or occur for no known reason. You will find more on how these abnormalities develop on pages 144–47.

CHROMOSOME ABNORMALITIES

These abnormalities are due to a fault in the baby's complement of 23 pairs of chromosomes, because there are too many, too few, or one chromosome is abnormal. The most common chromosomal disorder, Down syndrome, is discussed in detail on page 147.

Trisomies

PATAU'S SYNDROME (TRISOMY 13)

Patau syndrome occurs in 1 in 10,000 live births and three copies of chromosome 13 are present. The majority of affected babies miscarry early, but of the 20 percent that are liveborn, most die within a few days of birth. Those that survive have severe mental disabilities. Microcephaly and severe facial abnormalities, too many fingers and toes, HEART AND KIDNEY DEFECTS and EXOMPHALOS are characteristic and are usually identified during pregnancy on an ultrasound.

EDWARD SYNDROME (TRISOMY 18)

In Edward syndrome, three copies of chromosome 18 are present; this occurs in 1 in 7,000 live births. The physical malformations include strawberry-shaped head, HEART AND KIDNEY DEFECTS, CHOROID PLEXUS CYSTS, INTRAUTERINE GROWTH RESTRICTION (IUGR), DIAPHRAGMATIC HERNIA, EXOMPHALOS, low-set ears,

small receding jaw, shortened limbs, clenched hands, and feet with a curved sole. All this may be seen on the 20-week scan. The integrated screening test (see p.138) identifies 60 percent of babies with Edward syndrome. Severe developmental disability is usual and most die within the first year of life.

Triploidy (69XXY or XYY)

An extra set of 23 chromosomes, called triploidy, may result from an egg being fertilized by more than one sperm or the fertilized egg not dividing. It occurs in 2 percent of conceptions, but these usually miscarry (20 percent of all chromosomal miscarriages are triploidies). When the extra set of chromosomes comes from the father, the embryo does not develop but the placental tissues grow rapidly in an uncontrolled manner and the pregnancy rarely continues beyond 20 weeks (see HYDATIDIFORM MOLES p.423).

When the extra set comes from the mother, the pregnancy may go into the third trimester. Although the placenta is normal, the fetus usually has severe asymmetrical growth retardation. Triploidy is not influenced by maternal age.

Translocation

Translocation occurs when a piece of one chromosome becomes attached to the end of another. Individuals with a balanced translocation appear normal because the normal chromo-some counteracts the effect of the abnormal one. If he or she becomes a parent, there are three possible outcomes: the baby may have normal chromosomes; it can inherit the balanced translocation; or it may inherit an unbalanced translocation, which invariably leads to miscarriage or a severe abnormality in the baby. Translocations may be reciprocal or Robertsonian and are a cause of RECURRENT MISCARRIAGE. New translocations can arise that are not inherited from either parent. Sex chromosome abnormalities TURNER SYNDROME (45XO) This occurs in 1 in 2,500 live-borns; one of the two X chromosomes is completely missing. These girls have normal intelligence, but their growth is seriously affected, and they have no menstrual periods and are therefore infertile. Other physical characteristics are webbing of the neck and cubitus valgus (a wide carrying angle at the elbows). A high percentage of fetuses with a single X chromosome miscarry or are identified by prenatal diagnosis. Abnormalities detected on ultrasound include cystic hygroma (a large fluid-filled sac behind the neck), CARDIAC DEFECTS, particularly coarctation of the aorta, HYDROPS FETALIS, and horseshoe kidneys. Turner syndrome can also exist in mosaic form, meaning that the girl has two different cell lines—46XX and 45XO. If her egg cells contain a normal set of chromosomes, she will not necessarily be infertile.

KLINEFELTER SYNDROME (47XXY)

When boys have an extra X sex chromosome, it is known as Klinefelter syndrome and occurs in 1 in 1,000 live births. In adult life these men tend to be tall with a reduced head circumference and lower intelligence level, but are not usually considered developmentally disabled. Men with Klinefelter syndrome are infertile and more susceptible to autoimmune disease, malignancy, and cardiovascular disease in adult life.

TRIPLE X (47XXX)

Women with an extra X sex chromosome have normal fertility and there is a wide variation in mental capacity. Some may have lower intelligence than normal, but rarely suffer developmental disability. This occurs in 1 in 1,000 live births.

SUPERMALES (47XYY)

Some boys have an extra Y chromosome. They have normal appearance, mental development, and fertility. As adults, they have an increased incidence of language and reading difficulties, hyperactivity, and impulsive and aggressive behavior. This occurs in 1 in 1,000 live births.

Dominant genetic disorders

FAMILIAL HYPERCHOLESTEROLEMIA

More than 60 million people suffer from this genetic disorder, although men tend to be more severely affected than women. High cholesterol levels and narrowing of major blood vessels lead to heart attacks at an early age and fatty skin deposits around the eyelids. In parents who have a history of early heart disease, blood from their baby's umbilical cord can be tested for high cholesterol levels at birth so diagnosis can be made at birth, offering the possibility of preventative lifestyle measures to reduce the severity of the disease.

MARFAN'S SYNDROME

This syndrome affects 1 in 5,000 people and is inherited as an autosomal dominant so there is a 50 percent chance of the baby being affected. A mutation of the fibrillin protein gene on chromosome 15 causes variably weak and stretchy connective tissue in the eye, heart, blood vessels, joints, and skin. A person with Marfan's is tall, thin, with very long digits, flat feet, spinal curvature, hypermobile joints, and abnormal aortic heart valve. In

pregnancy, which needs to be managed by specialists, the biggest fear is that the dilated root of the aortic valve ruptures or dissects, which is fatal in 50 percent of cases. There is also a risk of miscarriage, cervical weakness, preterm labor, and postpartum hemorrhage.

VON WILLEBRANDS DISEASE

The most common inherited bleeding disorder (prevalence of 1 percent) is due to a reduction in the quantity or quality of von Willebrands factor in blood. Although pregnancy usually increases the levels of factor, this does not compensate sufficiently in severe cases and pregnant women are at risk of postpartum hemorrhage, which is managed with IV desmopressin treatment.

HUNTINGDON'S DISEASE

This dominant disease affects 1 in 20,000 and has an insidious onset in midlife, starting with personality changes progressing to uncontrolled movements, aggressive sexual behavior, and dementia. This disease has full penetrance so children of an affected parent have a 50 percent chance of getting the disease, which never skips a generation. Affected individuals and their families often try to hide the early onset of symptoms, but the sufferer is usually very aware of the problems ahead. Recent studies show that the abnormal gene is on chromosome 4 and analysis of fetal DNA obtained from CVS samples provides accurate prenatal diagnosis and prediction of late onset disease. Genetic counseling for affected families is essential.

Recessive genetic disorders

TAY-SACHS DISEASE

This is a fatal recessive disorder common in Ashkenazi Jewish families and French Canadians. It is caused by a deficiency in the enzyme hexosaminidase A, which results in fatty material being deposited in the nerve cells of the brain. Babies with the disease appear normal at first but by six months have started to develop progressive motor weakness and mental disability. The child becomes blind, deaf, unable to swallow, and suffers increasingly severe seizures before dying at three to five years old. Carriers can be identified by a simple blood test before or during pregnancy. If both parents are carriers, there is a 25 percent chance in each pregnancy of having an affected baby. Amniocentesis or chorionic villus sampling will be offered to confirm the diagnosis. Because there is no treatment for this disease, many couples choose to terminate an affected pregnancy.

CYSTIC FIBROSIS (CF)

This is the most common recessive genetic disease in Caucasians, affecting 1 in 2,500 live births. It is due to an abnormality in sodium transportation, which makes the secretions of the lungs, digestive system, and sweat glands thick and sticky. Mucus accumulates in the lungs, leading to severe chest infections and as the pancreas and liver are also affected, the normal flow of digestive enzymes in the gut is compromised. This causes malnutrition if lifelong daily enzyme supplements are not started promptly. The severity is variable, ranging from death in the first year of life to poor health in middle age. Regular physical therapy can help improve the lung problems. Males with cystic fibrosis are infertile, due to blockage of their sperm transport tubes (vas deferens).

One in 22 Caucasians are carriers for the most common gene mutation responsible for CF (DF508), which is located on chromosome 7, making it possible to screen prospective parents and get prenatal diagnosis on DNA samples from a fetus at risk. However, there are many different mutations in the CF gene and current screening tests can only identify 85 percent of carriers. Carrier detection for CF is offered to Caucasians and individuals with a history of CF, partners of identified CF carriers, parents of a fetus with echogenic bowel detected on ultrasound, and sperm donors. It is important that couples who have screening receive detailed genetic counseling and are aware of the limitations of the tests.

PHENYLKETONURIA (PKU)

This recessive genetic disease is present in 1 in 14,000 births. It is caused by a defective gene that results in a deficiency of the enzyme that converts the essential amino acid phenylalanine into tyrosine. The buildup of high levels of phenylalanine in the bloodstream is toxic to the developing brain. If a special diet with low phenylalanine is started in the first few weeks of life the irreversible brain damage and learning disabilities can be avoided. All babies are tested for PKU after birth (see p.389).

SICKLE-CELL ANEMIA AND THALASSEMIA

If you are of Mediterranean or African origin, you will be offered a special electrophoresis test of your hemoglobin to determine whether you have sickle-cell or thalassemia traits (see p.425). If you carry the sickle-cell trait, it is important that your partner's sickle-cell status is established early in your pregnancy, because there is a chance that your baby could inherit a double dose of the trait and develop full-blown sickle-cell disease. This leads to severe anemia, infections, pain, and eventually heart and kidney failure. Similarly, if you carry the A or B thalassemia trait, your partner should be tested, too. A baby with full-blown thalassemia suffers from severe anemia and iron overload, leading to multiple organ failure. If both parents are carriers they may choose to have an invasive test such as chorionic villus sampling or amniocentesis to find out whether the baby has inherited the disease. Prenatal diagnosis is also offered when both parents carry the alpha or beta thalassemia trait.

Sex-linked disorders

DUCHENNE'S MUSCULAR DYSTROPHY (DMD)

This is the most common sex-linked disorder and affects 1 in 4,000 boys. He may appear normal in infancy, but, between four and 10 years, he loses his ability to walk due to muscular weakness and is usually confined to a wheelchair soon after. Identifying female carriers of DMD used to depend on finding increased blood levels of the muscle enzyme creatinine kinase. This was an unreliable test and most couples carrying a male fetus were therefore offered a termination of pregnancy. The gene has now been identified. In approximately two-thirds of families, a deletion is present on the short arm of chromosome X. So, most female carriers can be identified before pregnancy. Prenatal diagnosis can also be done on fetal DNA samples during pregnancy, to establish whether the baby is affected by DMD.

HEMOPHILIA

This X-linked recessive disorder affects 1 in 10,000 males and is due to a deficiency in blood coagulation factors that makes the blood clot too slowly. There are two types of hemophilia. The most common is hemophilia A, in which low levels of factor VIII are present. In hemophilia B, there is a deficiency in factor IX. The symptoms of both are prolonged bleeding from wounds and into joints, muscles, and other tissues following minor trauma. The severity of the disease depends on how little of the clotting factor is present in the blood.

Both hemophilia A and B can be treated by injections or transfusions of plasma containing the missing clotting factors so sufferers can lead a more normal life. Since carrier females may have normal or low levels of the clotting factors, diagnosis was unreliable before DNA testing. In families with a history of hemophilia, female carriers can be identified before pregnancy and fetal DNA testing in pregnancy can establish whether a male baby is affected.

FRAGILE X SYNDROME

This X-linked disorder is the most common form of inherited developmental disability. Mental impairment is variable in carrier females, but DNA testing can confirm the suspected diagnosis of Fragile X or carrier status. Genetic counseling should be offered to all women and their families with a history of developmental disability, since as many as 1 in 200 women carry the gene mutation.

OTHER CONGENITAL ABNORMALITIES

This section includes fetal abnormalities for which there is no known specific genetic cause although some, such as neural tube defects, tend to run in families. The list includes most of the abnormalities that sometimes become apparent on ultrasound scans.

Neural tube defects (NTDs)

These are one of the most common and serious congenital abnormalities and are due to a combination of genetic and environmental factors. In the absence of prenatal screening about 1 in every 400 babies will be affected. The embryonic neural tube fails to close properly during the first four weeks of pregnancy, resulting in incomplete development of the brain and spinal cord and varying degrees of permanent neurological damage. The most severe forms are anencephaly (the skull bones are incomplete and the brain is underdeveloped) and encephalocele (brain tissue projects through a hole in the skull). These babies are rarely liveborn. In SPINA BIFIDA (myelomeningocele) the spinal cord is not protected by the bony spinal column and may be closed (covered by protective membranes) or open (no covering membranes). The degree of paralysis, weakness, and sensory disability is variable depending on the level of spinal defect. This can range from a wheelchair existence and complete lack of bladder and bowel function to mild walking difficulties. However, babies with open spina bifida are often severely disabled, requiring frequent surgical procedures and prolonged hospitalization. The majority of severe cases develop HYDROCEPHALUS, resulting in developmental disability and learning difficulties. The mildest form of NTD is spina bifida occulta, a lesion in the lowest part of the sacral spine that usually goes unnoticed and is present in 5 percent of healthy babies. Prenatal screening for open spina bifida has been greatly improved by routine ultrasound scanning. In addition to the bony defect in the vertebral column, most babies with a myelomeningocele have scalloping of the frontal skull bones (the lemon sign on an ultrasound) and a cerebellum that looks more like a banana than the normal dumbbell-shaped structure.

Closed spina bifida has a better prognosis because the defect is more easily treated surgically after delivery, but difficult to detect prenatally. SPINA BIFIDA can run in families, but 95 percent of these babies are born to women with no family history. It is linked to poor diet and the risk of recurrence is 1 in 20 after a previously affected baby. Taking folic acid supplements for three months before pregnancy and in the first trimester prevents 75 percent of cases. Women with a previous NTD or on epilepsy drugs should take high doses of folic acid before conception (see p.51).

Hydrocephalus

This condition (often referred to as water on the brain) is caused by too much cerebrospinal fluid. Hydrocephalus is usually due to a blockage in the circulation of the fluid or to an overproduction or reduced absorption of the fluid. It is often associated with SPINA BIFIDA or follows a brain hemorrhage in a premature baby. If the problem is present before birth it can be seen on an ultrasound scan. As the head swells, the brain tissue is compressed, the skull bones become thinner, the sutures of the head widen, and the fontanelles bulge. After delivery, if the hydrocephalus is due to a blockage it may be possible to insert a tube to drain the fluid from the ventricles of the brain into the abdominal cavity or heart. Occasionally, the hydrocephalus is inherited as a sex-linked recessive disorder in males and these families need genetic counseling.

Microcephaly

In these babies the bony skull and brain size are smaller than average. The babies are nearly always severely intellectually impaired. Recognized causes include rubella infection in the first trimester, CYTOMEGALOVIRUS, TOXOPLASMOSIS, and SYPHILIS infections, severe irradiation, and maternal heroin and alcohol addiction. A few cases are inherited as a recessive genetic disorder, but in many cases no obvious cause can be found. More recently, there has been some speculation around the association of microcephaly and the Zika virus, though this has not been conclusively proven.

Choroid plexus cysts (CPC)

These cysts in the ventricles of the baby's brain are usually bilateral and seen in as many as 1 percent of all 20-week screening anatomy ultrasounds. Most CPCs are now thought to be benign structures and usually disappear by 24 weeks. Therefore, if seen in isolation, they are no longer commented on. However, if other anatomical anomalies are detected, detailed counseling will be needed to help parents decide whether they should expose their pregnancy to the risk of an invasive diagnostic test, since CPCs are associated with TRISOMY 18.

Gut abnormalities

DUODENAL ATRESIA

In this disorder, the small gut between the bottom of the stomach and ileum is absent and is usually diagnosed by the "double bubble sign" on an ultrasound

scan (bubble 1 is the normal stomach, bubble 2 results from the duodenum being unable to empty into the lower gut, which can also cause POLYHYDRAMNIOS). The blockage can be resolved by surgery immediately after delivery, but in one-third of cases this abnormality is associated with Down syndrome.

ESOPHAGEAL ATRESIA

In this disorder the tube between the throat and stomach is partially absent, leading to vomiting and excess drooling immediately after delivery. It is often associated with a fistula (passage) between the esophagus and trachea (windpipe). Hence POLYHYDRAMNIOS and the "double bubble sign" may not be evident on ultrasounds. If a fistula is present there is a serious risk of food entering the lungs and choking the baby. Esophageal atresia requires immediate surgery and sometimes multiple operations, but responds well to surgery if it is an isolated problem.

HYPERECHOIC FETAL BOWEL

If the bowel is seen to be very echogenic (as white as bone) on a prenatal ultrasound scan, this may be secondary to a major chromosomal abnormality, CYSTIC FIBROSIS, bowel obstruction, fetal infection, or GROWTH RESTRICTION. It can also occur in entirely normal fetuses or if the mother had experienced some bleeding earlier in the pregnancy.

Diaphragmatic hernia

This serious congenital abnormality affects 1 in 3,000 babies and can be diagnosed on the 20-week scan. The muscular diaphragm separates the chest (heart and lungs) from the abdomen (liver, stomach, spleen, intestines). If a defect occurs in the diaphragm, varying amounts of these abdominal organs can herniate into the chest cavity. In about 50 percent of fetuses there are associated chromosomal abnormalities, genetic syndromes and other structural abnormalities. At birth the baby will need immediate intensive care and ventilation before undergoing multiple operations to replace the abdominal organs and reconstruct the diaphragm. In utero surgery may be tried in special centers to give the affected lung a better chance of developing normally.

Abdominal wall defects

OMPHALOCELE (EXOMPHALOS)

Omphalocele occurs in 1 in 5,000 babies and is due to a defect in the abdominal wall underneath the umbilicus, through which varying amounts of small bowel and liver protrude covered by the peritoneal membrane. Most of these babies are identified on ultrasounds during pregnancy and some 50 percent have associated chromosomal, cardiac, or bladder abnormalities. Karyotyping (chromosome analysis) and ultrasound scans are needed to assess the problem. If it is an isolated defect, surgical correction after birth has a good prognosis, although it may require multiple operations.

GASTROSCHISIS

In this condition loops of bowel protrude through the abdominal wall defect and are not covered by peritoneum. In most cases the abnormality is isolated and there is no increased incidence of chromosomal abnormalities. The abdominal wall defect is usually small and can be repaired easily. Early delivery helps avoid bowel damage and complications.

Cardiac abnormalities

Structural heart disorders are the most common severe congenital abnormality in newborn babies, affecting 8 per 1,000 live births. They are an important cause of perinatal and childhood death. The incidence of cardiac abnormalities is increased in babies who are premature, have Down syndrome, are infected with RUBELLA virus, or who have mothers with congenital HEART DISEASE, DIABETES, EPILEPSY, or families with a history of cardiac abnormalities. In 30 percent of babies with a cardiac defect another structural abnormality is found and in 20 percent a chromosomal disorder is present.

Many heart defects can be diagnosed on ultrasound scan during pregnancy. This is why the 20-week anatomy scan routinely aims to obtain a view of the four chamber heart. If an abnormality is suspected, more scans will be arranged. The option of fetal karyotyping (chromosome analysis) should be offered to parents since the results may influence the plan for management at the time of delivery. Some cardiac lesions may require immediate surgical correction, whereas others can be dealt with later on.

SEPTAL DEFECTS

These "holes in the heart" account for 50 percent of all congenital heart defects. A hole in the septum (wall) dividing the two upper (atrial) or the two lower (ventricular) chambers of the heart, results in oxygenated and deoxygenated blood being mixed instead of separated. An atrial septal defect may result in few symptoms, but ventricular septal defects give rise to a loud heart murmur, and since the heart has to pump harder, it becomes enlarged. Untreated, the problem can be very serious particularly when complicated by CYANOSIS (blue baby).

CYANOTIC DEFECTS (BLUE BABY)

These disorders account for 25 percent of all congenital heart disease and require expert surgical and medical care. The outlook is often poor, depending on the severity of the lesion. In Tetralogy of Fallot (one of the most common forms of complex congenital heart defects), there is a large septal defect, abnormal aorta, and narrowing of the pulmonary valve. When the aorta and pulmonary artery are connected the wrong way around most of the blood receives no oxygen from the lungs and any oxygenated blood is sent back to the lungs rather than being pumped around the body.

PATENT DUCTUS ARTERIOSUS

Ten percent of cases of congenital heart disease are caused by failure of the duct between the heart and lungs to close after birth. This is more common in premature babies. The duct usually closes on its own given time but may require indomethacin drug treatment and occasionally surgery.

HYPOPLASTIC LEFT HEART

This is present in 10 percent of babies with congenital heart disease and accounts for 25 percent of all neonatal cardiac deaths. The left side of the heart is so under-developed that when the ductus closes at birth, the baby cannot obtain oxygenated blood. The postpartum prognosis is poor, and, when diagnosed prenatally, some parents choose to terminate.

Hydrops fetalis

This is present in a wide variety of fetal, maternal and placental disorders. It is diagnosed on an ultrasound scan when widespread accumulation of fluid in the body leads to edema of the skin and effusions (pools of fluid) around the heart, lungs, and abdominal organs. Immune hydrops occurs in severe RHESUS SENSITIZATION (see p.128). Nonimmune hydrops may be seen with fetal chromosomal, heart, lung, blood, and metabolic disorders, some congenital infections, and malformations of the placenta and umbilical cord. The overall outlook is poor, with a mortality rate of 80–90 percent depending on the cause. Despite investigations, the abnormality remains unexplained in one-third of cases.

Renal abnormalities

The fetal kidneys and bladder are visible at the 20-week scan. Severe renal problems are usually accompanied by OLIGOHYDRAMNIOS or reduced amniotic fluid because urine production is compromised.

POTTER SYNDROME

The fetal kidneys are absent (agenesis) or poorly formed and the lungs are underdeveloped as a result of the lack of amniotic fluid. The baby usually has facial abnormalities including widely spaced eyes, down-turned nose, low-set ears, and a small jaw. At birth, no urine is passed and the baby dies within a few hours from respiratory failure. The problem is rare but more common in boys.

HYDRONEPHROSIS

Enlarged kidneys are identified on ultrasound scan in 2 percent of all fetuses during the second trimester. They are usually due to narrowing or blockage of one or both of the ureters. When isolated, the finding is often insignificant. However, it may be associated with chromosomal abnormalities, in particular Down syndrome. Severe hydronephrosis leads to kidney damage due to the pressure of retained urine on the normal kidney tissue and may need to be drained by inserting a ureteric shunt to prevent kidney failure.

POLYCYSTIC KIDNEYS

This is a recessive genetic disorder with variable expression. Some fetuses will appear to have normal kidneys on prenatal ultrasound scan and present with renal failure as a teenager. Others will be found at the 20-week anatomy scan because of absent liquid and grossly enlarged kidneys. Adult polycystic kidney disease is a dominant genetic disease which can present in utero with enlarged cystic kidneys. One of the parents usually has kidney cysts.

FETAL SURGERY

A few fetal abnormalities may be amenable to corrective treatment before birth. These procedures are only done by pioneering fetal surgeons. The most successful techniques to date are those that involve inserting needles or fine tubes through the mother's abdomen into the uterine cavity or the fetus under ultrasound guidance. Intrauterine blood transfusions may be needed for severe Rhesus incompatibility, and drugs to correct irregularities of the fetal heartbeat or destroy life-threatening tumors can be directly injected into the fetus. Occasionally, fetal tissue samples may need to be obtained to make the diagnosis of a rare genetic disorder. Other examples of ultrasound-guided fetal surgery are the insertion of drainage tubes into severe cases of HYDROCEPHALUS or HYDRONEPHROSIS.

Open fetal surgery is experimental and is usually only considered in situations where there is nothing to lose. The mother's abdomen is opened and an incision in the uterus is made to gain access to the fetus and conduct the surgery. Great care is taken to keep the baby warm, replace the amniotic fluid, and avoid damaging the placenta. Even if the surgery is successful, the baby remains at risk of premature labor, infection, and leakage of amniotic fluid. Diaphragmatic hernias and fetal tumors may be treated in this way.

Problems in pregnancy and labor

Although the majority of pregnancies and births are straightforward, there are inevitably instances when something does not go as planned. Some of the problems that you may encounter are summarized here with information on the up-to-date methods of managing and treating them.

Ectopic pregnancy

An ectopic pregnancy develops outside the uterine cavity. This occurs in 1 in 50 pregnancies. Although some ectopics "miscarry" without complication (tubal miscarriage), the pregnancy may continue to grow and rupture the walls of the fallopian tube. The vast majority of these pregnancies occur in the fallopian tubes, but some may be found on the ovary or in the abdominal cavity.

The general symptoms are the same as those of early pregnancy, accompanied by a positive pregnancy test but lower abdominal pain, which almost always starts before any vaginal bleeding has occurred. If your doctor suspects an ectopic pregnancy you will have an ultrasound. This will show that there is no pregnancy sac within the uterine cavity, although the lining of the uterus may be thickened. The ectopic may be surgically removed. More hospitals are developing the expertise to do this by laparoscopic surgery, avoiding the need for open surgery. If the ectopic has not ruptured and hormone HCG levels are low, treatment with methotrexate may be advised, but if the patient is asymptomatic and the HCG hormone levels continue to drop, conservative management is possible.

Hydatidiform mole

Hydatidiform moles are the most common type of placental tumor. There are complete moles and partial moles. Complete moles are rare in Caucasian women (1 in every 1,200–2,000 pregnancies), but are much more common in women from Southeast Asia.

Moles are derived entirely from the cells of the father, due to an accident that occurs at the time of fertilization. No embryo is present in the pregnancy sac, but the placental tissues develop rapidly in an uncontrolled fashion, resembling a bunch of grapes on an ultrasound. Persistent bleeding and severe nausea are often associated with the presence of moles, and the size of the uterus is usually larger than would be expected for the menstrual dates. A complete mole may develop into an invasive cancer in a small percentage of cases, so special treatment is needed.

Partial moles are more common and usually mimic the appearance of an inevitable or incomplete miscarriage. The partial mole contains a fetus/embryo which has three sets of chromosomes instead of two (TRIPLOIDY). The placental cells swell and proliferate but not to the same degree as in a complete mole. A partial mole may be distinguished from a miscarriage only when the pathologist examines tissue removed from the uterus.

Fibroids

Fibroids are benign growths of the uterine muscle wall and can vary in size from a small pea to a large melon. The causes are not understood, but they tend to run in families and are more common in African-American women. Most women will have no problems with their fibroids, but if the embryo implants over a fibroid protruding into the uterine cavity the risk of early miscarriage increases.

Fibroids usually increase in size during pregnancy because of high estrogen levels and increased uterine blood supply. If they undergo red degeneration (the fibroid's blood supply is cut off, causing it to turn red and die), late miscarriage or premature labor may result. Large fibroids distorting the uterine cavity may lead to abnormal presentations and positions. Occasionally, fibroids obstruct the birth canal, preventing a vaginal delivery, but they usually shrink in size after delivery.

Incompetent cervix

The cervix usually remains tightly shut and sealed with a plug of mucus throughout pregnancy. An incompetent cervix starts to shorten and open during the fourth or fifth month of pregnancy, exposing the membranes to the risk of rupture and miscarriage. The condition is uncommon and may be caused by damage to the cervix during a previous labor, cervical surgery, or termination. If you are diagnosed with a weak cervix you may be advised to have a cervical cerclage or suture inserted at 12–14 weeks in your next pregnancy to keep your cervix tightly closed until the end of pregnancy. The suture is usually removed a few weeks before term to allow a vaginal delivery.

Venous thromboembolism

Women are more likely to develop a blood clot or thrombosis in a pelvic or leg vein during pregnancy and the postpartum period. This is due to increased levels of clotting factors and decreased levels of anticoagulant factors that protect pregnant women from uncontrolled uterine bleeding during pregnancy and after delivery. Venous thromboembolism (VTE) occurs in less than 1 in 1,000 births, but there are several important risk factors that increase the likelihood of VTE in pregnancy. These factors include being over 35 years old, immobility, smoking, obesity,

surgical delivery, previous VTE, family history of VTE, severe varicose veins, PREECLAMPSIA, dehydration, SICKLE-CELL DISEASE, maternal illness, and infection.

The thrombosis usually starts in the deep veins of the lower leg but may have extended up into the femoral or pelvic veins before detection. The danger is that part of the blood clot may break off and be swept into the lungs where it blocks one of the major blood vessels. This is called pulmonary embolus (PE) and although the incidence is only about 1 in 6,000 births, it is potentially very serious. Any pregnant woman with signs of a DVT or PE should be started on anticoagulation treatment as a matter of urgency even before the diagnosis is confirmed.

The signs of a DVT are pain and swelling in the calf or thigh muscles, with localized redness and tenderness of the leg—inability to put your heel to the ground when walking is diagnostic. Once the DVT is proven, an elastic compression stocking should be fitted and you will be confined to bed with your leg elevated until fully heparinized and until the leg stops being tender.

The symptoms of a pulmonary embolus are shortness of breath, chest pain, coughing up blood, faintness, collapse, together with all the signs of a DVT. A chest X-ray and examination may show evidence of a PE and a ventilation perfusion scan of the lung and bilateral Doppler ultrasound leg studies should be done immediately since two-thirds of deaths from PE occur within the first two to four

hours. Women with a proven PE will be given warfarin treatment after delivery for three to six months. If you have had a previous VTE in pregnancy or the postpartum period, or have other risk factors, you may be offered prophylactic heparin injections before and after the delivery in any subsequent pregnancy.

Obstetric cholestasis

This is rare but can lead to complications in late pregnancy, including stillbirth. The main symptom is severe itching without a rash, usually most intensely on the palms of the hands and soles of the feet, caused by bile salts deposited under the skin. A small percentage of women become jaundiced. The lower levels of bile leads to a reduction in the absorption of vitamin K, which increases the chances of mother or baby bleeding.

Ursodeoxycholic acid treatment helps reduce the itching and liver function abnormalities and vitamin K improves blood clotting. You may be advised to have an induction of labor at 37–38 weeks to reduce the risks of late pregnancy complications.

Anemia

The red cells in your blood contain hemoglobin (a complex of four protein chains attached to iron), which carries the oxygen around your body. During pregnancy women often become mildly anemic, because hemoglobin levels fall due to the demands of the growing baby and also because of increased fluid content of the mother's blood, which "dilutes" the hemoglobin level. If your levels drop below 11g/dl you will look pale, feel tired, breathless, or faint, and require iron and folic acid supplements. Mild iron-deficiency anemia in pregnancy is not harmful to the baby, who will continue helping itself to all the iron it requires from your reserves. More serious anemia puts you at risk of having a preterm or low birthweight baby, so if your hemoglobin levels do not improve after a few weeks of treatment, more blood tests will be done to check for rarer causes of anemia. Occasionally, it may be necessary to give iron infusions or a blood transfusion.

SICKLE-CELL ANEMIA

Sickle cell anemia (see p.418) is an inherited abnormality in the production of the protein chains that make hemoglobin. This results in a change in the shape of the red blood cells making it difficult for them to navigate around blood vessels smoothly. As a result they are easily damaged. The breakdown of the damaged cells results in hemolytic anemia. Cell debris clogs the blood vessels leading to strokes, infection, and pain in the bones, limbs, chest, or abdomen. Pregnant women with sickle-cell disease are at constant risk of a sickle-cell crisis. This can be life threatening for the mother and compromises placental function and fetal growth. They need to be specialized treatment.

THALASSEMIA

The thalassemias (see p.418) are inherited hemoglobin abnormalities. Alpha thalassemia is common in Southeast Asia, and beta thalassemia is usually found among people living around the Mediterranean or in the Middle East, India, and Pakistan. Carriers of the trait are likely to develop more severe anemia during pregnancy. Beta thalassemia major results in severe anemia and a life-long problem of trying to get rid of excess iron in the circulation. This is why iron supplements are never given to thalassemics but folate treatment is often prescribed.

ABO incompatibility

ABO incompatibility may occur in babies with blood type A, B or AB born to mothers with blood type O. Type O women routinely have antibodies to blood types A and B, but they are too large to cross the placenta. During pregnancy, a few fetal red cells can enter the mother's circulation and stimulate the formation of a smaller sized anti-A or B antibody, which may cross back to the baby's circulation and attack the baby's red cells. If many red cells are destroyed this leads to jaundice after birth, which may need treatment with phototherapy or an exchange transfusion.

Maternal red cell antibodies

At your first appointment, your blood group is checked and the presence of any atypical antibodies to your red blood cells is noted and you will be given a special card with details of them which should be shown to other doctors. Red cell antibodies usually develop as a result of a blood transfusion or pregnancy, but can occur naturally. They are not related to any illness or infection and are not harmful to your health. But, it is important to know about these antibodies during pregnancy, because if you need a blood transfusion, other blood groups as well as ABO and rhesus have to be taken into account during the cross matching. In addition, red cell antibodies may occasionally attack the baby's red blood cells leading to jaundice (see ABO incompatibility).

Rhesus disease

The rhesus factor is present on the surface of red blood cells. It is made up of three paired parts—C, D, and E—of which D is the most important because it can lead to rhesus isoimmunization (see p.128). About 85 percent of Caucasians carry the D antigen and are known as rhesus D positive, while the 15 percent who lack it are rhesus negative. If you are Rh-negative, problems may arise if you are carrying a baby who is Rh-positive because of a risk that you could develop antibodies that can cross the placenta and attack and destroy the baby's red blood cells. Although this is rarely a problem in a first pregnancy, you may develop antibodies as a result of exposure to your baby's Rh-positive blood at the delivery, which can cause problems in a subsequent pregnancy.

Rhesus disease can largely be prevented by giving injections of anti-D to Rh-negative women during pregnancy and after delivery (see p.128) to mop up

BLOOD PRESSURE PROBLEMS

Uncontrolled high blood pressure in pregnancy can result in serious problems for both mother and baby. The most common forms are pregnancy-induced hypertension (PIH) or preeclampsia (PET). Depending on history and risk factors (such as preexisting hypertension, renal disease, obesity, and maternal age), you may be offered low-dose aspirin from 12 weeks until 36 weeks gestation. PIH may be a reflection of undiagnosed chronic hypertension and is diagnosed on the basis of isolated high blood pressure in pregnancy. PET, however, is diagnosed when protein is detected in the urine in combination with high blood pressure.

PREECLAMPSIA

Preeclampsia (PET) complicates 5–8 percent of all pregnancies. Most cases are mild and usually occur in first-time mothers during the second half of their pregnancy, and are resolved soon after the birth. Severe cases can occur earlier in pregnancy. Others may not arise until you are in labor or make a first appearance only after delivery with little warning.

Pregnant women are checked for symptoms of impending preeclampsia at every prenatal doctor visit, during labor, and after the birth. The classical signs are raised blood pressure, peripheral edema (swelling of the hands, feet, and legs), and protein in the urine. The only cure for preeclampsia is delivery of the baby. However, if the baby is very immature, it may be possible to buy some more time for the baby to continue growing in utero by giving the mother treatment.

Mild preeclampsia may have no significant effect on the growth and well-being of the baby. However, when placental blood flow and function are reduced there is always the risk of the baby developing IUGR and oxygen shortage (hypoxia). Regular ultrasound and Doppler blood flow monitoring of fetal growth in pregnancies complicated by preeclampsia is now commonplace and plays an important role in deciding on the best time to deliver. In mothers with severe PET, the baby is also at risk of premature delivery, placental abruption, and intrauterine death.

Normal blood pressure in a nonpregnant woman is less than 140/90mm of mercury. However, blood pressure measurements in pregnancy vary between individuals and at different stages of pregnancy. Therefore, it is more useful to compare the prenatal visit readings with the blood pressure measurements made at the first prenatal visit when assessing the risk of preeclampsia.

Mild PET— the blood pressure (BP) rises to 140/90, with mild edema and proteinuria. The woman usually feels well but may require oral antihypertensive treatment if the BP is consistently raised.

Moderate PET—the BP exceeds 140/100 and is accompanied by more significant proteinuria and marked edema. Most women will be admitted to the hospital to bring the BP under control and assess the baby's well-being.

Severe PET—the BP exceeds 160/110 and heavy proteinuria is present. Sudden and intense swelling of the face and limbs may occur with marked weight gain. Immediate treatment is needed to lower the BP and prevent the onset of convulsions, which usually requires intravenous antihypertensive and sedative drugs, followed by immediate delivery of the baby, generally by cesarean section. The causes of preeclampsia are not clearly understood. There is no doubt that there is a genetic component since the problem tends to run in families. Preeclampsia is more common in first-time mothers, twin and diabetic pregnancies, and in women with preexisting hypertension or kidney disease. There is increasing evidence that PET is linked to a poor cardiovascular status and women who have had preeclampsia are at an increased risk of cardiovascular disease in later life.

ESSENTIAL HYPERTENSION

Essential hypertension complicates 1–3 percent of pregnancies and is more frequent in women over the age of 35. It is defined as persistently raised blood pressure (BP) greater than 140/90 before 20 weeks and may be diagnosed before your pregnancy or identified at your first prenatal doctor visit. Most women will already be taking an antihypertensive drug and need to be supervised by a physician in order to modify the dosages during pregnancy.

ECLAMPSIA

The signs of eclampsia are coma and convulsions and usually occur at the final stage of severe untreated PET or essential hypertension with superimposed PET. Eclampsia is now rare in the developed world, although it remains a potentially life-threatening obstetric emergency for the mother and baby, because all of the maternal blood vessels go into spasm leading to kidney, liver, and brain dysfunction together with a dramatic shortage of blood flow and oxygen to the fetus. Immediate measures are needed to sedate the mother's irritable brain, stabilize her blood pressure, and deliver the baby, which invariably requires a cesarean section delivery.

any Rh-positive fetal blood cells that may have entered your circulation and stop development of destructive maternal antibodies. However, if antibodies are detected in your blood tests in a subsequent pregnancy, you will need special care. You will be given more blood tests every four weeks, and your baby will be monitored for signs of anemia or heart failure. Severely affected babies may need multiple blood transfusions in utero to allow the pregnancy to continue until it is safe to deliver. At birth the baby will have tests for hemoglobin, ABO, and rhesus blood grouping, bilirubin levels, and a Coombs test (which detects the maternal rhesus antibodies). Jaundice often develops in the first 48 hours of life, requiring immediate treatment.

Amniotic fluid problems

POLYHYDRAMNIOS

This excess of amniotic fluid usually becomes noticeable when the volume is more than two liters. You will develop a tense abdomen through which it is difficult to feel fetal parts clearly and in severe cases, heartburn, breathlessness, and abdominal discomfort develop particularly when the onset is sudden. Polyhydramnios can be caused by increased production of amniotic fluid because the placental surface is large (twins) or the fetal urine production is increased (poorly controlled diabetes). It may also result because a fetal malformation prevents the fetus from swallowing fluid or absorbing it. Polyhydramnios almost always occurs in pregnancies affected by HYDROPS, because the fetus

develops heart failure or severe anemia, but in many cases no specific cause is identified. Polyhydramnios increases the risk of premature labor, CORD PROLAPSE and ABNORMAL PRESENTATIONS and may be relieved by amniocentesis to drain some of the fluid.

OLIGOHYDRAMNIOS

A reduced amniotic fluid pool is most likely to be due to INTRAUTERINE GROWTH RESTRICTION (IUGR) or ruptured membranes, but can also occur in healthy post-mature pregnancies. The amount of amniotic fluid in late pregnancy is a good measure of fetal well-being; ultrasound measurements that show reduced fluid near to term usually prompt a decision to deliver the baby. Rarely, oligohydramnios is noted on the 20-week scan, which usually means that the fetus has an abnormality of the renal tract. Oligohydramnios in early pregnancy results in poor lung development and limb pressure deformities in the baby.

Gestational diabetes

Between 1 and 3 percent of pregnant women develop gestational diabetes (glucose intolerance). The risk is greater in women who are obese, over 35, have a history of gestational diabetes, large babies, or intrauterine death or stillbirth, or a family history of diabetes. During pregnancy the placenta produces hormones that block the effects of insulin. This insulin resistance usually begins at 20–24

weeks and increases until delivery. If your pancreas cannot produce sufficient insulin to counteract this effect you will develop hyperglycemia (high sugar levels) and will be diagnosed as having gestational diabetes. All pregnant women are offered a glucose tolerance test between 24 and 28 weeks (see p.212). Most women diagnosed with gestational diabetes can be treated with dietary measures, but 10–20 percent will also require oral hypoglycemic agents and/or insulin treatment before the end of pregnancy. Careful monitoring of the pregnancy by a multidisciplinary team consisting of a dietician, an obstetrician, and an endocrinologist improves the outcome for mother and baby.

Gestational diabetes is not associated with increased risk of miscarriage or congenital abnormality because the glucose intolerance starts later in pregnancy. However, later complications are common because the fetal pancreas produces more insulin to manage the high maternal sugar levels that cross the placenta. This may lead to abnormal presentations, macrosomia (fat baby), and POLYHYDRAMNIOS, all of which increase the risk of premature labor and complications during labor and birth, particularly shoulder dystocia. Therefore, you will be given regular fetal growth ultrasounds and estimations of fluid volume and may be advised to undergo an induction of labor before term. Some 30 percent of gestational diabetics will develop overt diabetes or high blood pressure in later life.

Postpartum hemorrhage

Postpartum hemorrhage (APH) is best defined as significant vaginal bleeding after the 24th week of pregnancy. Before this date the bleeding is called a threatened miscarriage. After 24 weeks the baby has a chance of survival, so it is particularly important to diagnose or exclude bleeding from the placenta (due to placenta previa or placental abruption), which may require immediate delivery to protect you and your baby. Occasionally, the bleeding is from a cervical erosion or polyp. If you experience bleeding you and your baby need to be assessed at the hospital immediately.

Placental abruption

In this condition the placenta starts to separate from the wall of the uterus. The underlying cause is often unclear, but it is more common in multiparous women, cigarette smokers, cocaine and crack users, and women with poor nutrition, high blood pressure, or thrombophilia (pro-clotting tendency). The bleeding that results may be "revealed" if some of the blood is able to escape from the uterus into the vagina or "concealed" when the blood is trapped between the wall of the uterus and placenta.

Placental abruption is invariably painful, because blood seeps into the muscles of the uterus, causing irritation and contractions. If you experience a minor abruption, the baby is not distressed and your own condition is stable, it may be possible for you to go home after a few days

observation in the hospital. If the bleeding is severe, the buildup of blood behind the placenta leads to severe pain and further detachment of the placenta. If, upon examination, the uterus feels "woody" and is tender to touch an emergency delivery will be done, usually by cesarean section.

Placenta previa

One in 200 pregnancies at term are complicated by placenta previa where the placenta is implanted in the lower segment of the uterus and lies in front of (previa) the presenting part of the baby. If the placenta completely covers the internal part of the cervix, the cervical os (complex previa), the only option for delivery is cesarean section. In a low-lying placenta, where the baby's head is able to descend past the lower edge of the placenta, a vaginal birth may be possible.

It is common for the placenta to be reported as a partial previa or low-lying at the 20-week scan. However, by 32 weeks, the lower segment of the uterus has started to extend downward and previously low-lying placentas now appear to be placed higher in the uterus. Placenta previa is responsible for 20 percent of cases of postpartum hemorrhage and is more common in multiparous women. The bleeding is characteristically painless, usually recurrent, and can be very severe, necessitating emergency delivery and a blood transfusion. If surgical removal of the placenta is not possible, a hysterectomy is often needed to prevent life-threatening hemorrhage.

Abnormally adherent placenta

The placenta usually separates from the uterine wall minutes after delivery. Occasionally it has invaded too deeply into the endometrial lining and the uterine muscle (placenta accreta—1 in 1,500 births) or penetrates through the muscle wall (placenta percreta—1 in 2,500 births). It cannot spontaneously separate and attempts to remove it manually may lead to a major postpartum hemorrhage (see p.335). If all means to stop the bleeding fail, the woman may have to undergo a hysterectomy. Placenta accreta or percreta are more common in women with a placenta that has implanted in the lower uterine segment or who have a uterine scar. If surgical removal is not possible, and the patient remains stable, the placenta may be left to slough off.

Uterine rupture

Uterine rupture usually follows an obstructed labor, the inappropriate use of oxytocic drugs and the rupture of a previous cesarean section or myomectomy (fibroid) scar, but occasionally occurs before labor in women with a uterine scar. A classical cesarean section scar is far more likely to rupture during labor than a lower segment scar, which is why elective cesarean delivery before term is advised for these women. The rupture may be silent and painless, or be accompanied by severe pain and shock due to intra-abdominal bleeding combined with acute fetal distress. An emergency hysterectomy may be required.

FETAL GROWTH IN PREGNANCY

INTRAUTERINE GROWTH RESTRICTION (IUGR)

In approximately 3–5 percent of pregnancies the baby suffers from growth restriction, which is also referred to as small for dates, small for gestational age, or placental insufficiency. The best definition of IUGR is a baby whose birthweight is below the 5th percentile for growth at that gestational age. IUGR is the third most important cause of perinatal mortality (after prematurity and congenital abnormalities). Consequently, trying to identify babies at risk of IUGR is part of routine prenatal care.

CAUSES OF IUGR

• General factors include racial variations, low socioeconomic status, high parity (many previous births), poor prenatal education, and a previous baby with unexplained IUGR.
• Maternal health issues include low maternal weight before pregnancy, weight gain of less than 22lb (10kg) during pregnancy, and poor nutrition. Smoking is an important and preventable cause of IUGR worldwide. Alcohol, amphetamines, heroin, and cocaine all have a powerful adverse effect on fetal growth. IUGR has a high risk of recurrence, suggesting that the problem runs in families.
• About 5 percent of IUGR babies have a chromosomal abnormality such as Down syndrome or a structural congenital abnormality of the heart, kidney, or skeleton. Symmetrical growth IUGR should always raise the suspicion of a fetal infection with RUBELLA, CYTOMEGALOVIRUS, SYPHILIS, or TOXOPLASMOSIS. Syphilis and toxoplasmosis can be

treated with antibiotics during pregnancy to reduce the damage to the fetus.
• Any disorder that reduces placental function or blood flow will result in IUGR because the supply of oxygen and nutrients to the fetus is reduced. This may be due to abnormalities in the early development of the placenta or because the placenta becomes less efficient in later pregnancy following placental bleeding or abruption. Women with poor nutrition or an underlying medical disorder are more likely to develop placental insufficiency. IUGR develops in 20 percent of all twin pregnancies but is more likely in identical twins where there is a risk that one baby will receive a reduced share of the placental blood flow (see twin-to-twin transfusion syndrome, p.346).

SCREENING FOR IUGR

Awareness of the causes of IUGR and those mothers who are at greatest risk is an important part of screening for IUGR. Abdominal examinations will miss some 30 percent of babies with IUGR, and menstrual dates are inaccurate in at least 1 in 4 pregnancies. This is why a dating scan in the first trimester of pregnancy is such a useful baseline measurement, particularly in pregnancies that later become complicated. Sequential ultrasound scans during the second and third trimester of pregnancy (see p.214 and p.257) are the best way to identify fetal growth problems and will distinguish between symmetrical and asymmetrical IUGR.

If a fetal chromosomal abnormality or infection is suspected, the mother will be advised to have an amniocentesis or infective screen to confirm the

diagnosis. The inclusion of Doppler blood flow measurements of the fetal cerebral, umbilical, and uterine vessels can help doctors assess the severity of the IUGR. Combined with other tests of fetal well-being they will influence how the rest of your pregnancy is treated.

MANAGING AN IUGR PREGNANCY

If consecutive ultrasound scans show that your baby's growth has been static or that the fluid volume or blood flow to the baby is now reduced, it may be necessary to consider delivering your baby early. Of course, this will only be an option if your baby has reached a viable age and if your doctors feel that the baby would be better taken care of in the outside world rather than being left in utero.

Babies with IUGR are more likely to develop fetal distress and asphyxia during labor and be born with low Apgar scores which is why you may be offered steroids prior to delivery. In severe cases, an elective cesarean section may be the preferred method of delivery. Babies with moderate IUGR may require an induction of labor, particularly when the amniotic fluid volume is reduced. If carefully managed, babies with mild IUGR are often delivered vaginally.

IUGR babies are more likely to experience postpartum complications and a pediatrician may be on hand at the time of delivery to assess the severity of the IUGR and the support that the newborn baby may require. IUGR babies usually catch up in growth after delivery unless the IUGR is due to a congenital abnormality or was very severe during early pregnancy.

Umbilical cord problems

CORD PROLAPSE

If the umbilical cord is positioned below the baby and the membranes rupture, the cord can slip or prolapse through the cervix. This occurs in 1 in 300 pregnancies and is more common in premature babies, breech presentations, transverse or oblique lies, and when there is POLYHYDRAMNIOS. Prolapsed cord is an obstetric emergency, because once exposed to cold air the umbilical blood vessels will go into spasm, cutting off the oxygen supply to the baby. Prompt delivery is needed.

CORD COMPRESSION

Mild, intermittent cord compression during a contraction occurs in about 10 percent of labors. The CTG may show signs of mild fetal stress, but usually the baby has sufficient energy reserves to recover quickly from the temporary lack of oxygen. Cord compression is often seen in the second stage of labor, particularly when the cord is short or wrapped around the baby's neck. Cord compression is more likely to lead to fetal distress and asphyxia in a baby that is already at risk because of IUGR or post-maturity, particularly if OLIGOHYDRAMNIOS is also present.

SINGLE UMBILICAL CORD ARTERY

The cord usually contains three blood vessels: two arteries and one vein. However, in about 5 percent of babies only one artery and one vein are present, which may be detected on an ultrasound scan. In 15 percent of affected babies the abnormality is associated with other congenital abnormalities and IUGR, and consequently the finding will prompt more tests. The condition is more common in twins.

VELAMENTOUS CORD INSERTION

When the cord runs through the membranes before entering the placenta (vasa previa) there is a risk that the vessels will be damaged when the membranes rupture, resulting in fetal bleeding. This condition is present in 1 percent of term pregnancies and is more common in twins.

Abnormal fetal lie and presentation

A transverse lie, an oblique lie, or a shoulder presentation is more common in women who have already had a baby because the uterus is more lax. These abnormal presentations are also associated with prematurity, multiple pregnancy, FIBROIDS, uterine malformations, POLYHYDRAMNIOS and PLACENTA PREVIA. It may be possible to perform a gentle cephalic version (if placenta previa is not present) but the baby often returns to the previous presentation. In late pregnancy the risk of cord prolapse may require admission to the hospital, and delivery may have to be by cesarean section.

Face presentations occur in 1 in 500 deliveries, usually by chance but occasionally because the baby is anencephalic or has a swelling in the neck or shortening of the neck muscles. There is little point in trying to diagnose the problem before labor or in early labor, because the face presentation may resolve spontaneously when it passes through the pelvis. The labor is usually prolonged and the facial swelling can be severe and take several days to settle down.

Brow presentation is the least common (1 in 1,500 births) and most unfavorable of all abnormal presentations, as the presenting part is too broad to be delivered vaginally. It can be associated with fetal abnormalities, particularly HYDROCEPHALY.

Shoulder dystocia

This is an obstetric emergency because the head has already delivered but the shoulders do not follow because they are wedged in the pelvis. Swift action is required to prevent the baby from developing asphyxia. Placing the mother flat and her legs in the McRoberts' position (flexion and abduction of the maternal hips) allows the doctor to apply firm axial traction on the baby's head and neck to encourage the anterior shoulder under the pubic symphysis. An extensive episiotomy, and pressure applied above the pubic bone by an assistant, helps the delivery. Shoulder dystocia is more common in obese and diabetic women, in babies weighing more than 9lb (4kg), and following a prolonged labor or instrumental delivery. If there is a history of shoulder dystocia an experienced obstetrician should always be present at the next delivery.

Miscarriage

Miscarriage—the spontaneous loss of a pregnancy before the fetus is able to survive outside the mother's uterus—is the most common complication of pregnancy. It occurs in 15 percent of recognized pregnancies but we now know that some 50 percent of fertilized eggs are lost, many of which never reach the stage of being visible on the ultrasound scan. The vast majority of random miscarriages occur early in pregnancy and are due to fetal chromosome abnormalities that are incompatible with further development. Miscarriage after 12 weeks of gestation is uncommon and affects only 1–2 percent of pregnancies.

Miscarriage is a process, not a single event, so if you experience vaginal bleeding or pain in pregnancy there are several possible outcomes. In a threatened miscarriage there is no obvious problem on ultrasound scan and the bleeding will stop after a few days or recur again, but the cervix remains closed. If the process continues and the cervix starts to open, usually accompanied by cramping abdominal pain, the miscarriage becomes inevitable. It may go on to be complete, in which case the uterus empties its contents entirely, or incomplete if some pregnancy tissues are left behind.

Removal of the remaining tissues using drugs or surgery to evacuate the uterus is usually advised to prevent hemorrhage and infection. Sometimes the pregnancy stops developing but there are no obvious signs of a problem (missed miscarriage) until an ultrasound scan demonstrates that the fetus has died (missed miscarriage) or there is no evidence of a fetal pole within the early pregnancy sac (blighted ovum).

The chance of suffering from a miscarriage increases with the maternal age and also if the woman has experienced previous miscarriages. Recurrent miscarriage—usually defined as three or more consecutive pregnancy losses—is not common, affecting only 1 percent of couples. Most couples who experience this distressing problem want to have investigations done sooner rather than later in order to establish whether or not they have an underlying cause for the repeated miscarriages. In the majority of cases, no cause is found and these couples should be reassured that their prognosis for a successful pregnancy the next time is excellent. Nonetheless, referral to a unit that specializes in miscarriage is very valuable for these couples, since they will be offered the latest tests and opportunities to participate in new treatment trials, not to mention gaining the psychological benefits of knowing that they have done all they can to help prevent the miscarriages from recurring.

I cannot overstate the value of joining a research program if you have suffered from recurrent miscarriages. Repeatedly, studies have shown that the future pregnancy outcome is improved for women who are taken care of by a dedicated team of specialists.

RECOVERY AND COUNSELING

Losing a baby at any stage in pregnancy can be a devastating experience. We all react to and deal with grief in different ways, but when it comes to pregnancy loss the stages of acute grief, mourning, and healing can be lengthy processes and there are no shortcuts that can be taken. The first stage invariably includes shock, disbelief, numbness, confusion, and sometimes denial. Next comes the angry stage, often complicated by guilt, despair, depression, and physical symptoms of anxiety, such as insomnia, poor sleep, and loss of appetite.

With time, acute grief gives way to a deep sadness and later will be replaced by feelings of mourning, regret, and yearning for the baby you have lost. Eventually, you will become resigned and accepting of what has happened although this never completely takes away the emotional pain, but you will be able to deal with it in a more controlled way. You will need support during the recovery period and this may come from many different sources: family, friends, the hospital medical team, your physician, other parents who have had similar experiences, and local or national support groups. Most maternity units can put you in touch with special counselors who have experience in helping couples cope with the aftermath of early and late pregnancy loss.

Stillbirth and neonatal death

Stillbirth (SB) follows the death of a baby in utero after 20 weeks. It may be anticipated because a serious congenital abnormality was identified during pregnancy, but 50 percent of all stillbirths occur without warning.

The diagnosis may be suspected if the mother reports a lack of fetal movements. This can be confirmed on an ultrasound scan that reveals the absence of a fetal heart beat. Labor usually starts spontaneously a few days after the baby's death, but you may prefer to have labor induced at the earliest opportunity and occasionally some women may choose to deliver their baby by cesarean section. If delivery has not occurred within seven days you will be advised to undergo an induction of labor, because after this time there is a chance that you could develop a serious blood-clotting abnormality if the fetal tissues remain in utero.

The risk of stillbirth is increased in high-risk pregnancies. However, the stillbirth rate has fallen dramatically as a result of improved maternal health, nutrition, and prenatal monitoring for problems such as high blood pressure, diabetes, growth restriction, cholestasis, and rhesus disease.

Stillbirth in labor is now rare (1 in 1,000), due to improved intrapartum monitoring, but it still occasionally follows a massive placental abruption. Even when a detailed postmortem is

performed, the cause of the stillbirth is not always understood, which is particularly distressing for the parents.

Neonatal death (NND) is the loss of a baby within four weeks of birth and affects 3–4 per 1,000 babies, most of them dying within the first week of life. In 25 percent of neonatal deaths the baby has either a severe genetic or chromosomal abnormality or a structural problem, most commonly affecting the heart. Neonatal death is also associated with early premature birth, and occasionally follows infection during pregnancy, or fetal distress and asphyxia during labor. If the baby dies after four weeks of life, it is called an infant death.

Sudden infant death syndrome (SIDS)—previously called crib death—is rare, affecting 1 in 1,600 babies, but is more common among premature babies, severely growth-restricted babies, boys, and multiple births.

Termination (early and late)

The decision to undergo a termination of pregnancy for a fetal abnormality is never an easy one (see p.138). If you find yourself in this distressing situation you will need some information about your options and what is likely to happen, and my patients tell me that this is a subject that most pregnancy books fail to include. Before 12 weeks of pregnancy the termination can be performed surgically or medically, but after 12 weeks it is usually safer to induce labor using

medicine and deliver the fetus vaginally. The surgical approach involves clearing out the uterus using suction under general anesthesia, which has the advantage of being quick and physically painless since the mother is unconscious. You should expect to have some vaginal bleeding for up to a week after the procedure and you will probably be prescribed a course of antibiotics to prevent an infection.

The medical approach to a termination involves a combination of two medicines; the first is prostaglandins and then sometimes Pitocin. The prostaglandin can be given orally, but this route can be associated with nausea and stomach upsets. The vaginal tablets are repeated at regular intervals until the termination is complete, the number of doses required being related to how far advanced the pregnancy is. However, vaginal bleeding and cramping abdominal pains usually start before or soon after the first vaginal tablet and in the majority of cases the fetus is expelled within 24 hours. You will be given painkilling medication to help you deal with any discomfort and you can expect to have light vaginal bleeding for up to a week after the termination.

If it is suspected that some of the pregnancy tissues have been retained, you will be advised to have a surgical curettage under anesthesia, but in the vast majority of cases this will not be necessary.

Concerns after birth

The days and weeks after birth are often dogged with minor problems. I have included general advice on many of these in Life After Birth (see pp.370–405), so I am concentrating here on maternal complications that may arise and also problems that may be present in newborns.

MATERNAL PROBLEMS

It is important to seek advice promptly if you notice symptoms of any of the following conditions that sometimes occur in the days and weeks after the birth. Most are only temporary and can be resolved speedily with appropriate treatment, although a few may need special help.

Postpartum pyrexia

Postpartum or puerperal pyrexia is defined as a rise in maternal temperature to 100.4°F (38°C) or more, from day 1 to day 10 after delivery and is usually caused by infection. Thanks to improved hygiene, obstetric care, and infection control in hospitals, the incidence of postpartum infection has now fallen to about 3 percent and is rarely life threatening. The most likely site of infection is the uterine cavity (endometritis) or perineum, but urinary tract and breast infections are also common. THROMBOEMBOLISM may cause a postpartum pyrexia and is more likely to develop after a cesarean section, as are chest or incision infections.

UTERINE INFECTION

Most cases are caused by infections ascending from the cervix or vagina. The organisms infect the placental bed and any pieces of retained placenta or membranes left in the endometrial cavity. If the lochia begins to smell offensive or you start to develop lower abdominal pain and tenderness, it is likely that you have developed endometritis.

This is important to diagnose and treat promptly to avoid complications such as damage to the fallopian tubes, resulting in future difficulties becoming pregnant.

Your doctor will examine you internally and take some vaginal swabs for testing. If the examination suggests that there may be tissues left within the uterine cavity (cervix partly open and the uterus enlarged, tender, and boggy), you will be started on a course of antibiotics and advised to undergo a uterine evacuation to remove the tissues. Since no one wants to give a woman who has just delivered a baby general anesthesia unless absolutely necessary, an ultrasound scan is often performed to confirm the examination findings.

URINARY TRACT INFECTION

Urinary tract infections are particularly common in women who have been catheterized during labor or have had a difficult delivery. Any rise in temperature after delivery should prompt your doctor to send a sample of urine to the laboratory to be tested and antibiotics are usually started immediately. It is important to check that the urinary tract infection has been treated effectively by testing another urine sample after the treatment has been completed.

MASTITIS

Almost every mother develops some degree of breast engorgement as the milk starts to come into her breasts. The breasts become swollen, hard, and sore, and this commonly causes a slight rise in her temperature. Fortunately, the problem usually disappears spontaneously in a day or two as breastfeeding becomes established. However, if you become feverish and start to feel sick, your breasts need to be examined for signs of localized or patchy redness and induration (pitting hardness). This is called mastitis and can be extremely painful because one of the milk ducts has become blocked and the stagnant pool of milk in the blocked duct can quickly become infected. The organism is usually a staphylococcus present on your or your baby's skin, which then enters a cracked or sore nipple and spreads into the breast tissues.

Caught early, mastitis responds promptly to antibiotics, mild analgesics, and keeping the breast milk flowing by continuing to breastfeed, or expressing the milk to relieve the pressure. However, if mastitis is not identified and treated quickly, it can develop into a full-blown breast abscess. In addition to feeling generally sick with a high temperature, you will have a hot, firm lump in one of your breasts that contains pus, which will have to be surgically incized (opened) and drained at the hospital.

Perineal problems

Approximately 69 percent of women who give birth vaginally require some sort of stitching. If at any stage in the few weeks after the delivery your perineum starts to throb, appears inflamed, or produces any sort of discharge, consult your obstetrician. You may have developed an incision infection, and this is usually treated simply and effectively with antibiotics. Sometimes it is judged best to remove one or two of the stitches to release the pressure in the inflamed area and make it easier to access the

incision to clean it thoroughly. Occasionally, after a traumatic vaginal delivery, a hematoma (collection of blood) can swell up in the walls of the vagina. This is removed surgically to relieve the pain, tie off the bleeding point, and keep infection from developing.

Some women continue to experience problems with their episiotomy or perineal wounds for several weeks, but there is no need to suffer in silence. Your physician will be able to make sure that there is no sign of infection. She may suggest that you consider having some physical therapy using ultrasound to help relieve the discomfort. Your maternity unit will also be able to refer you to the obstetric physical therapists, who have expertise in helping women with postpartum problems involving the perineum, bowel, bladder, and vagina.

Stress incontinence

Women who have had a vaginal delivery often suffer mild, temporary urinary incontinence. This is because the bladder neck has been stretched and pulled downward by the pressure of the baby's head passing through the birth canal. Usually this manifests itself as stress incontinence, where urine seeps out when you laugh, cough, sneeze, or move quickly.

Kegel exercises will improve your ability to regain full bladder control—the sooner you start the Kegel exercises, the quicker the benefits of them will be. However if, despite regularly performing these exercises, you are still

suffering from urinary or urge incontinence that disrupts your life (leaking persistently, unable to go out of the house because of the fear of accidents, or not being able to find a toilet quickly enough), then arrange for more specialized help through your physician.

Fecal incontinence

After a vaginal delivery, particularly those involving a prolonged second stage of labor and an extensive episiotomy or tear, some women may lose a degree of control over their bowels. This usually disappears soon after giving birth, with the help of some Kegel exercises. However, on the rare occasions when a woman develops full fecal incontinence and is simply unable to control her bowel movements, a specialist's help is needed, because this usually means that the anal sphincter and rectal skin have been torn.

Anemia

Symptomatic anemia after the birth may be due to acute blood loss (more common after prolonged labor, operative delivery, and postpartum hemorrhage) or because the mother's iron reserves have been depleted during pregnancy because of poor nutrition, difficulties in absorbing iron, twin pregnancies, or several pregnancies in quick succession. A blood transfusion may be required in severe cases, but iron and folic acid replacement is usually sufficient. Treatment should be started promptly.

PROBLEMS IN BABIES

Most of these problems are detected only after birth, usually during the baby's checkup in the hospital and at six weeks. However, cleft lip and palate is sometimes seen on ultrasound and fetal alcohol syndrome may be suspected if your baby has growth problems during pregnancy.

Cerebral palsy

Cerebral palsy (CP) describes a range of abnormalities in movement, muscle tone, posture, speech, vision, and hearing in young children, caused by damage to one or more areas of the brain. Cerebral palsy affects 1 in 400 children and is more common in pregnancies complicated by premature delivery, IUGR, and infection. There are three types of cerebral palsy depending on the area of the brain affected, and affected children usually have a combination of two or more types. There is no prenatal test that can detect cerebral palsy.

Fetal alcohol syndrome

Regular alcohol consumption during pregnancy may result in teratogenic (early) and toxic (later) damage to the fetus depending on how much alcohol is consumed. The main features of fetal alcohol syndrome (FAS) are intrauterine growth restriction (IUGR), failure to thrive after birth, damage to the nervous system, and poor childhood growth. Attention deficit disorder, language delays, and mild to moderate developmental disabilities gradually become evident. The characteristic facial appearance includes microcephaly, flattened nasal bridge, underdeveloped middle of the face, short upturned nose, and thin upper lip. Newborn babies are irritable, hyperactive, and have poor muscle tone. FAS affects at many as 1.5 in 1,000 babies in the US and is an important and preventable cause of learning difficulties.

Pathological jaundice

Occasionally, neonatal jaundice (see p.388) may be a sign of a more serious underlying condition such as anemia caused by blood group incompatibility, liver or thyroid disease, or an inherited enzyme disorder that weakens the red cells making them break down more easily. These rarer forms of jaundice are referred to as pathological jaundice and usually necessitate treatment with phototherapy and possibly even a blood transfusion. In severe cases, or when the baby is very premature, drugs to stimulate the liver to get rid of the excess bilirubin may be needed.

Cleft lip and palate

The development of the upper lip and palate (roof of the mouth) in the fetus involves the joining together of tissues in the midline of the face. When this is incomplete, as occurs in 1 in 750 babies, a split in the lip (harelip) and/or the palate results. The defect can be seen on prenatal ultrasound scans. Babies with a cleft palate may have difficulty feeding and are at risk of choking, because the absence of a bony roof to the mouth interferes with sucking and swallowing. They are also likely to suffer recurrent ear infections and hearing difficulties. Corrective surgery is done after birth. A cleft lip is usually closed at about three months but surgery to close the palate may be delayed until at least 12 months, to ensure it has developed fully.

Pyloric stenosis

Pyloric stenosis affects 1 in 500 newborn babies and is more common in boys. It is caused by thickening of the pylorus muscle between the bottom of the stomach and the small intestine. As food builds up, the stomach contracts strongly in an attempt to force food into the upper intestine. The problem starts soon after birth with persistent projectile vomiting during or right after being fed. As a result, the baby becomes hungry and irritable, and quickly becomes dehydrated and loses weight. Pyloric stenosis is diagnosed by feeling the contracted muscle in an abdominal examination and confirmed with an ultrasound scan or barium swallow X-ray. Prompt surgery to loosen the muscle cures the condition.

Umbilical hernia

This is caused by a weakness in the muscles of the abdominal wall at the point where the umbilical cord entered your baby's abdomen. A small bulge around the baby's umbilicus containing a section of intestine is common, occurring in about 10 percent of all babies, with a higher incidence in African-American children. It usually closes up of its own accord in time.

Inguinal hernia

This weakness in the lower abdominal wall in the groin region is caused by failure of the inguinal canal to close after birth. It occurs in about 3 percent of newborn babies and is often bilateral. During pregnancy the testes of a baby boy pass through the inguinal canal to reach the scrotum. Although girls do not have testes, they do have an inguinal canal so they too are susceptible to hernias.

Hernias are more likely in premature babies, babies with CYSTIC FIBROSIS, and baby boys with UNDESCENDED TESTES. As long as the contents of the bulge can be pushed back into the abdominal cavity, there is little cause for concern. However, occasionally, a loop of intestine becomes trapped in the hernia, obstructing the bowel. These strangulated hernias are a surgical emergency and need to be dealt with promptly to save the bowel and repair the abdominal wall defect.

Hypospadias

This common abnormality is found in 1 in 500 baby boys. The external opening of the urethra is positioned on the underside of the penis instead of at the end. The penis may curve downward and the foreskin or prepuce appears hooded. Sometimes the urethral opening is far back in the scrotum or on the upper side of the penis (epispadias). Curative surgery is performed at one year.

Undescended testes

This condition is found in 1 in 125 newborn boys. In 15 percent of cases both testes are undescended. The majority will have descended spontaneously after nine months but if the problem persists after this time, your pediatrician may recommend a pediatric surgeon. If the testes remain undescended it can lead to testicular cancers, abnormal sperm production, and infertility.

Imperforate anus

The anus is closed, either because it is sealed by a thin membrane of skin over the external opening or because the passage between the rectum and anal canal has not developed (anal atresia). The baby's lower bowel becomes distended and swollen toward the end of the pregnancy, which is visible on an ultrasound scan. All babies are examined at the time of delivery and will undergo immediate corrective surgery if necessary.

Dislocated hips

This congenital abnormality is identified in as many as 1 in 200 babies at the routine postpartum checkup (see p.387). It is more common in girls, in the left hip, in multiple pregnancies, and in babies that are born breech, or have another abnormality such as Down syndrome or a NEURAL TUBE DEFECT.

If the hip is dislocated, the hip joint is unstable and will make a clicking sound when the knees are bent up toward the hips and the legs rotated outward. A dislocated hip is usually curable with the use of orthopedic manipulation and splints to hold the hip in the correct position during the first few months of life. Occasionally surgery may be required.

Talipes

This is a condition in which the baby's feet are turned inward (equino varus) so that the soles are facing each other. Less commonly, the baby's feet are turned outward (calcaneo valgus). Talipes can be diagnosed on an ultrasound scan during pregnancy and often the problem runs in families.

The mildest form of talipes is caused by abnormal positioning of the feet during pregnancy and usually corrects itself during the first few months after the birth. However, if the feet cannot be easily manipulated into the correct position, regular physical therapy will be required and the baby will probably need to wear corrective splints for many months to make sure that his walking develops normally. Very severe forms of talipes may necessitate surgical correction at various points over several years.

Useful contacts

Alexander Graham Bell Association for the Deaf
3417 Volta Place, NW Washington, D.C. 20007
Tel: (202) 337-5220
www.agbell.org

American Academy of Pediatrics
345 Park Boulevard
Itasca, IL 60143
Tel: (800) 433-9016
www.aap.org
Educational brochures and fact sheets on children's health

American Chiropractic Association
1701 Clarendon Blvd. Suite 200
Arlington, VA 22209
Tel: (703) 276-8800
www.amerchiro.org

American College of Emergency Physicians (ACEP)
4950 West Royal Lane
Irving, TX 75063
Tel: (800) 798-1822
www.acep.org

American College of Nurse-Midwives
8403 Colesville Road, Suite 1500
Silver Springs, MD 20910
Tel: (240) 485-1800
www.midwife.org

American College of Obstetricians and Gynecologists
409 12th Street SW
Washington, D.C. 20024
Tel: (800) 673-8444
www.acog.org
Wide-ranging advice on health matters

American Diabetes Association
2451 Crystal Drive, Suite 900
Arlington, VA 22202
Tel: (800) 342-2383
www.diabetes.org

American Institute of Homeopathy
10418 Whitehead Street
Fairfax, VA 22030
Tel: (888) 445-9988
www.homeopathyusa.org

American Osteopathic Association
142 East Ontario Street
Chicago, IL 60611
Tel: (800) 621-1773
www.osteopathic.org

American Pregnancy Helpline
Tel: (866) 942-6466
www.thehelpline.org
A hotline providing information on unplanned pregnancy, parenting, and maternity facilities

American Sudden Infant Death Syndrome (SIDS) Institute
528 Raven Way
Naples, FL 110
Tel: (239) 431-5425
www.sids.org

Association for Children with Down Syndrome
4 Fern Place
Plainview, NY 11803
Tel: (516) 933-4700
www.acds.org

Asthma and Allergy Foundation of America
8201 Corporate Drive Suite 1000
Landover, MD 20785
Tel: (800) 727-8462
www.aafa.org

Attention Deficit Disorder Association
9930 Johnnycake Ridge Road
Mentor, OH 44060
Tel: (440) 350-9595
www.add.org
Support and resources for parents of children with ADD

Autism Society of America
4340 East-West Highway, Suite 350
Bethesda, MD 20814
Tel: (800) 328-8476
www.autism-society.org

Birth Defect Research for Children
976 Lake Baldwin Lane, Suite 104
Orlando, FL 32814
Tel: (407) 895-0802
www.birthdefects.org

Compassionate Friends
P.O. Box 3696
Oak Brook, IL 60522 Tel: (630) 990-0010
www.compassionatefriends.org
Grief support for the death of a child

Postpartum Support International
6706 SW 54th Avenue
Portland, OR 97219
Tel: (800) 944-4773
www.postpartum.net

Federation for Children with Special Needs
529 Main Street, Suite 1M3
Boston, MA 02129
Tel: (617) 236-7210
www.fcsn.org

Healthy Mothers, Healthy Babies Coalition
121 North Washington Street
Alexandria, VA 22314
www.hmhb.org

Hydrocephalus Association
4340 East-West Highway, Suite 905
Bethesda, MD 20814
Tel: (888) 598-3789
www.hydroassoc.org

International Cesarean Awareness Network
1304 Kingsdale Drive
Redondo Beach, CA 90278
Tel: (888) 686-4226
www.ican-online.org
Support and information for women who have had a cesarean section

International Childbirth Education Association
110 Horizon Drive, Suite 210
Raleigh, NC 27615
Tel: (919) 674-4183
www.icea.org
Information, pamphlets, and booklets for expectant and new parents

Internet Resource for Special Children
www.isaac-online.org
Internet resource providing information, news articles, and online communities on special-needs children

La Leche League International
1400 N. Meacham Road, Schaumburg, IL
60173 Tel: (847) 519-7730
www.lllusa.org
Breast-feeding advice and support

March of Dimes/Birth Defects Foundation
1275 Mamaroneck Avenue White Plains,
NY 10605 Tel: (888) MO-DIMES
TTY: (914) 997-4764
www.modimes.org

MEND (Mommies Enduring Neonatal Disease)
P.O.Box 631566 Irving, TX 75063 Tel:
(972) 506-9000 www.mend.org

American Association of Birth Centers
3123 Gottschall Road
Perkiomenville, PA 18074
Tel: (215) 234-8068
www.birthcenters.org

National Association for Sickle Cell Disease
3700 Koppers Street Ste. 570, Baltimore,
MD 21227 Tel: (800) 421-8453
www.sicklecelldisease.org

National Autism Hotline, Autism Service Center
929 4th Avenue, P.O. Box 507
Huntington, WV 25710
www.autismservicescenter.org

National Cystic Fibrosis Foundation
4550 Montgomery Avenue, Suite 1100N
Bethesda, MD 20814
Tel: (212) 986-8783 www.cff.org

National Organization on Disability
77 Water Street Suite 204, New York,
NY 0005 Tel: (646) 505-1191
www.nod.org

North American Registry of Midwives (NARM)
5257 Rosestone Drive, Lilburn, GA
30047 Tel: (888) 842-4784
www.narm.org

Parents Helping Parents
1400 Parkmoor Avenue, Suite 100
San Jose, CA 95126 Tel: (408) 727-5775
www.php.com
Resources and support for parents of children with special needs

Parents Without Partners
1650 South Dixie Highway, Suite 510
Boca Raton, FL 33432
Tel: (561) 391-8833
www.parentswithoutpartners.org

Preeclampsia Foundation
6905 N. Wickham Road, Suite 302
Melbourne, FL 32940
Tel: (800) 665-9341
www.preeclampsia.org

Thalassemia Action Group/ Cooley's Anemia Foundation, Inc.
129-09 26th Avenue
Flushing, NY 11354
Tel: (800) 522-7222
www.thalassemia.org

Index

Page numbers in **bold** refer to the illustrations

Acknowledgments

Author's acknowledgments

Updated edition 2019

Revising and updating this new edition of *I'm Pregnant!* has been a labor of love. Above all, I want to thank my colleagues at St. Mary's, Miss Shankari Arulkumaran, the hospital's lead consultant obstetrician and gynecologist for prenatal services. Her help has been invaluable and this revised edition benefits especially from her expertise and research in preterm labor, high-risk obstetrics, and postnatal care.

First edition 2005

Writing this book has been an exciting and rewarding challenge. It had a lengthy gestation period, during which I enjoyed the pleasure of working with some very talented individuals. I want to acknowledge their contribution and thank them for their expertise, encouragement, guidance, and practical support. There are too many to include each by name but a few need special mention. Maggie Pearlstine persuaded me to embark on the original project and convinced me that it could, should, and would be written. Debbie Beckerman, author and mother, devoted many hours to ensuring that we included all the issues that other pregnancy books had failed to address. Esther Ripley contributed more than editorial skills to the project. Her enthusiasm for the subject matter was only surpassed by her ability to remain calm, patient, and encouraging at all times, even when I was late meeting a deadline. Additional thanks must go to the creative team at Dorling Kindersley who have guided me. I am also grateful to my medical colleagues May Backos, Lorin Lakasing, and Lorna Phelan, who read through and sense- checked the original and revised manuscripts. Thanks also to my experienced midwifery colleagues at St. Mary's for their help and advice, and the many enthusiastic patients who have so generously shared their feelings, thoughts, anxieties, fears, and triumphs with me over the years. I hope I have done justice to their requests for a comprehensive and fresh pregnancy bible.

Publisher's acknowledgments

DK would like to thank Arunesh Talapatra and Vikas Sachdeva for design help; Devangana Ojha for editorial assistance; Vishal Ghavri for image research; and Umesh Rawat and Anurag Trivedi for technical checks.

For the 2005 edition

Project editors Esther Ripley, Angela Baynham
Art editor Nicola Rodway
Designers Briony Chappell, Alison Gardner
DTP designers Karen Constanti, Jackie Plant
Picture researchers Sarah Duncan, Anna Bedewell
Illustrator Philip Wilson
Production controller Shwe Zin Win
Managing editor Liz Coghill
Managing art editors Glenda Fisher, Emma Forge
Art director Carole Ash
Publishing manager Anna Davidson
Publishing director Corinne Roberts

Thanks to the following for their editorial support:
Julia Halford, Katie Dock, Isabella Jones
Additional photography Ruth Jenkinson
Additional illustrator Debbie Maizels
Additional DTP design Julian Dams, Grahame Kitto
Picture research administrator Carlo Ortu
Additional picture research Franziska Marking
Picture librarian Romaine Werblow
Proofreader Constance Novis
Indexer Hilary Bird

Picture credits

Most of the images in this book are of the embryo and fetus live in utero, pictured using endoscopic and ultrasound technology. When this has not been possible, images have been taken by reputable medical professionals as part of research or to promote educational awareness.

Dorling Kindersley would like to thank the following for their kind permission to reproduce their photographs: (abbreviations key: a=above, b=below/bottom, c=center, f=far, l=left, r=right, t=top)

1 Prof. J.E. Jirasek MD, DSc.: CRC Press/Parthenon. **2–3 Getty Images:** Blend Images - Kidstock. **4 Corbis:** I.WA-Dann Tardif (b); **Mother & Baby Picture Library:** Paul Mitchell (t); **OSF:** (m). **7 Photonica:** Henrik Sorensen. **8 Science Photo Library:** Edelmann (tl). **9 Corbis:** Susan Solie Patterson (tr). **10–11 Getty Images:** David Oliver. **12 Getty Images:** Bill Ling. **13 Mother & Baby Picture Library:** Ian Hooton. **14 Science Photo Library:** Prof. P. Motta/Dept. Of Anatomy/University "La Sapienza," Rome (cla); D. Phillips (crb); VVG (clb). **15 Science Photo Library:** Edelmann (c); Prof. P. Motta/Dept. Of Anatomy/University "La Sapienza," Rome (cra). **Wellcome Library, London:** Yorgos Nikas (cfr). **16 Science Photo Library:** Richard Rawlins/Custom Medical Stock Photo. **18 Science Photo Library:** Professors P.M. Motta & J. Van Blerkom (bl); Prof. P. Motta/Dept. Of Anatomy/University "La Sapienza," Rome (br). **19 Science Photo Library:** Dr Yorgos Nikas (bl); D. Phillips (br). **20 Science Photo Library:** Edelmann (all). **24 Mother & Baby Picture Library:** Ruth Jenkinson. **26 Alamy Images:** Camera Press Ltd. **31 Mother & Baby Picture Library:** Ian Hooton. **33 Science Photo Library:** CNRI (tr); Moredun Scientific Ltd (crb); Dr Gopal Murti (cr). **36 Bubbles:** Lucy Tizard. **37 Mother & Baby Picture Library:** Ian Hooton. **38 Mother & Baby Picture Library:** Ian Hooton. **44 Getty Images:** Tom Mareschal (bl). **46 Getty Images:** Chris Everard (bl). **Prof. J.E. Jirasek MD, DSc.:** CRC Press/Parthenon (br). **47 Alamy Images:** foodfolio (bl); **Science Photo Library:** Ian Hooton (tl); TISSUEPIX (br). **51 Getty Images:** Anthony Johnson. **57 Mother & Baby Picture Library:** Ruth Jenkinson. **58 Dreamtime.com:** Kirill Ryzhov. **60 Punchstock:** Blend Images **64 Getty Images:** Chronoscope. **66–67 Prof. J.E. Jirasek MD, DSc.:** CRC Press/Parthenon (all). **68–69 Prof. J.E. Jirasek MD, DSc.:** CRC Press/Parthenon (both). **70–71 Prof. J.E. Jirasek MD, DSc.:** CRC Press/Parthenon (all). **74 Science Photo Library:** Zephyr. **75 Getty Images:** Peter Correz. **76 Corbis:** Ariel Skelley. **80 Professor Lesley Regan. 84 Science Photo Library:** Ian Hooton. **88 Depositphotos Inc.:** SimpleFoto (l). **89 Science Photo Library:** Eddie Lawrence (r). **90 Mother & Baby Picture Library:** Ian Hooton. **92–93 Prof. J.E. Jirasek MD, DSc.:** CRC Press/ Parthenon (both). **Life Issues Institute** (bl); **Science Photo Library:** Edelmann (bc) (br). **96 Bubbles:** Jennie Woodcock (cl); **Mediscan:** Medical-On-Line (bl). **99 Getty Images:** Ericka McConnell. **105 Mother & Baby Picture Library:** Ian Hooton. **106–107 Science Photo Library:** Edelmann (both). **108 LOGIQlibrary** (cl). **109 Science Photo Library:** GE Medical Systems (b). **110 LOGIQlibrary. 111 Science Photo Library:** BSIP (cla). **115 Getty Images:** Daniel Bosler. **118 Mother & Baby Picture Library:** Ian Hooton. **121 Mother & Baby Picture**

Library: Ian Hooton. **123 Science Photo Library:** Garo / Phanie (br). **124 LOGIQlibrary. 125 LOGIQlibrary** (cl) (cr); **Patients and staff of St. Mary's Hospital, Fetal Medicine Unit** (br). **126 Dreamstime.com:** Roman Kosolapov (tl). **127 Getty Images:** Jose Luis Pelaez Inc. (br). **135 Getty Images:** Terry Vine (r). **136 Alamy Stock Photo:** Lakov Filimonov (t). **138 Dreamstime.com:** Alexander Raths (crb). **139 Professor Lesley Regan** (both). **140 Alamy Stock Photo:** Phanie (l). **141 Science Photo Library:** Saturn Stills (cra). **142 Science Photo Library:** Dr. P. Boyer (crb); Saturn Stills (cb). **148 Mother & Baby Picture Library:** Ian Hooton. **149 Mother & Baby Picture Library:** Paul Mitchell. **150 Getty Images:** Steve Allen (l); **Prof. J.E. Jirasek MD, DSc.:** CRC Press/ Parthenon (b). **150–151 Getty Images:** Ranald Mackechnie. **151 Prof. J.E. Jirasek MD, DSc.:** CRC Press/Parthenon (br); **Life Issues Institute** (tr). **152–153 Getty Images:** Steve Allen (both). **154 Science Photo Library:** Professor P.M. Motta & E. Vizza (cl); VVG (tl). **155 Science Photo Library:** GE Medical Systems (b). **158 Science Photo Library:** CNRI (bc); Edelmann (bl). **159 Alamy Images:** Janine Wiedel (bc); **Photonica:** Henrik Sorensen (bl). **162 iStockphoto.com:** Voyata (clb). **163 Alamy Images:** Camera Press Ltd. **166–167 Science Photo Library:** Neil Bromhall (both). **168 OSF:** (bl); **Science Photo Library:** Neil Bromhall / Genesisi Films (br). **171 Science Photo Library:** DR P. Marazzi (cr); **Science Photo Library:** (crb). **175 LOGIQlibrary** (l) (tl); **Professor Lesley Regan** (cr) (c). **176 Mother & Baby Picture Library:** Ian Hooton. **177 Alamy Images:** Camera Press Ltd. **179 Alamy Images:** Bubbles Photolibrary. **180–181 Prof. J.E. Jirasek MD, DSc.:** CRC Press/Parthenon (both). **182–183 Prof. J.E. Jirasek MD, DSc.:** CRC Press/Parthenon (all). **185 iStockphoto. com:** Dirima (r). **187 Corbis:** Cameron. **189 LOGIQlibrary. 190 Mother & Baby Picture Library. 193 Alamy Images:** Bill Bachmann. **195 SuperStock. 197 Alamy Images:** Dan Atkin. **198 Corbis:** Jim Craigmyle. **199 Science Photo Library:** Ian Hooton. **200 LOGIQlibrary** (l) (c); **Science Photo Library:** Dr Najeeb Layyous (b). **200–201 Getty Images:** Jim Craigmyle. **201 LOGIQlibrary** (tr) (br). **202–203 Life Issues Institute** (both). **204 Science Photo Library:** BSIP, MARIGAUX. **205 Professor Lesley Regan** (bl). **211 Mother & Baby Picture Library:** Ian Hooton. **218 Oppo. 221 Mother & Baby Picture Library:** Ian Hooton. **223 Mother & Baby Picture Library:** Ian Hooton. **225 Alamy Images:** Camera Press Ltd. **227 Corbis:** Roy McMahon. **228 Getty Images:** Ross Whitaker. **230–231 Science Photo Library:** GE Medical Systems (both). **232 LOGIQlibrary** (br); **Science Photo Library:** GE Medical Systems (bl). **233 Alamy Images:** Nick Veasey x-ray. **236 Mother & Baby Picture Library:** Ian Hooton. **243 Alamy Images:** Stock Image. **246 Science Photo Library:** Colin Cuthbert. **248 Getty Images:** BSIP (bl). **249 Alamy Stock Photo:** Lev Dolgachov (tr). **253 Dreamstime.com:** Vadimgozhda (br). **256 Mother & Baby Picture Library:** Caroline Molloy. **257 Professor Lesley Regan** (both). **260–261 Science Photo Library:** GE Medical Systems (both). **263 Science Photo Library:** Mehau Kulyk. **265 Mother & Baby Picture Library:** Ian Hooton. **271 Mother & Baby Picture Library:** Ian Hooton. **278–279 Alamy Images:** plainpicture/Kirch, S.. **280 Alamy Stock Photo:** Science Photo Library. **281 Corbis:** Jules Perrier. **282 Mother & Baby Picture Library:** Ian Hooton. **290 Science Photo Library:** BSIP, Laurent. **291 SuperStock:** Science Photo Library (tr). **293 Mother & Baby Picture Library:** Ruth Jenkinson. **297 Mother & Baby Picture Library:** Ruth Jenkinson. **309 Mother & Baby Picture Library:** Ian Hooton. **314 Alamy Images:** Janine Wiedel. **322 Wellcome Library, London:** Anthea Sieveking. **324 Angela Hampton/Family Life Picture Library. 329 Wellcome Library, London:** Anthea Sieveking. **332 Mother & Baby Picture Library:** Moose Azim. **334 Science Photo Library:** CNRI. **336 Alamy Images:** Peter Usbeck. **337 Corbis:** Annie Griffiths Belt. **342 Corbis:** Tom Stewart. **344 Corbis:** ER Productions. **345 Professor Lesley Regan. 348 Mother & Baby Picture Library:** Indira Flack. **360 Alamy Images:** Yoav Levy. **368 Alamy Images:** Janine Wiedel. **370–371 Getty Images:** Kaz Mori. **372 Getty Images:** Rubberball Productions. **373 Alamy Images:** plainpicture/Kirch, S.. **374 Alamy Images:** Bubbles Photolibrary. **380 Alamy Images:** Shout (cl); **Mother & Baby Picture Library:** Ruth Jenkinson (tl). **381 PunchStock:** Brand X Pictures. **382 SuperStock. 390 Bubbles. 394 Getty Images:** Logan Mock-Bunting (cl). **399 Bubbles:** Loisjoy Thurstun. **402 Mother & Baby Picture Library:** Ian Hooton. **404 Alamy Images:** Janine Wiedel. **405 Alamy Images:** Peter Usbeck (l); **Science Photo Library:** Joseph Nettis (r). **406–407 Corbis:** Norbert Schaefer.

All other images © Dorling Kindersley
For further information see: www.dkimages.com

Fourth Edition
Senior Editor Dawn Bates
Designer Philippa Nash
Jacket Designer Nicola Powling
Senior Producer, Pre-Production Tony Phipps
Producer Ché Creasey
Managing Editor Dawn Henderson
Managing Art Editor Marianne Markham
Art Director Maxine Pedliham
Publishing Director Mary-Clare Jerram

DK INDIA
Senior Editor Arani Sinha
Editor Madhurika Bhardwaj
Managing Editor Soma B. Chowdhury
Senior DTP Designer Tarun Sharma
CTS Manager Sunil Sharma
Picture Research Coordinator Sumita Khatwani
Picture Research Manager Taiyaba Khatoon

A WORLD OF IDEAS:
SEE ALL THERE IS TO KNOW

www.dk.com